Svetlin G. Georgiev, İnci M. Erhan
Numerical Analysis on Time Scales

Also of Interest

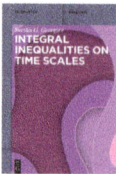

Integral Inequalities on Time Scales
Svetlin G. Georgiev, 2018
ISBN 978-3-11-070550-8, e-ISBN (PDF) 978-3-11-070555-3

Functional Analysis with Applications
Svetlin G. Georgiev, Khaled Zennir, 2019
ISBN 978-3-11-065769-2, e-ISBN (PDF) 978-3-11-065772-2

Data Science
Time Complexity, Inferential Uncertainty, and Spacekime Analytics
Ivo D. Dinov, Milen Velchev Velev, 2021
ISBN 978-3-11-069780-3, e-ISBN (PDF) 978-3-11-069782-7

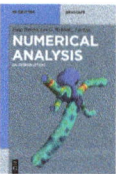

Numerical Analysis
An Introduction
Timo Heister, Leo G. Rebholz, Fei Xue, 2019
ISBN 978-3-11-057330-5, e-ISBN (PDF) 978-3-11-057332-9

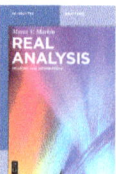

Real Analysis
Measure and Integration
Marat V. Markin, 2019
ISBN 978-3-11-060097-1, e-ISBN (PDF) 978-3-11-060099-5

Svetlin G. Georgiev, İnci M. Erhan

Numerical Analysis on Time Scales

—

DE GRUYTER

Mathematics Subject Classification 2010
Primary: 34N05, 39A10, 41A05; Secondary: 65D05, 65L05

Authors
Prof. Dr. Svetlin G. Georgiev
Kliment Ohridski University of Sofia
Department of Differential Equations
Faculty of Mathematics and Informatics
1126 Sofia
Bulgaria
svetlingeorgiev1@gmail.com

Prof. Dr. İnci M. Erhan
Atılım University
Department of Mathematics
06830 Ankara
Turkey
inci.erhan@atilim.edu.tr

ISBN 978-3-11-078725-2
e-ISBN (PDF) 978-3-11-078732-0
e-ISBN (EPUB) 978-3-11-078734-4

Library of Congress Control Number: 2022936834

Bibliographic information published by the Deutsche Nationalbibliothek
The Deutsche Nationalbibliothek lists this publication in the Deutsche Nationalbibliografie;
detailed bibliographic data are available on the Internet at http://dnb.dnb.de.

© 2022 Walter de Gruyter GmbH, Berlin/Boston
Cover image: maxkabakov / iStock / Getty Images Plus
Typesetting: VTeX UAB, Lithuania
Printing and binding: CPI books GmbH, Leck

www.degruyter.com

Preface

Numerical analysis is an extremely important field in mathematics and other natural sciences. Almost all real life problems that are modeled mathematically do not have exact solutions. Moreover, the mathematical models often have a nonlinear structure which makes them even more difficult to solve analytically. In this sense, the development and construction of efficient numerical methods gain a big significance. Motivated by this fact, the studies related to the development of new powerful numerical methods or improvement of the existing ones are still continuing.

In particular, numerical solutions of differential equations are of great importance since many processes in nature are time dependent and their mathematical models are usually described by partial or ordinary differential equations and often, by difference equations. The theory of time scales and dynamic equations, on the other hand, unifies the continuous and discrete models, thus providing a more general view to the subject.

Dynamic equations, which describe how quantities change across the time or space, arise in any field of study where measurements can be taken. Most realistic mathematical models cannot be solved using the traditional analytical methods for dynamic equations on time scales. They must be handled with computational methods that deliver approximate solutions.

Until recently, there were very few studies related to numerical methods on time scales. In the last few years, some initial results on the subject have been published, which initiated the development of numerical analysis on time scales.

This book is devoted to designing, analyzing, and applying computational techniques for dynamic equations on time scales. The book provides material for a typical first course. This book is an introduction to numerical methods for initial value problems for dynamic equations on time scales.

The book contains 12 chapters. In Chapter 1, the Lagrange, σ-Lagrange, Hermite, and σ-Hermite polynomial interpolations are introduced. From these interpolations, approximations for the delta derivative of continuously delta-differentiable functions are deducted. In Chapter 2, formulae for numerical integration on time scales are derived and the associated approximation errors are estimated. In Chapter 3, linear interpolating splines, linear interpolating σ-splines, cubic and Hermite splines are introduced. Chapter 4 is presented as a study of the Euler method. Chapters 5 and 6 consider the Taylor series methods of order-2 and order-p and analyze convergence of these methods. Linear multistep methods are investigated in Chapter 7. Chapter 8 contains the analysis of Runge–Kutta methods. Chapter 9 deals with the series solution method for fractional dynamic equations and dynamic equations on time scales. The Adomian polynomials method is investigated in Chapter 10. Chapter 11 is devoted to weak solutions and variational methods for some classes of linear dynamic equations on time scales. Nonlinear dynamic equations and variational methods are investigated in Chapter 12.

https://doi.org/10.1515/9783110787320-201

We presume that the readers are familiar with the basic notions on time scales such as forward and backward jump operators, graininnes function, right and left scattered, dense and isolated points, as well as with the basic calculus concepts on time scales such as the delta differentiation and integration and their properties, elementary functions on time scales, Taylor formula. For the readers who are studying the time scales for the first time, we suggest learning these basic notions and concepts from the numerous references given in this book and elsewhere.

The text material of this book is presented in a highly readable, mathematically solid format. Many practical problems are illustrated, displaying a wide variety of solution techniques. The authors welcome any suggestions for the improvement of the text.

Paris/Ankara, July 2022

Svetlin Georgiev

İnci Erhan

Contents

1 Polynomial interpolation

Polynomial interpolation is a very useful tool employed in many areas of science in which a given set of data needs to be represented by a function, in particular, by a polynomial taking the given values at the points of the given data set. From the classical numerical analysis we know that there is a unique polynomial satisfying the given conditions which can be constructed in different ways. The Lagrange interpolation polynomial is one of the most used. For more details we refer the reader to the books [5] and [16].

In some physical problems, together with the values of a certain quantity, the values of the rates of change of this quantity, that is, the derivative values at a given set of points are also known. In such applications the polynomial representing this data is also required to match the derivatives. The Hermite interpolation polynomial is used for such problems.

In this chapter, we consider the problem of polynomial interpolation on time scales. This problem involves finding a polynomial that agrees with some information that we have for a given real-valued function f of a single real variable x. We construct the Lagrange, σ-Lagrange, Hermite, and σ-Hermite interpolation polynomials for a given real-valued function f defined on an arbitrary time scale.

Throughout this chapter, we assume that $\mathbb{T}$ is a time scale with forward jump operator σ, delta differentiation operator Δ and graininess function μ.

1.1 Lagrange interpolation

In this section, we construct the Lagrange interpolation polynomial on an arbitrary time scale. We present the theoretical background of this construction and solve numerical examples.

Let $\mathcal{P}_n$, $n \in \mathbb{N}_0$, denote the set of all polynomials of degree $\leq n$ defined over the set $\mathbb{R}$ of real numbers. Let $n \in \mathbb{N}$ and $x_i \in \mathbb{T}$, $i \in \{0, 1, \ldots, n\}$, be distinct and y_i, $i \in \{0, 1, \ldots, n\}$, be given real numbers. We will find $p_n \in \mathcal{P}_n$ such that $p_n(x_i) = y_i$, $i \in \{0, 1, \ldots, n\}$. Below we introduce the form of a polynomial taking the given values y_i at the points x_i for $i \in \{0, 1, \ldots, n\}$.

Theorem 1.1. *Suppose that $n \in \mathbb{N}$. Then there exist polynomials $L_k \in \mathcal{P}_n$, $k \in \{0, 1, \ldots, n\}$, such that*

$$L_k(x_i) = \begin{cases} 1 & \text{if } i = k, \\ 0 & \text{if } i \neq k, \end{cases}$$

$i, k \in \{0, 1, \ldots, n\}$. Moreover,

https://doi.org/10.1515/9783110787320-001

$$p_n(x) = \sum_{k=0}^{n} L_k(x)y_k, \quad x \in \mathbb{T},$$

satisfies the condition $p_n(x_i) = y_i$, $i \in \{0, 1, \ldots, n\}$, $p_n \in \mathcal{P}_n$.

Proof. Define

$$L_k(x) = C_k \sum_{i=0, i\neq k}^{n} (x - x_i), \quad x \in \mathbb{T},$$

where $C_k \in \mathbb{R}$, $k \in \{0, 1, \ldots, n\}$, will be determined below. We have $L_k(x_i) = 0$, $i \in \{0, 1, \ldots, n\}$, $i \neq k$, and

$$L_k(x_k) = C_k \prod_{i=0, i\neq k}^{n} (x_k - x_i) = 1, \quad k \in \{0, 1, \ldots, n\}.$$

Thus,

$$C_k = \frac{1}{\prod_{i=0, i\neq k}^{n} (x_k - x_i)}, \quad k \in \{0, 1, \ldots, n\},$$

and

$$L_k(x) = \prod_{i=0, i\neq k}^{n} \frac{x - x_i}{x_k - x_i}, \quad x \in \mathbb{T}, \quad k \in \{0, 1, \ldots, n\}. \tag{1.1}$$

We have that $L_k \in \mathcal{P}_n$, $k \in \{0, 1, \ldots, n\}$, and $p_n \in \mathcal{P}_n$. This completes the proof. □

The uniqueness of the polynomial given in the previous theorem is proved next.

Theorem 1.2. *Assume that $n \in \mathbb{N}_0$. Let $x_i \in \mathbb{T}$, $i \in \{0, 1, \ldots, n\}$, be distinct and $y_i \in \mathbb{R}$, $i \in \{0, 1, \ldots, n\}$. Then there exists a unique polynomial $p_n \in \mathcal{P}_n$ such that*

$$p_n(x_i) = y_i, \quad i \in \{0, 1, \ldots, n\}.$$

Proof. The existence of the polynomial p_n follows by Theorem 1.1. Suppose that there exist two polynomials $p_n, q_n \in \mathcal{P}_n$ such that

$$p_n(x_i) = q_n(x_i) = y_i, \quad i \in \{0, 1, \ldots, n\}.$$

Then the polynomial $h_n = p_n - q_n$ has $n + 1$ distinct roots. Therefore, $h_n \equiv 0$ or $p_n \equiv q_n$. This completes the proof. □

Now, we formally define the polynomial in the above theorems.

Definition 1.3. Assume that $n \in \mathbb{N}_0$. Let $x_i \in \mathbb{T}$, $i \in \{0, 1, \ldots, n\}$, be distinct and $y_i \in \mathbb{R}$, $i \in \{0, 1, \ldots, n\}$. The polynomial

$$p_n(x) = \sum_{k=0}^{n} L_k(x) y_k, \quad x \in \mathbb{T},$$

where L_k, $k \in \{0, 1, \ldots, n\}$, are defined in (1.1), will be called the Lagrange interpolation polynomial of degree n with interpolation points (x_i, y_i), $i \in \{0, 1, \ldots, n\}$.

Definition 1.4. Assume that $n \in \mathbb{N}_0$. Let $x_i \in [a, b] \subset \mathbb{T}$, $i \in \{0, 1, \ldots, n\}$, be distinct and $f : [a, b] \to \mathbb{R}$ be a given function. The polynomial

$$p_n(x) = \sum_{k=0}^{n} L_k(x) f(x_k), \quad x \in \mathbb{T},$$

where L_k, $k \in \{0, 1, \ldots, n\}$, are defined in (1.1), will be called the Lagrange interpolation polynomial of degree n with interpolation points x_i, $i \in \{0, 1, \ldots, n\}$, for the function f.

Example 1.5. Let $\mathbb{T} = \mathbb{Z}$. We will construct the Lagrange interpolation polynomial for the set

$$\{(-3, 0), (-2, 2), (0, 1), (1, 1)\}.$$

Here

$$x_0 = -3, \quad x_1 = -2, \quad x_2 = 0, \quad x_3 = 1,$$
$$y_0 = 0, \quad y_1 = 2, \quad y_2 = 1, \quad y_3 = 1, \quad n = 3.$$

Then

$$L_0(x) = \prod_{i=1}^{3} \frac{x - x_i}{x_0 - x_i} = \frac{(x+2)x(x-1)}{-1 \cdot (-3) \cdot (-4)} = -\frac{(x-1)x(x+2)}{12},$$

$$L_1(x) = \prod_{i=0, i \neq 1}^{3} \frac{x - x_i}{x_1 - x_i} = \frac{(x+3)x(x-1)}{1 \cdot (-2) \cdot (-3)} = \frac{(x-1)x(x+3)}{6},$$

$$L_2(x) = \prod_{i=0, i \neq 2}^{3} \frac{x - x_i}{x_2 - x_i} = \frac{(x+3)(x+2)(x-1)}{3 \cdot 2 \cdot (-1)} = -\frac{(x-1)(x+2)(x+3)}{6},$$

$$L_3(x) = \prod_{i=0}^{2} \frac{x - x_i}{x_3 - x_i} = \frac{(x+3)(x+2)x}{4 \cdot 3 \cdot 1} = \frac{x(x+2)(x+3)}{12}, \quad x \in \mathbb{T}.$$

Hence,

$$p_3(x) = y_0 L_0(x) + y_1 L_1(x) + y_2 L_2(x) + y_3 L_3(x)$$

$$= 2\left(\frac{x(x-1)(x+3)}{6}\right) - \frac{(x-1)(x+2)(x+3)}{6} + \frac{x(x+2)(x+3)}{12}$$

$$= \frac{(x-2)(x-1)(x+3)}{6} + \frac{x(x+2)(x+3)}{12}$$

$$= \frac{(x+3)(2(x-1)(x-2) + x(x+2))}{12}$$

$$= \frac{(x+3)(3x^2 - 4x + 4)}{12}, \quad x \in \mathbb{T}.$$

Exercise 1.6. Let $\mathbb{T} = 2^{\mathbb{N}_0}$. Construct the Lagrange interpolation polynomial for the set

$$\{(1,-1), (4,0), (8,1), (16,2)\}.$$

Suppose that $n \in \mathbb{N}_0$ and $x_j \in \mathbb{T}, j \in \{0, 1, \ldots, n\}$, are distinct points. For $x \in \mathbb{T}$, we define the polynomials

$$\pi_{n+1}(x) = \prod_{j=0}^{n}(x - x_j), \quad \Pi_{n+1}^k(x) = \pi_{n+1}^{\Delta^k}(x), \quad k \in \mathbb{N}_0,$$

which will be employed in the error analysis of polynomial interpolation.

Example 1.7. Let $\mathbb{T} = 2^{\mathbb{N}_0}$, $x_0 = 1$, $x_1 = 2$. Here

$$n = 1, \quad \sigma(x) = 2x, \quad x \in \mathbb{T}.$$

Then

$$\pi_2(x) = (x-1)(x-2) = x^2 - 3x + 2, \quad x \in \mathbb{T},$$

and

$$\Pi_2^1(x) = \pi_2^{\Delta}(x) = \sigma(x) + x - 3 = 3x - 3,$$
$$\Pi_2^2(x) = \pi_2^{\Delta^2}(x) = 3, \quad x \in \mathbb{T}.$$

Example 1.8. Let $\mathbb{T} = \mathbb{Z}$ and

$$x_0 = 1, \quad x_1 = 2, \quad x_2 = 3.$$

Then

$$\pi_3(x) = (x-1)(x-2)(x-3)$$
$$= (x^2 - 3x + 2)(x-3)$$
$$= x^3 - 6x^2 + 11x - 6, \quad x \in \mathbb{T}.$$

We have

$$\sigma(x) = x + 1, \quad x \in \mathbb{T},$$

and

$$\Pi_3^1(x) = \pi_3^\Delta(x)$$
$$= (\sigma(x))^2 + x\sigma(x) + x^2 - 6(\sigma(x) + x) + 11$$
$$= (x + 1)^2 + x(x + 1) + x^2 - 6(x + 1 + x) + 11$$
$$= 3x^2 - 9x + 6,$$
$$\Pi_3^2(x) = \pi_3^{\Delta^2}(x)$$
$$= 3(\sigma(x) + x) - 9$$
$$= 6x - 6,$$
$$\Pi_3^3(x) = \pi_3^{\Delta^3}(x) = 6, \quad x \in \mathbb{T}.$$

Exercise 1.9. Let $\mathbb{T} = (\frac{1}{4})^{\mathbb{N}_0}$,

$$x_0 = \frac{1}{64}, \quad x_1 = \frac{1}{16}, \quad x_2 = \frac{1}{4}, \quad x_3 = 1.$$

Find $\pi_4(x)$, $\Pi_4^1(x)$, $\Pi_4^2(x)$, $\Pi_4^3(x)$, $\Pi_4^4(x)$, $x \in \mathbb{T}$.

The following theorem gives the error in approximating a function f by a Lagrange polynomial.

Theorem 1.10. *Suppose that $n \in \mathbb{N}_0$, $a, b \in \mathbb{T}$, $a < b$, $x_j \in [a, b]$, $j \in \{0, 1, \ldots, n\}$, are distinct and $f : [a, b] \rightarrow \mathbb{R}$, $f^{\Delta^k}(x)$ exist for any $x \in [a, b]$ and $k \in \{1, \ldots, n + 1\}$. Then for any $x \in [a, b]$ there exists $\xi = \xi(x) \in (a, b)$ such that*

$$f(x) - p_n(x) = \frac{f^{\Delta^{n+1}}(\xi)}{\Pi_{n+1}^{n+1}(\xi)} \pi_{n+1}(x), \quad x \in [a, b],$$

or

$$F_{\min,n+1}(\xi) \leq \frac{f(x) - p_n(x)}{\pi_{n+1}(x)} \leq F_{\max,n+1}(\xi), \quad x \in [a, b],$$

where

$$F_{\max,n+1}(\xi) = \max\left\{ \frac{f^{\Delta^{n+1}}(\xi)}{\Pi_{n+1}^{n+1}(\xi)}, \frac{f^{\Delta^{n+1}}(\rho(\xi))}{\Pi_{n+1}^{n+1}(\rho(\xi))} \right\},$$

$$F_{\min,n+1}(\xi) = \min\left\{ \frac{f^{\Delta^{n+1}}(\xi)}{\Pi_{n+1}^{n+1}(\xi)}, \frac{f^{\Delta^{n+1}}(\rho(\xi))}{\Pi_{n+1}^{n+1}(\rho(\xi))} \right\}.$$

Proof. Let p_n be the Lagrange interpolation polynomial for the function f with interpolation points x_j, $j \in \{0, 1, \ldots, n\}$. Define the function

$$\phi(t) = f(t) - p_n(t) - \frac{f(x) - p_n(x)}{\pi_{n+1}(x)}\pi_{n+1}(t), \quad t \in [a, b].$$

Then

$$\begin{aligned}\phi(x_j) &= f(x_j) - p_n(x_j) - \frac{f(x) - p_n(x)}{\pi_{n+1}(x)}\pi_{n+1}(x_j) \\ &= f(x_j) - f(x_j) \\ &= 0, \quad j \in \{0, 1, \dots, n\},\end{aligned}$$

and $\phi(x) = 0$. Thus, $\phi : [a, b] \to \mathbb{R}$ has at least $n + 2$ generalized zeros (GZs). Hence, by Rolle's theorem (see Theorem A.6 of Appendix A), it follows that $\phi^{\Delta^{n+1}}$ has at least one GZ on (a, b). Therefore, there exists an $\xi = \xi(x) \in (a, b)$ such that

$$\phi^{\Delta^{n+1}}(\xi) = 0 \quad \text{or} \quad \phi^{\Delta^{n+1}}(\rho(\xi))\phi^{\Delta^{n+1}}(\xi) < 0.$$

Note that

$$\phi^{\Delta^{n+1}}(t) = f^{\Delta^{n+1}}(t) - \frac{f(x) - p_n(x)}{\pi_{n+1}(x)}\pi_{n+1}^{\Delta^{n+1}}(t), \quad t \in [a, b].$$

We now consider each case separately.

1. Let $\phi^{\Delta^{n+1}}(\xi) = 0$. Then

$$f^{\Delta^{n+1}}(\xi) = \frac{f(x) - p_n(x)}{\pi_{n+1}(x)}\pi_{n+1}^{\Delta^{n+1}}(\xi),$$

or

$$f(x) - p_n(x) = \frac{f^{\Delta^{n+1}}(\xi)}{\pi_{n+1}^{\Delta^{n+1}}(\xi)}\pi_{n+1}(x) = \frac{f^{\Delta^{n+1}}(\xi)}{\Pi_{n+1}^{n+1}(\xi)}\pi_{n+1}(x).$$

2. Let

$$\phi^{\Delta^{n+1}}(\rho(\xi))\phi^{\Delta^{n+1}}(\xi) < 0.$$

Then

$$\begin{aligned}\phi^{\Delta^{n+1}}(\rho(\xi)) &= f^{\Delta^{n+1}}(\rho(\xi)) - \frac{f(x) - p_n(x)}{\pi_{n+1}(x)}\pi_{n+1}^{\Delta^{n+1}}(\rho(\xi)) \\ &= f^{\Delta^{n+1}}(\rho(\xi)) - \frac{f(x) - p_n(x)}{\pi_{n+1}(x)}\Pi_{n+1}^{n+1}(\rho(\xi)),\end{aligned}$$

and

$$\phi^{\Delta^{n+1}}(\xi) = f^{\Delta^{n+1}}(\xi) - \frac{f(x) - p_n(x)}{\pi_{n+1}(x)}\Pi_{n+1}^{n+1}(\xi).$$

Hence,

$$0 > \phi^{\Delta^{n+1}}(\rho(\xi))\phi^{\Delta^{n+1}}(\xi)$$

$$= \left(f^{\Delta^{n+1}}(\rho(\xi)) - \frac{f(x) - p_n(x)}{\pi_{n+1}(x)}\Pi_{n+1}^{n+1}(\rho(\xi)) \right)$$

$$\times \left(f^{\Delta^{n+1}}(\xi) - \frac{f(x) - p_n(x)}{\pi_{n+1}(x)}\Pi_{n+1}^{n+1}(\xi) \right)$$

$$= \left(\frac{f(x) - p_n(x)}{\pi_{n+1}(x)} \right)^2 \Pi_{n+1}^{n+1}(\rho(\xi))\Pi_{n+1}^{n+1}(\xi)$$

$$- \frac{f(x) - p_n(x)}{\pi_{n+1}(x)}\left(\Pi_{n+1}^{n+1}(\rho(\xi))f^{\Delta^{n+1}}(\xi) + \Pi_{n+1}^{n+1}(\xi)f^{\Delta^{n+1}}(\rho(\xi))\right)$$

$$+ f^{\Delta^{n+1}}(\rho(\xi))f^{\Delta^{n+1}}(\xi).$$

We conclude that

$$F_{\min,n+1}(\xi) \le \frac{f(x) - p_n(x)}{\pi_{n+1}(x)} \le F_{\max,n+1}(\xi).$$

This completes the proof. □

Note that, as stated in the next remark, the error vanishes if the number of data points increases to infinity.

Remark 1.11. Suppose that all the conditions of Theorem 1.10 hold. If

$$\lim_{n\to\infty} \max_{x\in[a,b]} \left(\frac{f^{\Delta^{n+1}}(\xi)}{\Pi_{n+1}^{n+1}(\xi)}\pi_{n+1}(x) \right) = 0$$

and

$$\lim_{n\to\infty} \max_{x\in[a,b]} \left(\frac{f^{\Delta^{n+1}}(\rho(\xi))}{\Pi_{n+1}^{n+1}(\rho(\xi))}\pi_{n+1}(x) \right) = 0,$$

then

$$\lim_{n\to\infty} \max_{x\in[a,b]} |f(x) - p_n(x)| = 0.$$

Example 1.12. Let $\mathbb{T} = \{0, \frac{1}{6}, \frac{1}{4}, \frac{1}{2}, 1, 2, 3, 4\}$. Let

$$a = x_0 = 0, \quad x_1 = \frac{1}{4}, \quad x_2 = 1, \quad x_3 = 3, \quad b = 4,$$

and

$$f(x) = \frac{x+1}{x^2 + x + 6}, \quad x \in \mathbb{T}.$$

We will construct the Lagrange interpolation polynomial p_3 of f for the points x_0, x_1, x_2, x_3 and compare the graphs of p_3 and f. First, note that

$$f(x_0) = f(0) = \frac{1}{6}, \quad f(x_1) = f\left(\frac{1}{4}\right) = \frac{20}{101},$$

$$f(x_2) = f(1) = \frac{1}{4}, \quad f(x_3) = f(3) = \frac{2}{9}.$$

Also, we compute

$$L_0(x) = \frac{(x - \frac{1}{4})(x - 1)(x - 3)}{(0 - \frac{1}{4})(0 - 1)(0 - 3)} = -\frac{4}{3}\left(x - \frac{1}{4}\right)(x - 1)(x - 3),$$

$$L_1(x) = \frac{(x - 0)(x - 1)(x - 3)}{(\frac{1}{4} - 0)(\frac{1}{4} - 1)(\frac{1}{4} - 3)} = \frac{64}{33}x(x - 1)(x - 3),$$

$$L_2(x) = \frac{(x - 0)(x - \frac{1}{4})(x - 3)}{(1 - 0)(1 - \frac{1}{4})(1 - 3)} = -\frac{2}{3}x\left(x - \frac{1}{4}\right)(x - 3),$$

$$L_3(x) = \frac{(x - 0)(x - \frac{1}{4})(x - 1)}{(3 - 0)(3 - \frac{1}{4})(3 - 1)} = \frac{2}{33}x\left(x - \frac{1}{4}\right)(x - 1), \quad x \in \mathbb{T}.$$

Then the Lagrange interpolation polynomial p_3 of f for the points $0, \frac{1}{4}, 1, 3$ is

$$p_3(x) = f(0)L_0(x) + f\left(\frac{1}{4}\right)L_1(x) + f(1)L_2(x) + f(3)L_3(x)$$

$$= \frac{1}{6}\left(-\frac{4}{3}\left(x - \frac{1}{4}\right)(x - 1)(x - 3)\right) + \frac{20}{101}\left(\frac{64}{33}x(x - 1)(x - 3)\right)$$

$$+ \frac{1}{4}\left(-\frac{2}{3}x\left(x - \frac{1}{4}\right)(x - 3)\right) + \frac{2}{9}\left(\frac{2}{33}x\left(x - \frac{1}{4}\right)(x - 1)\right)$$

$$= -\frac{2}{9}\left(x - \frac{1}{4}\right)(x - 1)(x - 3) + \frac{1280}{3333}x(x - 1)(x - 3)$$

$$- \frac{1}{6}x\left(x - \frac{1}{4}\right)(x - 3) + \frac{4}{297}x\left(x - \frac{1}{4}\right)(x - 1), \quad x \in \mathbb{T}.$$

The graphs of f and p_3 are compared in Figure 1.1. Moreover, the values of f and p_3 at noninterpolating points of the time scale are compared in Table 1.1. From both Table 1.1 and Figure 1.1, we observe a good approximation at noninterpolation points inside the interval $[a, b] = [0, 4]$.

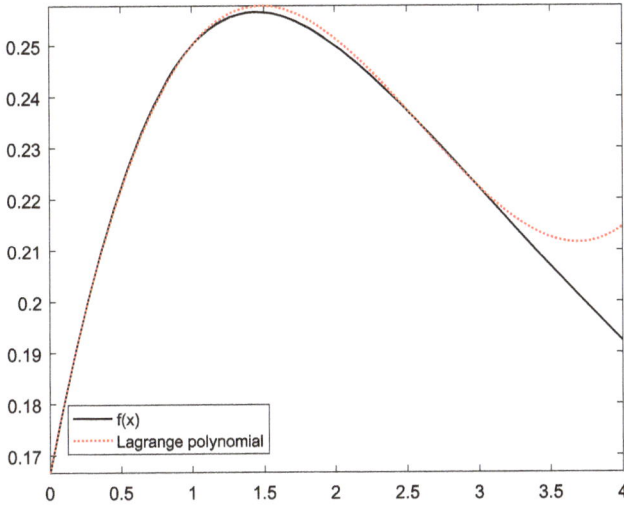

Figure 1.1: The graphs of f and the Lagrange polynomial p_3.

Table 1.1: The values of f and p_3 at noninterpolating points of $\mathbb{T}$.

x	$f(x)$	$p_3(x)$
1/6	0.1883	0.1884
1/2	0.2222	0.2218
2	0.2500	0.2513

1.2 σ-Lagrange interpolation

In this section, we will show that a given function can be approximated with the so-called σ-Lagrange polynomials. We will show that there are classes of time scales for which the Lagrange interpolation polynomials and the σ-Lagrange interpolation polynomials are different, and there are classes of time scales for which the Lagrange interpolation polynomials and the σ-Lagrange interpolation polynomials coincide.

In the following, we define the σ-polynomials. With $\mathcal{P}_n^\sigma$, $n \in \mathbb{N}_0$, we will denote the set of all functions in the form

$$g(x) = a_n\big(\sigma(x)\big)^n + a_{n-1}\big(\sigma(x)\big)^{n-1} + \cdots + a_1\sigma(x) + a_0, \quad x \in \mathbb{T},$$

where $a_j \in \mathbb{R}$, $j \in \{0, 1, \ldots, n\}$. Let $a, b \in \mathbb{T}$, $a < b$. A function $g \in \mathcal{P}_n^\sigma$ will be called a σ-polynomial. To define a σ-Lagrange polynomial, we have a requirement that the points in the data set should be σ-distinct.

Definition 1.13. Let $n \in \mathbb{N}_0$. The points $x_j \in [a, b)$, $j \in \{0, 1, \ldots, n\}$, will be called σ-distinct if $\sigma(x_n) \le b$ and

$$\sigma(x_0) < \sigma(x_1) < \cdots < \sigma(x_n).$$

Example 1.14. Let $\mathbb{T} = \{-1, 1\} \cup \{1 + (\frac{1}{2})^n : n \in \mathbb{N}_0\} \cup \{3, 4, 5\}$ and $a = -1$, $b = 5$. Take the points

$$x_0 = -1, \quad x_1 = 1, \quad x_2 = 3.$$

Then

$$\sigma(x_0) = 1, \quad \sigma(x_1) = 1, \quad \sigma(x_2) = 4.$$

Thus, the points $\{x_0, x_1, x_2\}$ are not σ-distinct.

Example 1.15. Let $\mathbb{T} = 2^{\mathbb{N}_0}$, $a = 1$, $b = 16$. Take the points

$$x_0 = 1, \quad x_1 = 2, \quad x_2 = 4.$$

Then

$$\sigma(x_0) = 2, \quad \sigma(x_1) = 4, \quad \sigma(x_2) = 8.$$

Therefore, $\{x_0, x_1, x_2\}$ are σ-distinct points.

As in the previous section, one can prove the following result.

Theorem 1.16. *Suppose that $n \in \mathbb{N}$ and $x_j \in \mathbb{T}$, $j \in \{0, 1, \ldots, n\}$, are σ-distinct. Then there exist unique σ-polynomials $L_{\sigma k} \in \mathcal{P}_n^\sigma$, $k \in \{0, 1, \ldots, n\}$, such that*

$$L_{\sigma k}(x_i) = \begin{cases} 1 & if\ i = k, \\ 0 & if\ i \neq k, \end{cases}$$

$i, k \in \{0, 1, \ldots, n\}$. Moreover,

$$p_{\sigma n}(x) = \sum_{k=0}^{n} L_{\sigma k}(x) y_k = \sum_{k=0}^{n} \left(\prod_{j=0, j \neq k}^{n} \frac{\sigma(x) - \sigma(x_j)}{\sigma(x_k) - \sigma(x_j)} \right) y_k, \quad x \in \mathbb{T},$$

satisfies the condition $p_{\sigma n}(x_i) = y_i$, $i \in \{0, 1, \ldots, n\}$, $p_{\sigma n} \in \mathcal{P}_n^\sigma$.

Based on the statement of the above theorem, we define the σ-Lagrange interpolation polynomials as follows.

Definition 1.17. Assume that $n \in \mathbb{N}_0$. Let $x_i \in \mathbb{T}$, $i \in \{0, 1, \ldots, n\}$, be σ-distinct and $y_i \in \mathbb{R}$, $i \in \{0, 1, \ldots, n\}$. The σ-polynomial

$$p_{\sigma n}(x) = \sum_{k=0}^{n} L_{\sigma k}(x) y_k, \quad x \in \mathbb{T},$$

where $L_{\sigma k}$, $k \in \{0, 1, \ldots, n\}$, are defined in Theorem 1.16, will be called the σ-Lagrange interpolation polynomial of degree n with σ-interpolation points (x_i, y_i), $i \in \{0, 1, \ldots, n\}$.

Definition 1.18. Assume that $n \in \mathbb{N}_0$. Let $x_i \in [a, b] \subset \mathbb{T}$, $i \in \{0, 1, \ldots, n\}$, be σ-distinct and $f : [a, b] \to \mathbb{R}$ be a given function. The σ-polynomial

$$p_{\sigma n}(x) = \sum_{k=0}^{n} L_{\sigma k}(x) f(x_k), \quad x \in \mathbb{T},$$

where $L_{\sigma k}$, $k \in \{0, 1, \ldots, n\}$, are defined in Theorem 1.16, will be called the σ-Lagrange interpolation polynomial of degree n with σ-interpolation points x_i, $i \in \{0, 1, \ldots, n\}$, for the function f.

In the following, we will compute the Lagrange and σ-Lagrange polynomials for a given set of data on a time scale and compare them.

Example 1.19. Let $\mathbb{T} = \{-2, -1, 0, 3, 7\}$,

$$a = -2, \quad x_0 = -2, \quad x_1 = 0, \quad b = 7,$$

and assume $f : \mathbb{T} \to \mathbb{R}$ is defined by

$$f(x) = x + 3, \quad x \in \mathbb{T}.$$

We will find the σ-Lagrange interpolation polynomial for the function f with σ-interpolation points $\{x_0, x_1\}$. We have

$$\sigma(x_0) = \sigma(-2) = -1,$$
$$\sigma(x_1) = \sigma(0) = 3,$$
$$L_{\sigma 0}(x) = \frac{\sigma(x) - \sigma(x_1)}{\sigma(x_0) - \sigma(x_1)} = -\frac{1}{4}(\sigma(x) - 3),$$
$$L_{\sigma 1}(x) = \frac{\sigma(x) - \sigma(x_0)}{\sigma(x_1) - \sigma(x_0)} = \frac{1}{4}(\sigma(x) + 1), \quad x \in \mathbb{T},$$
$$f(x_0) = f(-2) = 1,$$
$$f(x_1) = f(0) = 3.$$

Thus,

$$p_{\sigma 1}(x) = f(x_0) L_{\sigma 0}(x) + f(x_1) L_{\sigma 1}(x)$$
$$= -\frac{1}{4}(\sigma(x) - 3) + \frac{3}{4}(\sigma(x) + 1)$$
$$= \frac{1}{2}(\sigma(x) + 3), \quad x \in [-2, 7].$$

Note that

$$p_{\sigma 1}(0) = \frac{1}{2}(\sigma(0) + 3) = 3 = f(0),$$

$$p_{\sigma 1}(-2) = \frac{1}{2}(\sigma(-2) + 3) = 1 = f(-2).$$

Also, we have

$$p_{\sigma 1}(-1) = \frac{1}{2}(\sigma(-1) + 3) = \frac{3}{2}.$$

Now, we will find the Lagrange interpolation polynomial for the function f with interpolation points $\{x_0, x_1\}$. We have

$$L_0(x) = \frac{x - x_1}{x_0 - x_1} = -\frac{1}{2}x,$$

$$L_1(x) = \frac{x - x_0}{x_1 - x_0} = \frac{x + 2}{2}, \quad x \in [-2, 7].$$

Therefore,

$$p_1(x) = f(x_0)L_0(x) + f_1(x)L_1(x)$$
$$= -\frac{1}{2}x + 3\frac{x + 2}{2}$$
$$= x + 3, \quad x \in [-2, 7].$$

Then

$$p_1(0) = 3 = f(0), \quad p_1(-2) = 1 = f(-2).$$

We also have

$$p_1(-1) = -1 + 3 = 2 \neq p_{\sigma 1}(-1) = \frac{3}{2}.$$

Remark 1.20. In the above example we see that, in general, the Lagrange and σ-Lagrange interpolation polynomials for a function f are different.

Exercise 1.21. Let $\mathbb{T} = \{-1, -\frac{1}{4}, -\frac{1}{8}, 0, 2, 3, 7\}$,

$$a = x_0 = -1, \quad x_1 = -\frac{1}{8}, \quad x_2 = 3, \quad b = 7.$$

Find the σ-Lagrange and Lagrange interpolation polynomials for the function $f : \mathbb{T} \to \mathbb{R}$ defined by

$$f(x) = \frac{x + 1}{x^2 - x + 1} + 3x, \quad x \in [-1, 7],$$

with σ-interpolation and interpolation points x_0, x_1, x_2.

In the following example, we see that on some time scales, the Lagrange and σ-Lagrange polynomials are the same.

Example 1.22. Let $\mathbb{T} = \mathbb{Z}$,

$$a = x_0 = -1, \quad x_1 = 1, \quad x_2 = 3, \quad b = 4,$$

and $f : \mathbb{T} \to \mathbb{R}$ be defined by

$$f(x) = x^2 + x + 1, \quad x \in \mathbb{T}.$$

We will find the σ-Lagrange interpolation polynomial with σ-interpolation points x_0, x_1, x_2. We have $\sigma(x) = x + 1$, $x \in \mathbb{T}$, and

$$\sigma(x_0) = \sigma(-1) = 0,$$

$$\sigma(x_1) = \sigma(1) = 2,$$

$$\sigma(x_2) = \sigma(3) = 4.$$

Then

$$\sigma(x_0) < \sigma(x_1) < \sigma(x_2),$$

i. e., x_0, x_1, x_2 are σ-distinct points. We have

$$
\begin{aligned}
L_{\sigma 0}(x) &= \frac{(\sigma(x) - \sigma(x_1))(\sigma(x) - \sigma(x_2))}{(\sigma(x_0) - \sigma(x_1))(\sigma(x_0) - \sigma(x_2))} \\
&= \frac{(x + 1 - 2)(x + 1 - 4)}{(0 - 2)(0 - 4)} \\
&= \frac{(x - 1)(x - 3)}{8},
\end{aligned}
$$

$$
\begin{aligned}
L_{\sigma 1}(x) &= \frac{(\sigma(x) - \sigma(x_0))(\sigma(x) - \sigma(x_2))}{(\sigma(x_1) - \sigma(x_0))(\sigma(x_1) - \sigma(x_2))} \\
&= \frac{(x + 1 - 0)(x + 1 - 4)}{(2 - 0)(2 - 4)} \\
&= -\frac{(x - 3)(x + 1)}{4},
\end{aligned}
$$

$$
\begin{aligned}
L_{\sigma 2}(x) &= \frac{(\sigma(x) - \sigma(x_0))(\sigma(x) - \sigma(x_1))}{(\sigma(x_2) - \sigma(x_0))(\sigma(x_2) - \sigma(x_1))} \\
&= \frac{(x + 1 - 0)(x + 1 - 2)}{(4 - 0)(4 - 2)} \\
&= \frac{(x - 1)(x + 1)}{8}, \quad x \in [-1, 4],
\end{aligned}
$$

$$f(x_0) = f(-1) = 1,$$
$$f(x_1) = f(1) = 3,$$
$$f(x_2) = f(3) = 13.$$

Hence,

$$p_{\sigma 2}(x) = f(x_0)L_{\sigma 0}(x) + f(x_1)L_{\sigma 1}(x) + f(x_2)L_{\sigma 2}(x)$$
$$= \frac{(x-1)(x-3)}{8} - 3\frac{(x-3)(x+1)}{4} + 13\frac{(x-1)(x+1)}{8}$$
$$= \frac{x^2 - 4x + 3 - 6x^2 + 12x + 18 + 13x^2 - 13}{8}$$
$$= x^2 + x + 1, \quad x \in [-1, 4],$$

is the σ-Lagrange interpolation polynomial for the function f.

Now, we will find the Lagrange interpolation polynomial for the function f with interpolation points x_0, x_1, x_2. We have

$$L_0(x) = \frac{(x-x_1)(x-x_2)}{(x_0-x_1)(x_0-x_2)} = \frac{(x-1)(x-3)}{8},$$
$$L_1(x) = \frac{(x-x_0)(x-x_2)}{(x_1-x_0)(x_1-x_2)} = -\frac{(x+1)(x-3)}{4},$$
$$L_2(x) = \frac{(x-x_0)(x-x_1)}{(x_2-x_0)(x_2-x_1)} = \frac{(x+1)(x-1)}{8}, \quad x \in [-1, 4].$$

Hence,

$$p_2(x) = f(x_0)L_0(x) + f(x_1)L_1(x) + f(x_2)L_2(x)$$
$$= \frac{(x-1)(x-3)}{8} - 3\frac{(x+1)(x-3)}{4} + 13\frac{(x+1)(x-1)}{8}$$
$$= \frac{x^2 - 4x + 3 - 6x^2 + 12x + 18 + 13x^2 - 13}{8}$$
$$= x^2 + x + 1, \quad x \in [-1, 4].$$

Moreover,

$$p_2(x) = p_{\sigma 2}(x) = f(x), \quad x \in [-1, 4].$$

Exercise 1.23. Let $\mathbb{T} = 2^{\mathbb{N}_0}$,

$$a = 1 = x_0, \quad x_1 = 2, \quad x_2 = 8, \quad x_3 = 32 = b,$$

and $f : \mathbb{T} \to \mathbb{R}$ be defined by

$$f(x) = x^3 + x^2 - x + 4, \quad x \in \mathbb{T}.$$

Find the σ-Lagrange interpolation polynomial for the function f with σ-interpolation points x_0, x_1, x_2.

Now, we will describe the classes of time scales for which the σ-Lagrange and Lagrange interpolation polynomials coincide.

Theorem 1.24. *Let $\mathbb{T}$ be a time scale such that $\sigma(t) = ct + d$ for any $t \in \mathbb{T}$ and some constants c, d. Let also, $n \in \mathbb{N}$ and*

$$a = x_0 < x_1 < \cdots < x_n = b, \quad x_j \in \mathbb{T}, \quad j \in \{0, 1, \ldots, n\},$$

be σ-interpolation and interpolation points. Then

$$L_k(x) = L_{\sigma k}(x), \quad x \in [a, b], \quad k \in \{0, 1, \ldots, n\}.$$

Proof. Since $\sigma(t) = ct + d$ for any $t \in \mathbb{T}$, we get

$$L_{\sigma k}(x) = \prod_{j=0, j \neq k}^{n} \frac{\sigma(x) - \sigma(x_j)}{\sigma(x_k) - \sigma(x_j)}$$

$$= \prod_{j=0, j \neq k}^{n} \frac{(cx + d) - (cx_j + d)}{(cx_k + d) - (cx_j + d)}$$

$$= \prod_{j=0, j \neq k}^{n} \frac{x - x_j}{x_k - x_j}$$

$$= L_k(x), \quad k \in \{0, 1, \ldots, n\}, \quad x \in [a, b].$$

This completes the proof. □

Remark 1.25. From Theorem 1.24, it is clear that the uniqueness of interpolation polynomial of degree n is not violated. Indeed, if σ is not a linear function, the σ-Lagrange interpolation polynomial of a function f is not a polynomial of degree n or not a polynomial at all. For instance, if $\mathbb{T} = \mathbb{N}_0^2$, then $\sigma(t) = (\sqrt{t} + 1)^2$, $t \in \mathbb{T}$, and the σ-Lagrange interpolation polynomial is not a polynomial. Therefore, for any function f, or any data set on a time scale $\mathbb{T}$, there is a unique interpolation polynomial of degree n.

Suppose that $n \in \mathbb{N}_0$ and $x_j \in \mathbb{T}, j \in \{0, 1, \ldots, n\}$, are σ-distinct points. For $x \in \mathbb{T}$, define the σ-polynomials

$$\pi_{\sigma n+1}(x) = \prod_{j=0}^{n}(\sigma(x) - \sigma(x_j)), \quad \Pi_{\sigma n+1}^{k}(x) = \pi_{\sigma n+1}^{\Delta^k}(x), \quad x \in \mathbb{T}, \quad k \in \mathbb{N}_0.$$

These functions are employed in the error estimate for the σ-Lagrange interpolation.

Example 1.26. Let $\mathbb{T} = \{0, \frac{1}{6}, \frac{1}{3}, \frac{1}{2}, 1 + \frac{1}{2^n} : n \in \mathbb{N}_0\}$,

$$a = 0, \quad x_0 = 0, \quad x_1 = \frac{1}{3}, \quad b = 1.$$

We will compute $\pi_{\sigma 2}(x)$, $\pi_{\sigma 2}(\frac{1}{2})$, and $\Pi_{\sigma 2}^{1}(\frac{1}{2})$.
We have

$$\sigma(x_0) = \frac{1}{6}, \quad \sigma(x_1) = \frac{1}{2},$$

$$\sigma\left(\frac{1}{2}\right) = 1, \quad \sigma(1) = 1.$$

Hence,

$$\pi_{\sigma 2}(x) = (\sigma(x) - \sigma(x_0))(\sigma(x) - \sigma(x_1)) = \left(\sigma(x) - \frac{1}{6}\right)\left(\sigma(x) - \frac{1}{2}\right), \quad x \in \mathbb{T},$$

$$\pi_{\sigma 2}\left(\frac{1}{2}\right) = \left(\sigma\left(\frac{1}{2}\right) - \frac{1}{6}\right)\left(\sigma\left(\frac{1}{2}\right) - \frac{1}{2}\right) = \left(1 - \frac{1}{6}\right)\left(1 - \frac{1}{2}\right) = \frac{5}{12},$$

$$\Pi_{\sigma 2}^{1}\left(\frac{1}{2}\right) = \pi_{\sigma 2}^{\Delta}\left(\frac{1}{2}\right)$$

$$= \frac{(\sigma^2(\frac{1}{2}) - \frac{1}{6})(\sigma^2(\frac{1}{2}) - \frac{1}{2}) - (\sigma(\frac{1}{2}) - \frac{1}{6})(\sigma(\frac{1}{2}) - \frac{1}{2})}{\sigma(\frac{1}{2}) - \frac{1}{2}}$$

$$= \frac{(\sigma(1) - \frac{1}{6})(\sigma(1) - \frac{1}{2}) - (\sigma(\frac{1}{2}) - \frac{1}{6})(\sigma(\frac{1}{2}) - \frac{1}{2})}{\sigma(\frac{1}{2}) - \frac{1}{2}}$$

$$= \frac{(1 - \frac{1}{6})(1 - \frac{1}{2}) - (1 - \frac{1}{6})(1 - \frac{1}{2})}{1 - \frac{1}{2}} = 0.$$

Exercise 1.27. Let $\mathbb{T} = \{0, 2, 3, 5, 9, 16, 18\}$,

$$a = 0, \quad x_0 = 0, \quad x_1 = 3, \quad x_2 = 9, \quad b = 18.$$

Find $\Pi_{\sigma 2}^{2}(3)$.

The error in the σ-Lagrange interpolation is given in the next theorem.

Theorem 1.28. *Suppose that $n \in \mathbb{N}_0$, $a, b \in \mathbb{T}$, $a < b$, $x_j \in [a, b]$, $j \in \{0, 1, \ldots, n\}$, are σ-distinct and $f : [a, b] \to \mathbb{R}$, $f^{\Delta^k}(x)$ exist for any $x \in [a, b]$ and for any $k \in \{1, \ldots, n+1\}$. Then for any $x \in [a, b]$ there exists $\xi = \xi(x) \in (a, b)$ such that*

$$f(x) - p_{\sigma n}(x) = \frac{f^{\Delta^{n+1}}(\xi)}{\Pi_{\sigma n+1}^{n+1}(\xi)} \pi_{\sigma n+1}(x), \quad x \in [a, b],$$

or

$$F_{\sigma \min, n+1}(\xi) \leq \frac{f(x) - p_{\sigma n}(x)}{\pi_{\sigma n+1}(x)} \leq F_{\sigma \max, n+1}(\xi), \quad x \in [a, b],$$

where

$$F_{\sigma \max, n+1}(\xi) = \max\left\{ \frac{f^{\Delta^{n+1}}(\xi)}{\Pi_{\sigma n+1}^{n+1}(\xi)}, \frac{f^{\Delta^{n+1}}(\rho(\xi))}{\Pi_{\sigma n+1}^{n+1}(\rho(\xi))} \right\},$$

$$F_{\sigma \min, n+1}(\xi) = \min\left\{ \frac{f^{\Delta^{n+1}}(\xi)}{\Pi_{\sigma n+1}^{n+1}(\xi)}, \frac{f^{\Delta^{n+1}}(\rho(\xi))}{\Pi_{\sigma n+1}^{n+1}(\rho(\xi))} \right\}.$$

Proof. Let $p_{\sigma n}$ be the σ-Lagrange interpolation polynomial for the function f with σ-interpolation points x_j, $j \in \{0, 1, \ldots, n\}$. Define the function

$$\phi(t) = f(t) - p_{\sigma n}(t) - \frac{f(x) - p_{\sigma n}(x)}{\pi_{\sigma n+1}(x)} \pi_{\sigma n+1}(t), \quad t \in [a, b].$$

From here, the proof repeats that of Theorem 1.10 and we omit it. □

Remark 1.29. Suppose that all conditions of Theorem 1.28 hold. If

$$\lim_{n \to \infty} \max_{x \in [a, b]} \left(\frac{f^{\Delta^{n+1}}(\xi)}{\Pi_{\sigma n+1}^{n+1}(\xi)} \pi_{\sigma n+1}(x) \right) = 0$$

and

$$\lim_{n \to \infty} \max_{x \in [a, b]} \left(\frac{f^{\Delta^{n+1}}(\rho(\xi))}{\Pi_{\sigma n+1}^{n+1}(\rho(\xi))} \pi_{\sigma n+1}(x) \right) = 0,$$

then

$$\lim_{n \to \infty} \max_{x \in [a, b]} |f(x) - p_{\sigma n}(x)| = 0.$$

The last example clearly shows the difference between Lagrange and σ-Lagrange polynomials on a time scale whose forward jump operator is not a linear function. The computed polynomials are also compared graphically.

Example 1.30. Let $\mathbb{T} = \{\sqrt{2n+1}, n \in \mathbb{N}_0\} = \{1, \sqrt{3}, \sqrt{5}, \ldots\}$. Take

$$a = x_0 = \sqrt{3}, \quad x_1 = \sqrt{7}, \quad x_2 = \sqrt{11}, \quad x_3 = \sqrt{17}, \quad b = \sqrt{23},$$

and

$$f(x) = \frac{x^2 + 1}{2x^2 + 5}, \quad x \in \mathbb{T}.$$

We will construct the Lagrange interpolation polynomial p_3 and the σ-Lagrange interpolation polynomial $p_{\sigma 3}$ interpolating the function f at the points x_0, x_1, x_2, x_3. We will also compare the graphs of the two types interpolating polynomials for the function f. Notice that, on the given time scale we have

$$\sigma(x) = \sqrt{x^2 + 2}, \quad x \in \mathbb{T}.$$

Also,

$$f(x_0) = f(\sqrt{3}) = \frac{3+1}{6+5} = \frac{4}{11},$$

$$f(x_1) = f(\sqrt{7}) = \frac{7+1}{14+5} = \frac{8}{19},$$

$$f(x_2) = f(\sqrt{11}) = \frac{11+1}{22+5} = \frac{4}{9},$$

$$f(x_3) = f(\sqrt{17}) = \frac{17+1}{34+5} = \frac{6}{13}.$$

First, we compute $L_0(x)$, $L_1(x)$, $L_2(x)$, and $L_3(x)$ as follows:

$$L_0(x) = \frac{(x - \sqrt{7})(x - \sqrt{11})(x - \sqrt{17})}{(\sqrt{3} - \sqrt{7})(\sqrt{3} - \sqrt{11})(\sqrt{3} - \sqrt{17})},$$

$$L_1(x) = \frac{(x - \sqrt{3})(x - \sqrt{11})(x - \sqrt{17})}{(\sqrt{7} - \sqrt{3})(\sqrt{7} - \sqrt{11})(\sqrt{7} - \sqrt{17})},$$

$$L_2(x) = \frac{(x - \sqrt{3})(x - \sqrt{7})(x - \sqrt{17})}{(\sqrt{11} - \sqrt{3})(\sqrt{11} - \sqrt{7})(\sqrt{11} - \sqrt{17})},$$

$$L_3(x) = \frac{(x - \sqrt{3})(x - \sqrt{7})(x - \sqrt{11})}{(\sqrt{17} - \sqrt{3})(\sqrt{17} - \sqrt{7})(\sqrt{17} - \sqrt{11})}, \quad x \in \mathbb{T}.$$

Then the Lagrange interpolation polynomial p_3 interpolating the function f at the points $\sqrt{3}, \sqrt{7}, \sqrt{11}, \sqrt{17}$ is obtained as

$$p_3(x) = f(\sqrt{3})L_0(x) + f(\sqrt{7})L_1(x) + f(\sqrt{11})L_2(x) + f(\sqrt{17})L_3(x)$$

$$= \frac{4}{11} \frac{(x - \sqrt{7})(x - \sqrt{11})(x - \sqrt{17})}{(\sqrt{3} - \sqrt{7})(\sqrt{3} - \sqrt{11})(\sqrt{3} - \sqrt{17})}$$

$$+ \frac{8}{19}\frac{(x-\sqrt{3})(x-\sqrt{11})(x-\sqrt{17})}{(\sqrt{7}-\sqrt{3})(\sqrt{7}-\sqrt{11})(\sqrt{7}-\sqrt{17})}$$

$$+ \frac{4}{9}\frac{(x-\sqrt{3})(x-\sqrt{7})(x-\sqrt{17})}{(\sqrt{11}-\sqrt{3})(\sqrt{11}-\sqrt{7})(\sqrt{11}-\sqrt{17})}$$

$$+ \frac{6}{13}\frac{(x-\sqrt{3})(x-\sqrt{7})(x-\sqrt{11})}{(\sqrt{17}-\sqrt{3})(\sqrt{17}-\sqrt{7})(\sqrt{17}-\sqrt{11})}, \quad x \in \mathbb{T}.$$

Next, we compute $L_{\sigma 0}(x)$, $L_{\sigma 1}(x)$, $L_{\sigma 2}(x)$, and $L_{\sigma 3}(x)$ as follows:

$$L_{\sigma 0}(x) = \frac{(\sigma(x)-\sigma(\sqrt{7}))(\sigma(x)-\sigma(\sqrt{11}))(\sigma(x)-\sigma(\sqrt{17}))}{(\sigma(\sqrt{3})-\sigma(\sqrt{7}))(\sigma(\sqrt{3})-\sigma(\sqrt{11}))(\sigma(\sqrt{3})-\sigma(\sqrt{17}))}$$

$$= \frac{(\sqrt{x^2+2}-3)(\sqrt{x^2+2}-\sqrt{13})(\sqrt{x^2+2}-\sqrt{19})}{(\sqrt{5}-3)(\sqrt{5}-\sqrt{13})(\sqrt{5}-\sqrt{19})},$$

$$L_{\sigma 1}(x) = \frac{(\sigma(x)-\sigma(\sqrt{3}))(\sigma(x)-\sigma(\sqrt{11}))(\sigma(x)-\sigma(\sqrt{17}))}{(\sigma(\sqrt{7})-\sigma(\sqrt{3}))(\sigma(\sqrt{7})-\sigma(\sqrt{11}))(\sigma(\sqrt{7})-\sigma(\sqrt{17}))}$$

$$= \frac{(\sqrt{x^2+2}-\sqrt{5})(\sqrt{x^2+2}-\sqrt{13})(\sqrt{x^2+2}-\sqrt{19})}{(3-\sqrt{5})(3-\sqrt{13})(3-\sqrt{19})},$$

$$L_{\sigma 2}(x) = \frac{(\sigma(x)-\sigma(\sqrt{3}))(\sigma(x)-\sigma(\sqrt{7}))(\sigma(x)-\sigma(\sqrt{17}))}{(\sigma(\sqrt{11})-\sigma(\sqrt{3}))(\sigma(\sqrt{11})-\sigma(\sqrt{7}))(\sigma(\sqrt{11})-\sigma(\sqrt{17}))}$$

$$= \frac{(\sqrt{x^2+2}-\sqrt{5})(\sqrt{x^2+2}-3)(\sqrt{x^2+2}-\sqrt{19})}{(\sqrt{13}-\sqrt{5})(\sqrt{13}-3)(\sqrt{13}-\sqrt{19})},$$

$$L_{\sigma 3}(x) = \frac{(\sigma(x)-\sigma(\sqrt{3}))(\sigma(x)-\sigma(\sqrt{7}))(\sigma(x)-\sigma(\sqrt{11}))}{(\sigma(\sqrt{17})-\sigma(\sqrt{3}))(\sigma(\sqrt{17})-\sigma(\sqrt{7}))(\sigma(\sqrt{17})-\sigma(\sqrt{11}))}$$

$$= \frac{(\sqrt{x^2+2}-\sqrt{5})(\sqrt{x^2+2}-3)(\sqrt{x^2+2}-\sqrt{13})}{(\sqrt{19}-\sqrt{5})(\sqrt{19}-3)(\sqrt{19}-\sqrt{13})}, \quad x \in \mathbb{T}.$$

Hence, the σ-Lagrange polynomial $p_{\sigma 3}$ interpolating the function f at the points $\sqrt{3}, \sqrt{7}, \sqrt{11}, \sqrt{17}$ is obtained as

$$p_{\sigma 3}(x) = f(\sqrt{3})L_{\sigma 0}(x) + f(\sqrt{7})L_{\sigma 1}(x) + f(\sqrt{11})L_{\sigma 2}(x) + f(\sqrt{17})L_{\sigma 3}(x)$$

$$= \frac{4}{11}\frac{(\sqrt{x^2+2}-3)(\sqrt{x^2+2}-\sqrt{13})(\sqrt{x^2+2}-\sqrt{19})}{(\sqrt{5}-3)(\sqrt{5}-\sqrt{13})(\sqrt{5}-\sqrt{19})}$$

$$+ \frac{8}{19}\frac{(\sqrt{x^2+2}-\sqrt{5})(\sqrt{x^2+2}-\sqrt{13})(\sqrt{x^2+2}-\sqrt{19})}{(3-\sqrt{5})(3-\sqrt{13})(3-\sqrt{19})}$$

$$+ \frac{4}{9}\frac{(\sqrt{x^2+2}-\sqrt{5})(\sqrt{x^2+2}-3)(\sqrt{x^2+2}-\sqrt{19})}{(\sqrt{13}-\sqrt{5})(\sqrt{13}-3)(\sqrt{13}-\sqrt{19})}$$

$$+ \frac{6}{13}\frac{(\sqrt{x^2+2}-\sqrt{5})(\sqrt{x^2+2}-3)(\sqrt{x^2+2}-\sqrt{13})}{(\sqrt{19}-\sqrt{5})(\sqrt{19}-3)(\sqrt{19}-\sqrt{13})}, \quad x \in \mathbb{T}.$$

It is clear that $p_3(x) \neq p_{\sigma 3}(x), x \in \mathbb{T}$. Moreover, $p_{\sigma 3}$ is not a polynomial. Figure 1.2 shows the graphs of f and p_3 and Figure 1.3 presents the graphs of f and $p_{\sigma 3}$. In addition, we compare the graphs of f, p_3, and $p_{\sigma 3}$ in Figure 1.4. The values of f, p_3, and $p_{\sigma 3}$ at some noninterpolating points are compared in Table 1.2.

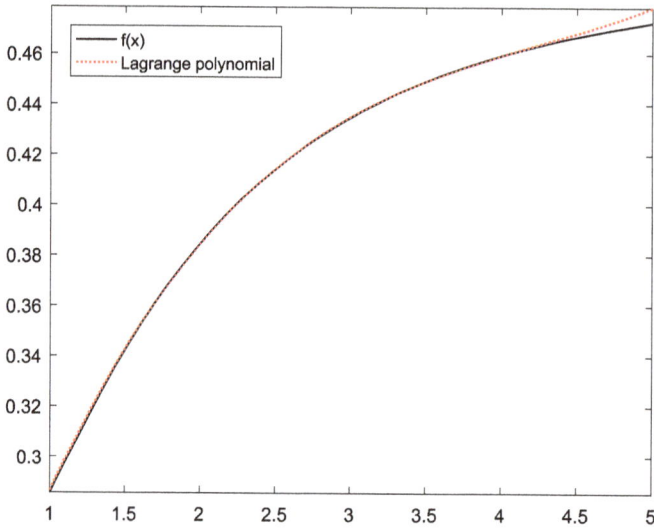

Figure 1.2: The graphs of f and the Lagrange polynomial p_3.

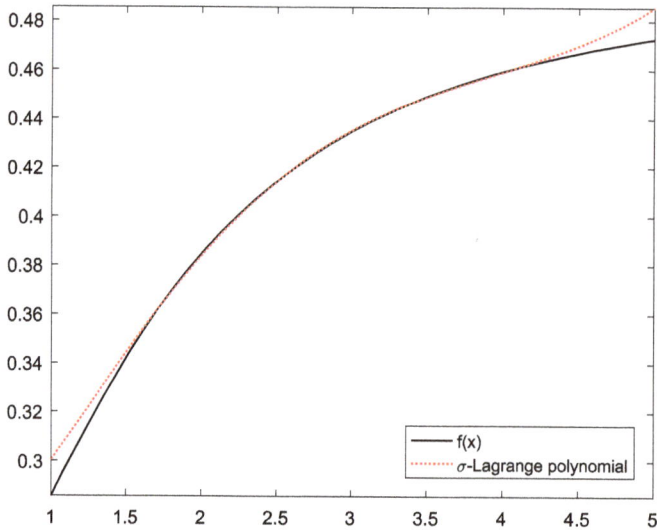

Figure 1.3: The graphs of f and the σ-Lagrange polynomial $p_{\sigma 3}$.

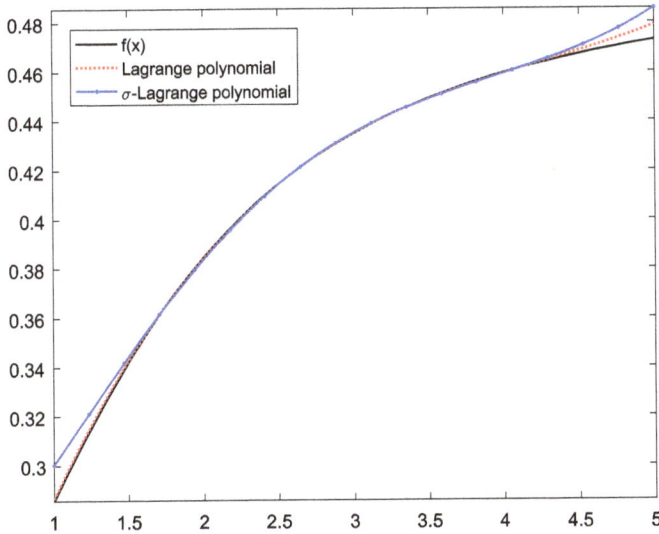

Figure 1.4: The graphs of the Lagrange polynomial p_3 and σ-Lagrange polynomial $p_{\sigma3}$.

Table 1.2: The values of f, p_3, and $p_{\sigma3}$ at some noninterpolating points of $\mathbb{T}$.

x	$f(x)$	$p_3(x)$	$p_{\sigma3}(x)$
$\sqrt{5}$	0.4000	0.3997	0.3993
3	0.4348	0.4349	0.4350
$\sqrt{15}$	0.4571	0.4569	0.4567
$\sqrt{19}$	0.4651	0.4657	0.4665
$\sqrt{21}$	0.4681	0.4699	0.4719

1.3 Hermite interpolation

The idea of Lagrange interpolation and σ-Lagrange interpolation can be generalized in various ways. Here, in this section, we consider one simple extension where a polynomial p is required to take given values and delta derivative values at the interpolation points. For given σ-distinct points x_j, $j \in \{0, 1, \ldots, n\}$, and two sets of real numbers y_j, z_j, $j \in \{0, 1, \ldots, n\}$, $n \in \mathbb{N}_0$, we need to find a polynomial $p_{2n+1} \in \mathcal{P}_{2n+1}$ satisfying the conditions

$$p_{2n+1}(x_j) = y_j, \quad p_{2n+1}^{\Delta}(x_j) = z_j, \quad j \in \{0, 1, \ldots, n\}.$$

The construction of such a polynomial is similar to that of the Lagrange interpolation polynomial and it is given in the following theorem.

Theorem 1.31 (Hermite interpolation theorem). *Let $n \in \mathbb{N}_0$ and let $a, b \in \mathbb{T}$, $a < b$, and $x_j \in [a, b] \subset \mathbb{T}$, $j \in \{0, 1, \ldots, n\}$, be σ-distinct and $x_j \neq \sigma(x_k)$ for all $j, k \in \{0, 1, \ldots, n\}$. Let* *also $y_j, z_j \in \mathbb{R}, j \in \{0, 1, \ldots, n\}$. Then there exists a unique polynomial $p_{2n+1} \in \mathcal{P}_{2n+1}$ such that*

$$p_{2n+1}(x_j) = y_j, \quad p_{2n+1}^{\Delta}(x_j) = z_j, \quad j \in \{0, 1, \ldots, n\}. \tag{1.2}$$

Proof. For $x \in [a, b]$, define the polynomial

$$M_k(x) = \prod_{j=0, j \neq k}^{n} \frac{x - \sigma(x_j)}{x_k - \sigma(x_j)}, \quad k \in \{0, 1, \ldots, n\}.$$

We have

$$M_k(\sigma(x_j)) = 0, \quad M_k(x_k) = 1, \quad j, k \in \{0, 1, \ldots, n\}, \quad j \neq k.$$

We will search for a polynomial $p_{2n+1} \in \mathcal{P}_{2n+1}$ in the following form:

$$p_{2n+1}(x) = \sum_{j=0}^{n} (y_j + (x - x_j)(\alpha_j y_j + \beta_j z_j)) M_j(x) L_j(x), \quad x \in [a, b],$$

where $\alpha_j, \beta_j \in \mathbb{R}, j \in \{0, 1, \ldots, n\}$, will be determined by conditions (1.2). We have

$$p_{2n+1}(x_k) = \sum_{j=0}^{n} (y_j + (x_k - x_j)(\alpha_j y_j + \beta_j z_j)) M_j(x_k) L_j(x_k) = y_k,$$

$$p_{2n+1}^{\Delta}(x) = \sum_{j=0}^{n} (\alpha_j y_j + \beta_j z_j) M_j(\sigma(x)) L_j(\sigma(x))$$

$$+ \sum_{j=0}^{n} (y_j + (x - x_j)(\alpha_j y_j + \beta_j z_j))(M_j^{\Delta}(x) L_j(x) + M_j(\sigma(x)) L_j^{\Delta}(x)),$$

$$p_{2n+1}^{\Delta}(x_k) = \sum_{j=0}^{n} (\alpha_j y_j + \beta_j z_j) M_j(\sigma(x_k)) L_j(\sigma(x_k))$$

$$+ \sum_{j=0}^{n} (y_j + (x_k - x_j)(\alpha_j y_j + \beta_j z_j))(M_j^{\Delta}(x_k) L_j(x_k) + M_j(\sigma(x_k)) L_j^{\Delta}(x_k))$$

$$= (\alpha_k y_k + \beta_k z_k) M_k(\sigma(x_k)) L_k(\sigma(x_k)) + y_k (M_k^{\Delta}(x_k) + M_k(\sigma(x_k)) L_k^{\Delta}(x_k))$$

$$= z_k$$

or, equivalently,

$$(\alpha_k y_k + \beta_k z_k) M_k(\sigma(x_k)) L_k(\sigma(x_k)) = z_k - y_k (M_k^{\Delta}(x_k) + M_k(\sigma(x_k)) L_k^{\Delta}(x_k)),$$

or

$$\alpha_k y_k + \beta_k z_k = \frac{z_k - y_k (M_k^{\Delta}(x_k) + M_k(\sigma(x_k)) L_k^{\Delta}(x_k))}{M_k(\sigma(x_k)) L_k(\sigma(x_k))},$$

and

$$
p_{2n+1}(x) = \sum_{j=0}^{n} \left(y_j + \frac{z_j - y_j(M_j^{\Delta}(x_j) + M_j(\sigma(x_j))L_j^{\Delta}(x_j))}{M_j(\sigma(x_j))L_j(\sigma(x_j))} (x - x_j) \right) M_j(x)L_j(x)
$$

$$
= \sum_{j=0}^{n} \left(\left(1 - \frac{M_j^{\Delta}(x_j) + M_j(\sigma(x_j))L_j^{\Delta}(x_j)}{M_j(\sigma(x_j))L_j(\sigma(x_j))} (x - x_j) \right) y_j \right.
$$

$$
\left. + \frac{z_j}{M_j(\sigma(x_j))L_j(\sigma(x_j))} (x - x_j) \right) M_j(x)L_j(x), \quad x \in [a, b].
$$

Now, suppose that there are two polynomials such that $p_{2n+1}, q_{2n+1} \in \mathcal{P}_{2n+1}$ and

$$
p_{2n+1}(x_k) = q_{2n+1}(x_k) = y_k, \quad p_{2n+1}^{\Delta}(x_k) = q_{2n+1}^{\Delta}(x_k) = z_k, \quad k \in \{0, 1, \ldots, n\}.
$$

Let $h_{2n+1} = p_{2n+1} - q_{2n+1}$. Then $h_{2n+1} \in \mathcal{P}_{2n+1}$ and it has at least $2n + 2$ GZs. Thus,

$$
h_{2n+1} \equiv 0 \quad \text{or} \quad p_{2n+1} \equiv q_{2n+1} \quad \text{on } [a, b].
$$

This completes the proof. □

Next, we give the definition and the general structure of a Hermite interpolation polynomial.

Definition 1.32. Let $n \in \mathbb{N}_0$ and let $a, b \in \mathbb{T}$, $a < b$, and $x_j \in [a, b] \subset \mathbb{T}, j \in \{0, 1, \ldots, n\}$, be σ-distinct. Let also $y_j, z_j \in \mathbb{R}, j \in \{0, 1, \ldots, n\}$. Then the polynomial

$$
p_{2n+1}(x) = \sum_{j=0}^{n} \left(\left(1 - \frac{M_j^{\Delta}(x_j) + M_j(\sigma(x_j))L_j^{\Delta}(x_j)}{M_j(\sigma(x_j))L_j(\sigma(x_j))} (x - x_j) \right) y_j \right.
$$

$$
\left. + \frac{z_j}{M_j(\sigma(x_j))L_j(\sigma(x_j))} (x - x_j) \right) M_j(x)L_j(x), \quad x \in [a, b],
$$

is called the Hermite interpolation polynomial for the set of values given in

$$
\{(x_j, y_j, z_j) : j \in \{0, 1, \ldots, n\}\}.
$$

In the next remark, we prove that the Hermite interpolation polynomial given in Definition 1.32 reduces to the classical Hermite interpolation polynomial whenever the time scale is the set of real numbers.

Remark 1.33. If $\mathbb{T} = \mathbb{R}$, then

$$
M_j^{\Delta}(x_j) = M_j'(x_j) = L_j'(x_j),
$$

$$
L_j^{\Delta}(x_j) = L_j'(x_j),
$$

$$
M_j(\sigma(x_j)) = L_j(\sigma(x_j)) = M_j(x_j) = L_j(x_j) = 1, \quad j \in \{0, 1, \ldots, n\}.
$$

Hence,

$$p_{2n+1}(x) = \sum_{j=0}^{n}\left((1 - 2L_j'(x_j)(x - x_j))y_j + z_j(x - x_j))(L_j(x))^2\right), \quad x \in [a, b].$$

Thus, we get the classical Hermite interpolation polynomial.

Example 1.34. Let $\mathbb{T} = 2^{\mathbb{N}_0}$,

$$x_0 = 1, \quad x_1 = 4, \quad y_0 = 1, \quad y_1 = -1, \quad z_0 = 1, \quad z_1 = 2.$$

Here $n = 1$, $\sigma(x) = 2x$, $x \in \mathbb{T}$. Then

$$L_0(x) = \frac{x - x_1}{x_0 - x_1} = -\frac{1}{3}(x - 4), \quad L_0^\Delta(x) = -\frac{1}{3}, \quad L_0(\sigma(x_0)) = L_0(2) = \frac{2}{3},$$

$$L_1(x) = \frac{x - x_0}{x_1 - x_0} = \frac{1}{3}(x - 1), \quad L_1^\Delta(x) = \frac{1}{3}, \quad L_1(\sigma(x_1)) = L_1(8) = \frac{7}{3},$$

$$M_0(x) = \frac{x - \sigma(x_1)}{x_0 - \sigma(x_1)} = -\frac{1}{7}(x - 8), \quad M_0^\Delta(x) = -\frac{1}{7}, \quad M_0(\sigma(x_0)) = M_0(2) = \frac{6}{7},$$

$$M_1(x) = \frac{x - \sigma(x_0)}{x_1 - \sigma(x_0)} = \frac{1}{2}(x - 2), \quad M_1^\Delta(x) = \frac{1}{2}, \quad M_1(\sigma(x_1)) = M_1(8) = 3, \quad x \in \mathbb{T}.$$

Hence,

$$p_3(x) = \left(y_0 + \frac{z_0 - y_0(M_0^\Delta(x_0) + M_0(\sigma(x_0))L_0^\Delta(x_0))}{M_0(\sigma(x_0))L_0(\sigma(x_0))}(x - x_0)\right)M_0(x)L_0(x)$$

$$+ \left(y_1 + \frac{z_1 - y_1(M_1^\Delta(x_1) + M_1(\sigma(x_1))L_1^\Delta(x_1))}{M_1(\sigma(x_1))L_1(\sigma(x_1))}(x - x_1)\right)M_1(x)L_1(x)$$

$$= \left(1 + \frac{1 - (-\frac{1}{7} + \frac{6}{7}(-\frac{1}{3}))}{\frac{6}{7}\cdot\frac{2}{3}}(x - 1)\right)\left(-\frac{1}{7}(x - 8)\right)\left(-\frac{1}{3}(x - 4)\right)$$

$$+ \left(-1 + \frac{2 + (\frac{1}{2} + 3\cdot\frac{1}{3})}{3\cdot\frac{7}{3}}(x - 4)\right)\left(\frac{1}{2}(x - 2)\right)\left(\frac{1}{3}(x - 1)\right)$$

$$= \frac{1}{21}\left(1 + \frac{5}{2}(x - 1)\right)(x - 4)(x - 8) + \frac{1}{6}\left(-1 + \frac{1}{2}(x - 4)\right)(x - 1)(x - 2)$$

$$= \frac{1}{42}(5x - 3)(x - 4)(x - 8) + \frac{1}{12}(x - 6)(x - 1)(x - 2), \quad x \in [1, 8].$$

Exercise 1.35. Let $\mathbb{T} = 2\mathbb{Z}$. Find the Hermite interpolation polynomial for the set

$$x_0 = -4, \quad x_1 = 0, \quad x_2 = 4,$$
$$y_0 = -1, \quad y_1 = 1, \quad y_2 = -1,$$
$$z_0 = 1, \quad z_1 = -1, \quad z_2 = 1.$$

In some applications, instead of a data set of points, function values, and delta derivative values at these points, a function itself may be given. The Hermite polynomial for this function, that is, the polynomial which takes the same values as the function and whose delta derivative takes the values of the delta derivative of the function at given points, is defined below.

Definition 1.36. Let $n \in \mathbb{N}_0$ and let $a, b \in \mathbb{T}$, $a < b$, and $x_j \in [a, b] \subset \mathbb{T}$, $j \in \{0, 1, \ldots, n\}$, be σ-distinct. Let also, $f : [a, b] \to \mathbb{R}$ be delta differentiable on $[a, b]$. Then the polynomial

$$p_{2n+1}(x) = \sum_{j=0}^{n} \left(\left(1 - \frac{M_j^{\Delta}(x_j) + M_j(\sigma(x_j))L_j^{\Delta}(x_j)}{M_j(\sigma(x_j))L_j(\sigma(x_j))}(x - x_j) \right) f(x_j) \right.$$

$$\left. + \frac{f^{\Delta}(x_j)}{M_j(\sigma(x_j))L_j(\sigma(x_j))}(x - x_j) \right) M_j(x)L_j(x), \quad x \in [a, b],$$

is called the Hermite interpolation polynomial for the function f.

Exercise 1.37. Let $\mathbb{T} = 3^{\mathbb{N}_0}$ and $f : \mathbb{T} \to \mathbb{R}$ be given by

$$f(x) = x^3 + 3x^2 + e_2(x, 1), \quad x \in \mathbb{T}.$$

Let also,

$$x_0 = 1, \quad x_1 = 9, \quad x_2 = 81.$$

Find the Hermite interpolation polynomial for the function f.

Let $n \in \mathbb{N}_0$ and $a, b \in \mathbb{T}$, $a < b$, and $x_j \in \mathbb{T}$, $j \in \{0, 1, \ldots, n\}$, be σ-distinct. Define the polynomials

$$\zeta_n(x) = \prod_{j=0}^{n} (x - \sigma(x_j)), \quad x \in [a, b].$$

The error in the Hermite interpolation is given in the following theorem.

Theorem 1.38. *Suppose that $n \in \mathbb{N}_0$, $a, b \in \mathbb{T}$, $a < b$, $x_j \in [a, b]$, $j \in \{0, 1, \ldots, n\}$, are σ-distinct and $f : [a, b] \to \mathbb{R}$, $f^{\Delta^k}(x)$ exist for any $x \in [a, b]$ and for any $k \in \{1, \ldots, 2n + 2\}$. Then for any $x \in [a, b]$ there exists $\xi = \xi(x) \in (a, b)$ such that*

$$f(x) - p_{2n+1}(x) = \frac{f^{\Delta^{2n+2}}(\xi)}{\pi_{n+1}(x)\zeta_{n+1}(x)}(\pi_{n+1}\zeta_{n+1})^{\Delta^{2n+2}}(\xi), \quad x \in [a, b],$$

or

$$G_{\min,2n+2}(\xi) \leq \frac{f(x) - p_{2n+1}(x)}{\pi_{n+1}(x)\zeta_{n+1}(x)} \leq G_{\max,2n+2}(\xi), \quad x \in [a, b],$$

where

$$G_{\max,2n+2}(\xi) = \max\left\{\frac{f^{\Delta^{2n+2}}(\xi)}{(\pi_{n+1}\zeta_{n+1})^{\Delta^{2n+2}}(\xi)}, \frac{f^{\Delta^{2n+2}}(\rho(\xi))}{(\pi_{n+1}\zeta_{n+1})^{\Delta^{2n+2}}(\rho(\xi))}\right\},$$

$$G_{\min,2n+2}(\xi) = \min\left\{\frac{f^{\Delta^{2n+2}}(\xi)}{(\pi_{n+1}\zeta_{n+1})^{\Delta^{2n+2}}(\xi)}, \frac{f^{\Delta^{2n+2}}(\rho(\xi))}{(\pi_{n+1}\zeta_{n+1})^{\Delta^{2n+2}}(\rho(\xi))}\right\}.$$

Proof. Let p_{2n+1} be the Hermite interpolation polynomial for the function f with interpolation points x_j, $j \in \{0, 1, \ldots, n\}$. Define the function

$$\psi(t) = f(t) - p_{2n+1}(t) - \frac{f(x) - p_{2n+1}(x)}{\pi_{n+1}(x)\zeta_{n+1}(x)}\pi_{n+1}(t)\zeta_{n+1}(t), \quad t \in [a, b].$$

Then

$$\psi(x_j) = f(x_j) - p_{2n+1}(x_j) - \frac{f(x) - p_{2n+1}(x)}{\pi_{n+1}(x)\zeta_{n+1}(x)}\pi_{n+1}(x_j)\zeta_{n+1}(x_j)$$

$$= f(x_j) - f(x_j)$$

$$= 0, \quad x \in [a, b], \quad j \in \{0, 1, \ldots, n\},$$

and $\psi(x) = 0$, $x \in [a, b]$. Thus, $\psi : [a, b] \to \mathbb{R}$ has at least $n + 2$ GZs. Hence, by Rolle's theorem (see Theorem A.6 in Appendix A), it follows that ψ^Δ has at least $n + 1$ GZs on (a, b) that do not coincide with x_j, $j \in \{0, 1, \ldots, n\}$. Next,

$$\psi^\Delta(t) = f^\Delta(t) - p_{2n+1}^\Delta(t) - \frac{f(x) - p_{2n+1}(x)}{\pi_{n+1}(x)\zeta_{n+1}(x)}(\pi_{n+1}\zeta_{n+1})^\Delta(t)$$

$$= f^\Delta(t) - p_{2n+1}^\Delta(t)$$

$$- \frac{f(x) - p_{2n+1}(x)}{\pi_{n+1}(x)\zeta_{n+1}(x)}(\pi_{n+1}^\Delta(t)\zeta_{n+1}(\sigma(t)) + \pi_{n+1}(t)\zeta_{n+1}^\Delta(t)), \quad t \in [a, b].$$

Hence,

$$\psi^\Delta(x_j) = f^\Delta(x_j) - p_{2n+1}^\Delta(x_j)$$

$$- \frac{f(x) - p_{2n+1}(x)}{\pi_{n+1}(x)\zeta_{n+1}(x)}(\pi_{n+1}^\Delta(x_j)\zeta_{n+1}(\sigma(x_j)) + \pi_{n+1}(x_j)\zeta_{n+1}^\Delta(x_j))$$

$$= 0, \quad j \in \{0, 1, \ldots, n\},$$

i. e., ψ^Δ has at least $n + 1$ GZs at x_j, $j \in \{0, 1, \ldots, n\}$. Therefore, ψ^Δ has at least $2n + 2$ GZs in $[a, b]$. Then $\psi^{\Delta^{2n+2}}$ has at least one GZ in (a, b) and there exists a $\xi = \xi(x) \in (a, b)$ such that

$$\psi^{\Delta^{2n+2}}(\xi) = 0 \quad \text{or} \quad \psi^{\Delta^{2n+2}}(\rho(\xi))\psi^{\Delta^{2n+2}}(\xi) < 0.$$

Observe that

$$\psi^{\Delta^{2n+2}}(t) = f^{\Delta^{2n+2}}(t) - \frac{f(x) - p_{2n+1}(x)}{\pi_{n+1}(x)\zeta_{n+1}(x)}(\pi_{n+1}\zeta_{n+1})^{\Delta^{2n+2}}(t), \quad t \in [a, b].$$

Now we consider each case separately.

1. Let $\psi^{\Delta^{2n+2}}(\xi) = 0$. Then

$$f^{\Delta^{2n+2}}(\xi) = \frac{f(x) - p_{2n+1}(x)}{\pi_{n+1}(x)\zeta_{n+1}(x)}(\pi_{n+1}\zeta_{n+1})^{\Delta^{2n+2}}(\xi), \quad x \in [a, b],$$

or

$$f(x) - p_{2n+1}(x) = \frac{f^{\Delta^{2n+2}}(\xi)}{(\pi_{n+1}\zeta_{n+1})^{\Delta^{2n+2}}(\xi)}\pi_{n+1}(x)\zeta_{n+1}(x), \quad x \in [a, b].$$

2. Let

$$\psi^{\Delta^{2n+2}}(\rho(\xi))\psi^{\Delta^{2n+2}}(\xi) < 0.$$

Then

$$\psi^{\Delta^{2n+2}}(\rho(\xi)) = f^{\Delta^{2n+2}}(\rho(\xi)) - \frac{f(x) - p_{2n+1}(x)}{\pi_{n+1}(x)\zeta_{n+1}(x)}(\pi_{n+1}\zeta_{n+1})^{\Delta^{2n+2}}(\rho(\xi)), \quad x \in [a, b],$$

and

$$\psi^{\Delta^{2n+2}}(\xi) = f^{\Delta^{2n+2}}(\xi) - \frac{f(x) - p_{2n+1}(x)}{\pi_{n+1}(x)\zeta_{n+1}(x)}(\pi_{n+1}\zeta_{n+1})^{\Delta^{2n+2}}(\xi), \quad x \in [a, b].$$

Then we have,

$$0 > \psi^{\Delta^{2n+2}}(\rho(\xi))\psi^{\Delta^{2n+2}}(\xi)$$

$$= \left(f^{\Delta^{2n+2}}(\rho(\xi)) - \frac{f(x) - p_{2n+1}(x)}{\pi_{n+1}(x)\zeta_{n+1}(x)}(\pi_{n+1}\zeta_{n+1})^{\Delta^{2n+2}}(\rho(\xi))\right)$$

$$\times \left(f^{\Delta^{2n+2}}(\xi) - \frac{f(x) - p_{2n+1}(x)}{\pi_{n+1}(x)\zeta_{n+1}(x)}(\pi_{n+1}\zeta_{n+1})^{\Delta^{2n+2}}(\xi)\right)$$

$$= \left(\frac{f(x) - p_{2n+1}(x)}{\pi_{n+1}(x)\zeta_{n+1}(x)}\right)^2 (\pi_{n+1}\zeta_{n+1})^{\Delta^{2n+2}}(\xi)(\pi_{n+1}\zeta_{n+1})^{\Delta^{2n+2}}(\rho(\xi))$$

$$- \frac{f(x) - p_{2n+1}(x)}{\pi_{n+1}(x)\zeta_{n+1}(x)}((\pi_{n+1}\zeta_{n+1})^{\Delta^{2n+2}}(\rho(\xi))f^{\Delta^{2n+2}}(\xi)$$

$$+ (\pi_{n+1}\zeta_{n+1})^{\Delta^{2n+2}}(\xi)f^{\Delta^{2n+2}}(\rho(\xi)))$$

$$+ f^{\Delta^{2n+2}}(\rho(\xi))f^{\Delta^{2n+2}}(\xi), \quad x \in [a, b].$$

Hence,

$$G_{\min,2n+2}(\xi) \le \frac{f(x) - p_{2n+1}(x)}{\pi_{n+1}(x)\zeta_{n+1}(x)} \le G_{\max,2n+2}(\xi), \quad x \in [a,b].$$

This completes the proof. □

Remark 1.39. Suppose that all the conditions of Theorem 1.38 hold. If

$$\lim_{n\to\infty}\max_{x\in[a,b]}\left(\frac{f^{\Delta^{2n+2}}(\xi)}{(\pi_{n+1}\zeta_{n+1})^{\Delta^{2n+2}}(\xi)}\pi_{n+1}(x)\zeta_{n+1}(x)\right) = 0$$

and

$$\lim_{n\to\infty}\max_{x\in[a,b]}\left(\frac{f^{\Delta^{2n+2}}(\rho(\xi))}{(\pi_{n+1}\zeta_{n+1})^{\Delta^{2n+2}}(\rho(\xi))}\pi_{n+1}(x)\zeta_{n+1}(x)\right) = 0,$$

then

$$\lim_{n\to\infty}\max_{x\in[a,b]}|f(x) - p_{2n+1}(x)| = 0.$$

In the following example, we compute the Hermite polynomial for a given function and compare it graphically with the function itself.

Example 1.40. Consider the time scale $\mathbb{T} = 2\mathbb{Z} = \{\ldots, -4, -2, 0, 2, 4, \ldots\}$. Let $a = x_0 = -4, x_1 = 0, x_2 = 4, b = 6$, and $f(x) = \frac{1}{x^2+1}$, $x \in \mathbb{T}$.

We will compute the Hermite polynomial $p_5(x)$ and compare the graphs of p_5 and f.

On this time scale, we have $\sigma(x) = x + 2$, $x \in \mathbb{T}$, so that $\sigma(x_0) = -2$, $\sigma(x_1) = 2$, and $\sigma(x_2) = 6$. We have

$$f^{\Delta}(x) = \frac{\frac{1}{(x+2)^2+1} - \frac{1}{x^2+1}}{x+2-x} = -\frac{2x+2}{(x^2+1)(x^2+4x+5)}, \quad x \in \mathbb{T}.$$

Hence,

$$y_0 = f(x_0) = f(-4) = \frac{1}{17},$$

$$y_1 = f(x_1) = f(0) = 1,$$

$$y_2 = f(x_2) = f(4) = \frac{1}{17},$$

$$z_0 = f^{\Delta}(x_0) = f^{\Delta}(-4) = \frac{6}{85},$$

$$z_1 = f^{\Delta}(x_1) = f^{\Delta}(0) = -\frac{2}{5},$$

$$z_2 = f^{\Delta}(x_2) = f^{\Delta}(4) = -\frac{10}{629}.$$

First, we compute L_i and M_i for $i = 0, 1, 2$,

$$L_0(x) = \frac{(x-0)(x-4)}{(-4-0)(-4-4)} = \frac{1}{32}(x^2 - 4x),$$

$$L_1(x) = \frac{(x+4)(x-4)}{(0+4)(0-4)} = -\frac{1}{16}(x^2 - 16),$$

$$L_2(x) = \frac{(x+4)(x-0)}{(4+4)(4-0)} = \frac{1}{32}(x^2 + 4x),$$

$$M_0(x) = \frac{(x-\sigma(0))(x-\sigma(4))}{(-4-\sigma(0))(-4-\sigma(4))} = \frac{(x-2)(x-6)}{(-4-2)(-4-6)} = \frac{1}{60}(x^2 - 8x + 12),$$

$$M_1(x) = \frac{(x-\sigma(-4))(x-\sigma(4))}{(0-\sigma(-4))(0-\sigma(4))} = \frac{(x+2)(x-6)}{(0+2)(0-6)} = -\frac{1}{12}(x^2 - 4x - 12),$$

$$M_2(x) = \frac{(x-\sigma(-4))(x-\sigma(0))}{(4-\sigma(-4))(4-\sigma(0))} = \frac{(x+2)(x-2)}{(4+2)(4-2)} = \frac{1}{12}(x^2 - 4), \quad x \in \mathbb{T}.$$

Let

$$h(x) = x^2, \quad x \in \mathbb{T}.$$

Then,

$$\left(h(x)\right)^\Delta = x + \sigma(x) = x + x + 2 = 2x + 2, \quad x \in \mathbb{T}.$$

We compute L_i^Δ and M_i^Δ for $i = 0, 1, 2$,

$$L_0^\Delta(x) = \frac{1}{32}(2x + 2 - 4) = \frac{1}{16}(x - 1),$$

$$L_1^\Delta(x) = -\frac{1}{16}(2x + 2) = -\frac{1}{8}(x + 1),$$

$$L_2^\Delta(x) = \frac{1}{32}(2x + 2 + 4) = \frac{1}{16}(x + 3),$$

$$M_0^\Delta(x) = \frac{1}{60}(2x + 2 - 8) = \frac{1}{30}(x - 3),$$

$$M_1^\Delta(x) = -\frac{1}{12}(2x + 2 - 4) = -\frac{1}{6}(x - 1),$$

$$M_2^\Delta(x) = \frac{1}{12}(2x + 2) = \frac{1}{6}(x + 1), \quad x \in \mathbb{T}.$$

The values involved in the Hermite polynomial are computed as

$$L_0(\sigma(x_0)) = L_0(-2) = \frac{1}{32}(4 + 8) = \frac{3}{8},$$

$$L_0^\Delta(x_0) = L_0^\Delta(-4) = \frac{1}{16}(-4 - 1) = -\frac{5}{16},$$

$$L_1(\sigma(x_1)) = L_1(2) = -\frac{1}{16}(4 - 16) = \frac{3}{4},$$

$$L_1^\Delta(x_1) = L_1^\Delta(0) = -\frac{1}{8}(0+1) = -\frac{1}{8},$$

$$L_2(\sigma(x_2)) = L_2(6) = \frac{1}{32}(36+24) = \frac{15}{8},$$

$$L_2^\Delta(x_2) = L_2^\Delta(4) = \frac{1}{16}(4+3) = \frac{7}{16},$$

and

$$M_0(\sigma(x_0)) = M_0(-2) = \frac{1}{60}(4+16+12) = \frac{8}{15},$$

$$M_0^\Delta(x_0) = M_0^\Delta(-4) = \frac{1}{30}(-4-3) = -\frac{7}{30},$$

$$M_1(\sigma(x_1)) = M_1(2) = -\frac{1}{12}(4-8-12) = \frac{4}{3},$$

$$M_1^\Delta(x_1) = M_1^\Delta(0) = -\frac{1}{6}(0-1) = \frac{1}{6},$$

$$M_2(\sigma(x_2)) = M_2(6) = \frac{1}{12}(36-4) = \frac{8}{3},$$

$$M_2^\Delta(x_2) = M_2^\Delta(4) = \frac{1}{6}(4+1) = \frac{5}{6}.$$

The Hermite polynomial $p_5(x)$ is computed as

$$p_5(x) = \left[\left(1 - \frac{-\frac{7}{30} - \frac{8}{15}\cdot\frac{5}{16}}{\frac{8}{15}\cdot\frac{3}{8}}(x+4)\right)\cdot\frac{1}{17} + \frac{\frac{6}{85}}{\frac{8}{15}\cdot\frac{3}{8}}(x+4)\right]L_0(x)M_0(x)$$

$$+ \left[\left(1 - \frac{\frac{1}{6} - \frac{4}{3}\cdot\frac{1}{8}}{\frac{4}{3}\cdot\frac{3}{4}}(x-0)\right)\cdot 1 + \frac{-\frac{2}{5}}{\frac{4}{3}\cdot\frac{3}{4}}(x-0)\right]L_1(x)M_1(x)$$

$$+ \left[\left(1 - \frac{\frac{5}{6} + \frac{8}{3}\cdot\frac{7}{16}}{\frac{8}{3}\cdot\frac{15}{8}}(x-4)\right)\cdot\frac{1}{17} + \frac{-\frac{10}{629}}{\frac{8}{3}\cdot\frac{15}{8}}(x-4)\right]L_2(x)M_2(x), \quad x \in \mathbb{T},$$

which after simplification becomes

$$p_5(x) = \left[\frac{1}{17} + \frac{8}{17}(x+4)\right]\frac{1}{32}(x^2-4x)\frac{1}{60}(x^2-8x+12)$$

$$+ \left[1 - \frac{2}{5}x\right]\frac{1}{16}(x^2-16)\frac{1}{12}(x^2-4x-12)$$

$$+ \left[\frac{1}{17} - \frac{84}{3145}(x-4)\right]\frac{1}{32}(x^2+4x)\frac{1}{12}(x^2-4), \quad x \in \mathbb{T}.$$

In Figure 1.5, the graphs of f and p_5 are given. It is clear that f and p_5 coincide at the points $-4, 0, 4$ and also at $\sigma(-4), \sigma(0), \sigma(4)$.

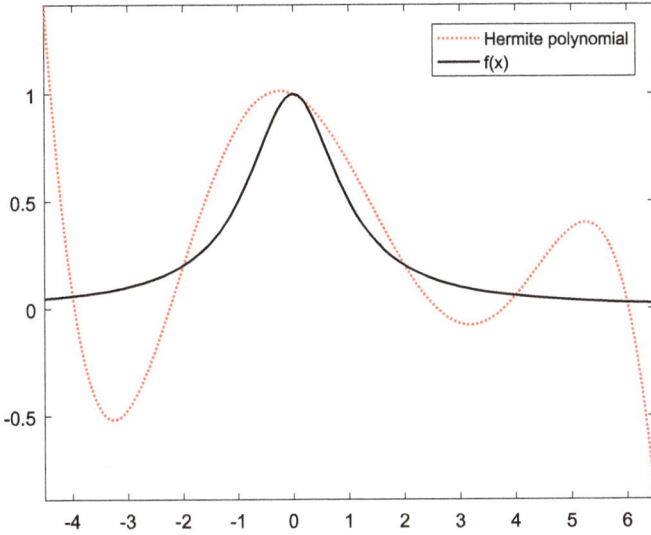

Figure 1.5: The graphs of the Hermite polynomial p_5 and the function f.

1.4 σ-Hermite interpolation

In this section, we will construct σ-Hermite interpolation polynomials. We will demonstrate the difference between Hermite and σ-Hermite interpolation polynomials. As was mentioned in the previous sections, the σ-interpolation polynomials provide an alternative way to interpolate a given set of data. They may coincide with the interpolation polynomials in certain cases and differ in others. The numerical examples presented in this section demonstrate these situations.

Theorem 1.41. *Let* $n \in \mathbb{N}_0$ *and let* $a, b \in \mathbb{T}, a < b,$ *and* $x_j \in [a, b] \subset \mathbb{T}, j \in \{0, 1, \ldots, n\},$ *be σ-distinct, the forward jump operator σ be delta differentiable on* $[a, b]$ *and* $\sigma^\Delta(x_j) \neq 0,$ $j \in \{0, 1, \ldots, n\}.$ *Let also,* $y_j, z_j \in \mathbb{R}, j \in \{0, 1, \ldots, n\}.$ *Then there exists a unique σ-polynomial* $p_{\sigma 2n+1} \in \mathcal{P}_{2n+1}^\sigma$ *such that*

$$p_{\sigma 2n+1}(x_j) = y_j, \quad p_{\sigma 2n+1}^\Delta(x_j) = z_j, \quad j \in \{0, 1, \ldots, n\}.$$

Proof. Let $M_k, k \in \{0, 1, \ldots, n\},$ be the polynomials as in the proof of Theorem 1.31. We will find a polynomial $p_{\sigma 2n+1} \in \mathcal{P}_{2n+1}^\sigma$ in the following form:

$$p_{\sigma 2n+1}(x) = \sum_{j=0}^{n} (y_j + (\sigma(x) - \sigma(x_j))(\alpha_j y_j + \beta_j z_j))M_j(x)L_j(x), \quad x \in [a, b],$$

where $\alpha_j, \beta_j \in \mathbb{R}, j \in \{0, 1, \ldots, n\},$ will be determined below. We have

$$p_{\sigma 2n+1}(x_k) = \sum_{j=0}^{n}(y_j + (\sigma(x_k) - \sigma(x_j))(\alpha_j y_j + \beta_j z_j))M_j(x_k)L_j(x_k) = y_k,$$

$$p_{\sigma 2n+1}^{\Delta}(x) = \sum_{j=0}^{n}\sigma^{\Delta}(x)(\alpha_j y_j + \beta_j z_j)M_j(\sigma(x))L_j(\sigma(x))$$

$$+ \sum_{j=0}^{n}(y_j + (\sigma(x) - \sigma(x_j))(\alpha_j y_j + \beta_j z_j))$$

$$\times (M_j^{\Delta}(x)L_j(x) + M_j(\sigma(x))L_j^{\Delta}(x)),$$

$$p_{\sigma 2n+1}^{\Delta}(x_k) = \sum_{j=0}^{n}\sigma^{\Delta}(x_k)(\alpha_j y_j + \beta_j z_j)M_j(\sigma(x_k))L_j(\sigma(x_k))$$

$$+ \sum_{j=0}^{n}(y_j + (\sigma(x_k) - \sigma(x_j))(\alpha_j y_j + \beta_j z_j))$$

$$\times (M_j^{\Delta}(x_k)L_j(x_k) + M_j(\sigma(x_k))L_j^{\Delta}(x_k))$$

$$= \sigma^{\Delta}(x_k)(\alpha_k y_k + \beta_k z_k)M_k(\sigma(x_k))L_k(\sigma(x_k))$$

$$+ y_k(M_k^{\Delta}(x_k) + M_k(\sigma(x_k))L_k^{\Delta}(x_k))$$

$$= z_k,$$

or, equivalently,

$$\sigma^{\Delta}(x_k)(\alpha_k y_k + \beta_k z_k)M_k(\sigma(x_k))L_k(\sigma(x_k)) = z_k - y_k(M_k^{\Delta}(x_k) + M_k(\sigma(x_k))L_k^{\Delta}(x_k)),$$

or

$$\alpha_k y_k + \beta_k z_k = \frac{z_k - y_k(M_k^{\Delta}(x_k) + M_k(\sigma(x_k))L_k^{\Delta}(x_k))}{\sigma^{\Delta}(x_k)M_k(\sigma(x_k))L_k(\sigma(x_k))},$$

and

$$p_{\sigma 2n+1}(x) = \sum_{j=0}^{n}\left(y_j + \frac{z_j - y_j(M_j^{\Delta}(x_j) + M_j(\sigma(x_j))L_j^{\Delta}(x_j))}{\sigma^{\Delta}(x_j)M_j(\sigma(x_j))L_j(\sigma(x_j))}(\sigma(x) - \sigma(x_j))\right)$$

$$\times M_j(x)L_j(x)$$

$$= \sum_{j=0}^{n}\left(\left(1 - \frac{M_j^{\Delta}(x_j) + M_j(\sigma(x_j))L_j^{\Delta}(x_j)}{\sigma^{\Delta}(x_j)M_j(\sigma(x_j))L_j(\sigma(x_j))}(\sigma(x) - \sigma(x_j))\right)y_j\right.$$

$$+ \frac{z_j}{\sigma^{\Delta}(x_j)M_j(\sigma(x_j))L_j(\sigma(x_j))}(\sigma(x) - \sigma(x_j))\left.\right)M_j(x)L_j(x), \quad x \in [a,b].$$

Now, suppose that there are two σ-polynomials such that $p_{\sigma 2n+1}, q_{\sigma 2n+1} \in \mathcal{P}_{2n+1}^{\sigma}$ and

$$p_{\sigma 2n+1}(x_k) = q_{\sigma 2n+1}(x_k) = y_k, \quad p_{\sigma 2n+1}^{\Delta}(x_k) = q_{\sigma 2n+1}^{\Delta}(x_k) = z_k, \quad k \in \{0, 1, \ldots, n\}.$$

Let $h_{\sigma 2n+1} = p_{\sigma 2n+1} - q_{\sigma 2n+1}$. Then $h_{\sigma 2n+1} \in P_{2n+1}^{\sigma}$ and it has at least $2n + 2$ GZs. Thus,

$$h_{\sigma 2n+1} \equiv 0 \quad \text{or} \quad p_{\sigma 2n+1} \equiv q_{\sigma 2n+1} \quad \text{on } [a, b].$$

This completes the proof. □

Taking into account the last theorem, the σ-Hermite interpolation polynomials for a given data set on a time scale and for a given function on an arbitrary time scale are defined as follows.

Definition 1.42. Let $n \in \mathbb{N}_0$ and let $a, b \in \mathbb{T}$, $a < b$, and $x_j \in [a, b] \subset \mathbb{T}$, $j \in \{0, 1, \ldots, n\}$, be σ-distinct, the forward jump operator σ be Δ-differentiable on $[a, b]$ and $\sigma^{\Delta}(x_j) \neq 0$, $j \in \{0, 1, \ldots, n\}$. Let also, $y_j, z_j \in \mathbb{R}$, $j \in \{0, 1, \ldots, n\}$. The polynomial

$$p_{\sigma 2n+1}(x) = \sum_{j=0}^{n} \left(\left(1 - \frac{M_j^{\Delta}(x_j) + M_j(\sigma(x_j))L_j^{\Delta}(x_j)}{\sigma^{\Delta}(x_j)M_j(\sigma(x_j))L_j(\sigma(x_j))}(\sigma(x) - \sigma(x_j)) \right) y_j \right.$$
$$\left. + \frac{z_j}{\sigma^{\Delta}(x_j)M_j(\sigma(x_j))L_j(\sigma(x_j))}(\sigma(x) - \sigma(x_j)) \right) M_j(x)L_j(x), \quad x \in [a, b],$$

will be called the σ-Hermite interpolation polynomial for the set

$$\{(x_j, y_j, z_j) : j \in \{0, 1, \ldots, n\}\}.$$

Definition 1.43. Let $n \in \mathbb{N}_0$ and let $a, b \in \mathbb{T}$, $a < b$, and $x_j \in [a, b] \subset \mathbb{T}$, $j \in \{0, 1, \ldots, n\}$, be σ-distinct, the forward jump operator σ be Δ-differentiable on $[a, b]$ and $\sigma^{\Delta}(x_j) \neq 0$, $j \in \{0, 1, \ldots, n\}$. Let also, $f : [a, b] \to \mathbb{R}$ be Δ-differentiable. The polynomial

$$p_{\sigma 2n+1}(x) = \sum_{j=0}^{n} \left(\left(1 - \frac{M_j^{\Delta}(x_j) + M_j(\sigma(x_j))L_j^{\Delta}(x_j)}{\sigma^{\Delta}(x_j)M_j(\sigma(x_j))L_j(\sigma(x_j))}(\sigma(x) - \sigma(x_j)) \right) f(x_j) \right.$$
$$\left. + \frac{f^{\Delta}(x_j)}{\sigma^{\Delta}(x_j)M_j(\sigma(x_j))L_j(\sigma(x_j))}(\sigma(x) - \sigma(x_j)) \right) M_j(x)L_j(x), \quad x \in [a, b],$$

will be called the σ-Hermite interpolation polynomial for the function f.

The next example demonstrates that on a general time scale, the Hermite and σ-Hermite polynomials may be different.

Example 1.44. Let $\mathbb{T} = \{-1, 1, 2, 5, 9, 10\}$,

$$a = -1, \quad n = 1, \quad x_0 = -1, \quad x_1 = 5, \quad b = 9, \quad y_0 = y_1 = 0, \quad z_0 = z_1 = 1.$$

We have

$$\sigma(x_0) = \sigma(-1) = 1, \qquad\qquad\qquad \sigma(x_1) = \sigma(5) = 9,$$

$$L_0(x) = \frac{x - x_1}{x_0 - x_1} = -\frac{1}{6}(x - 5), \qquad L_0(\sigma(x_0)) = L_0(1) = \frac{2}{3}, \quad L_0^\Delta(x) = -\frac{1}{6},$$

$$L_1(x) = \frac{x - x_0}{x_1 - x_0} = \frac{1}{6}(x + 1), \qquad L_1(\sigma(x_1)) = L_1(9) = \frac{5}{3}, \quad L_1^\Delta(x) = \frac{1}{6},$$

$$M_0(x) = \frac{x - \sigma(x_1)}{x_0 - \sigma(x_1)} = -\frac{1}{10}(x - 9), \quad M_0(\sigma(x_0)) = M_0(1) = \frac{4}{5}, \quad M_0^\Delta(x) = -\frac{1}{10},$$

$$M_1(x) = \frac{x - \sigma(x_0)}{x_1 - \sigma(x_0)} = \frac{1}{4}(x - 1), \qquad M_1(\sigma(x_1)) = M_1(9) = 2, \quad M_1^\Delta(x) = \frac{1}{4},$$

$$\sigma^\Delta(x_0) = \frac{\sigma(\sigma(x_0)) - \sigma(x_0)}{\sigma(x_0) - x_0} = \frac{\sigma(1) - 1}{1 - (-1)} = \frac{2 - 1}{2} = \frac{1}{2},$$

$$\sigma^\Delta(x_1) = \frac{\sigma(\sigma(x_1)) - \sigma(x_1)}{\sigma(x_1) - x_1} = \frac{\sigma(9) - 9}{9 - 5} = \frac{1}{4}.$$

Hence, using that $y_0 = y_1 = 0, z_0 = z_1 = 1$, we get

$$
\begin{aligned}
p_{\sigma 3}(x) &= \frac{z_0}{\sigma^\Delta(x_0)M_0(\sigma(x_0))L_0(\sigma(x_0))}(\sigma(x) - \sigma(x_0))M_0(x)L_0(x) \\
&\quad + \frac{z_1}{\sigma^\Delta(x_1)M_1(\sigma(x_1))L_1(\sigma(x_1))}(\sigma(x) - \sigma(x_1))M_1(x)L_1(x) \\
&= \frac{1}{\frac{1}{2} \cdot \frac{4}{5} \cdot \frac{2}{3}}(\sigma(x) - 1)\left(-\frac{1}{10}(x - 9)\right)\left(-\frac{1}{6}(x - 5)\right) \\
&\quad + \frac{1}{\frac{1}{4} \cdot 2 \cdot \frac{5}{3}}(\sigma(x) - 9)\left(\frac{1}{6}(x + 1)\right)\left(\frac{1}{4}(x - 1)\right) \\
&= \frac{1}{16}(\sigma(x) - 1)(x - 9)(x - 5) + \frac{1}{20}(\sigma(x) - 9)(x + 1)(x - 1), \quad x \in \mathbb{T}, \\
p_{\sigma 3}(2) &= \frac{1}{16}(\sigma(2) - 1)(2 - 9)(2 - 5) + \frac{1}{20}(\sigma(2) - 9)(2 + 1)(2 - 1) \\
&= \frac{1}{16} \cdot (5 - 1) \cdot (-7) \cdot (-3) + \frac{1}{20} \cdot (5 - 9) \cdot 3 \\
&= \frac{21}{4} - \frac{3}{5} = \frac{105 - 12}{20} = \frac{93}{20},
\end{aligned}
$$

and

$$
\begin{aligned}
p_3(x) &= \frac{z_0}{M_0(\sigma(x_0))L_0(\sigma(x_0))}(x - x_0)M_0(x)L_0(x) \\
&\quad + \frac{z_1}{M_1(\sigma(x_1))L_1(\sigma(x_1))}(x - x_1)M_1(x)L_1(x) \\
&= \frac{1}{\frac{2}{3} \cdot \frac{4}{5}}(x + 1)\left(-\frac{1}{10}(x - 9)\right)\left(-\frac{1}{6}(x - 5)\right) \\
&\quad + \frac{1}{2 \cdot \frac{5}{3}}(x - 5)\left(\frac{1}{6}(x + 1)\right)\left(\frac{1}{4}(x - 1)\right) \\
&= \frac{1}{32}(x + 1)(x - 9)(x - 5) + \frac{1}{80}(x - 5)(x + 1)(x - 1), \quad x \in \mathbb{T},
\end{aligned}
$$

$$p_3(2) = \frac{1}{32}(2+1)(2-9)(2-5) + \frac{1}{80}(2-5)(2+1)(2-1)$$

$$= \frac{63}{32} - \frac{9}{80} = \frac{315 - 18}{160} = \frac{297}{160}.$$

Consequently, we have

$$p_3(2) \neq p_{\sigma 3}(2).$$

Exercise 1.45. Let $\mathbb{T} = \{-4, -1, 0, \frac{1}{2}, \frac{5}{6}, 1, \frac{4}{3}, 2, 7\}$,

$$a = -4, \quad x_0 = -4, \quad x_1 = 0, \quad x_2 = \frac{5}{6}, \quad b = 7,$$

and let the function $f : \mathbb{T} \to \mathbb{R}$ be defined by

$$f(x) = \frac{x+1}{x^2 + 4x + 7} + 1, \quad x \in \mathbb{T}.$$

Find the σ-Hermite interpolation polynomial for the function f.

In the next theorem, we will give some criteria for the coincidence of the Hermite and σ-Hermite interpolation polynomials.

Theorem 1.46. *Assume that* $\mathbb{T}$ *is a time scale such that* $\sigma(t) = ct+d$, *for any* $t \in \mathbb{T}$ *and for some real constants* c, d. *Let* $n \in \mathbb{N}_0$ *and let* $a, b \in \mathbb{T}, a < b$, *and* $x_j \in \mathbb{T}, j \in \{0, 1, \ldots, n\}$, *be σ-distinct. Let also* $y_j, z_j \in \mathbb{R}, j \in \{0, 1, \ldots, n\}$. *Then*

$$p_{\sigma 2n+1} \equiv p_{2n+1}.$$

Proof. Let $\sigma(t) = ct+d$ for any $t \in \mathbb{T}$ and for some real constants c and d. Then $\sigma^{\Delta}(t) = c$, $t \in \mathbb{T}$, and

$$p_{\sigma 2n+1}(x) = \sum_{j=0}^{n} \left(\left(1 - \frac{M_j^{\Delta}(x_j) + M_j(\sigma(x_j))L_j^{\Delta}(x_j)}{\sigma^{\Delta}(x_j)M_j(\sigma(x_j))L_j(\sigma(x_j))}(\sigma(x) - \sigma(x_j))\right) y_j \right.$$

$$+ \frac{z_j}{\sigma^{\Delta}(x_j)M_j(\sigma(x_j))L_j(\sigma(x_j))}(\sigma(x) - \sigma(x_j))\bigg) M_j(x)L_j(x)$$

$$= \sum_{j=0}^{n} \left(\left(1 - \frac{M_j^{\Delta}(x_j) + M_j(\sigma(x_j))L_j^{\Delta}(x_j)}{cM_j(\sigma(x_j))L_j(\sigma(x_j))}(cx + d - (cx_j + d))\right) y_j \right.$$

$$+ \frac{z_j}{cM_j(\sigma(x_j))L_j(\sigma(x_j))}(cx + d - (cx_j + d))\bigg) M_j(x)L_j(x)$$

$$= \sum_{j=0}^{n} \left(\left(1 - \frac{M_j^{\Delta}(x_j) + M_j(\sigma(x_j))L_j^{\Delta}(x_j)}{M_j(\sigma(x_j))L_j(\sigma(x_j))}(x - x_j)\right) y_j \right.$$

$$+ \frac{z_j}{M_j(\sigma(x_j))L_j(\sigma(x_j))}(x - x_j)\Big)M_j(x)L_j(x)$$

$$= p_{2n+1}(x), \quad x \in [a, b].$$

This completes the proof. $\qquad\qquad\qquad\qquad\qquad\qquad\qquad\qquad\qquad\qquad\square$

As we have proved in Theorem 1.38, one can deduct the following result regarding the error in σ-Hermite interpolation.

Theorem 1.47. *Suppose that $n \in \mathbb{N}_0$, $a, b \in \mathbb{T}$, $a < b$, $x_j \in [a, b]$, $j \in \{0, 1, \dots, n\}$, are σ-distinct and $f : [a, b] \to \mathbb{R}$, $f^{\Delta^k}(x)$ exist for any $x \in [a, b]$ and for any $k \in \{1, \dots, 2n + 2\}$. Then for any $x \in [a, b]$ there exists $\xi = \xi(x) \in (a, b)$ such that*

$$f(x) - p_{\sigma 2n+1}(x) = \frac{f^{\Delta^{2n+2}}(\xi)}{\pi_{n+1}(x)\zeta_{n+1}(x)}(\pi_{n+1}\zeta_{n+1})^{\Delta^{2n+2}}(\xi), \quad x \in [a, b],$$

or

$$G_{\min,2n+2}(\xi) \le \frac{f(x) - p_{\sigma 2n+1}(x)}{\pi_{n+1}(x)\zeta_{n+1}(x)} \le G_{\max,2n+2}(\xi), \quad x \in [a, b],$$

where $G_{\min,2n+2}(\xi)$ and $G_{\max,2n+2}(\xi)$ are defined as in Theorem 1.38.

In the following example, we consider a time scale with a nonlinear forward jump operator. We show the difference between Hermite and σ-Hermite polynomials graphically.

Example 1.48. Let $\mathbb{T} = \mathbb{N}_0^2 = \{0, 1, 4, 9, 16, \dots\}$ and $[a, b] = [1, 49]$. Let

$$x_0 = 1, \quad x_1 = 9, \quad x_2 = 25.$$

We will find the Hermite interpolation polynomial $p_5(x)$ and the σ-Hermite interpolation polynomial $p_{\sigma 5}(x)$ for a function f satisfying

$$f(1) = 2, \quad f(9) = 4, \quad f(25) = 10,$$

and

$$f^{\Delta}(1) = 3, \quad f^{\Delta}(9) = 6, \quad f^{\Delta}(25) = 7.$$

On this time scale, we have

$$\sigma(x) = (\sqrt{x} + 1)^2,$$

$$\sigma^{\Delta}(x) = \frac{\sigma(\sigma(x)) - \sigma(x)}{\sigma(x) - x} = \frac{2\sqrt{x} + 3}{2\sqrt{x} + 1}, \quad x \in \mathbb{T}.$$

We first compute the polynomials L_k and M_k for $k = 0, 1, 2$,

$$L_0(x) = \frac{(x-9)(x-25)}{(-8)(-24)} = \frac{x^2 - 34x + 225}{192},$$

$$L_1(x) = \frac{(x-1)(x-25)}{(8)(-16)} = -\frac{x^2 - 26x + 25}{128},$$

$$L_2(x) = \frac{(x-1)(x-9)}{(24)(16)} = \frac{x^2 - 10x + 9}{384},$$

$$M_0(x) = \frac{(x-16)(x-36)}{(-15)(-35)} = \frac{x^2 - 52x + 576}{525},$$

$$M_1(x) = \frac{(x-4)(x-36)}{(5)(-27)} = -\frac{x^2 - 40x + 144}{135},$$

$$M_2(x) = \frac{(x-4)(x-16)}{(21)(9)} = \frac{x^2 - 20x + 64}{189}, \quad x \in [1, 49].$$

Let

$$g_1(x) = x^2, \quad g_2(x) = x, \quad x \in \mathbb{T}.$$

Then

$$g_1^\Delta(x) = x + \sigma(x) = 2x + 2\sqrt{x} + 1,$$
$$g_2^\Delta(x) = 1, \quad x \in \mathbb{T}.$$

We compute, for $x \in [1, 49]$,

$$L_0^\Delta(x) = \frac{2x + 2\sqrt{x} - 33}{192}, \quad L_1^\Delta(x) = -\frac{2x + 2\sqrt{x} - 25}{128}, \quad L_2^\Delta(x) = \frac{2x + 2\sqrt{x} - 9}{384},$$

$$M_0^\Delta(x) = \frac{2x + 2\sqrt{x} - 51}{525}, \quad M_1^\Delta(x) = -\frac{2x + 2\sqrt{x} - 39}{135}, \quad M_2^\Delta(x) = \frac{2x + 2\sqrt{x} - 19}{189}.$$

Then we have

$$L_0(\sigma(x_0)) = L_0(4) = \frac{35}{64}, \qquad L_0^\Delta(x_0) = L_0^\Delta(1) = -\frac{29}{192},$$

$$L_1(\sigma(x_1)) = L_1(16) = \frac{135}{128}, \qquad L_1^\Delta(x_1) = L_1^\Delta(9) = \frac{1}{128},$$

$$L_2(\sigma(x_2)) = L_2(36) = \frac{315}{128}, \qquad L_2^\Delta(x_2) = L_2^\Delta(25) = \frac{17}{128},$$

$$M_0(\sigma(x_0)) = M_0(4) = \frac{128}{175}, \qquad M_0^\Delta(x_0) = M_0^\Delta(1) = -\frac{47}{525},$$

$$M_1(\sigma(x_1)) = M_1(16) = \frac{16}{6}, \qquad M_1^\Delta(x_1) = M_1^\Delta(9) = \frac{1}{9},$$

$$M_2(\sigma(x_2)) = M_2(36) = \frac{640}{189}, \qquad M_2^\Delta(x_2) = M_2^\Delta(25) = \frac{41}{189}.$$

Using these values, for $x \in [1, 49]$, we compute

$$p_5(x) = \left(2 + \frac{17}{2}(x-1)\right)\left(\frac{x^2 - 34x + 225}{192}\right)\left(\frac{x^2 - 52x + 576}{525}\right)$$
$$+ \left(4 + \frac{44}{15}(x-9)\right)\left(\frac{x^2 - 26x + 25}{128}\right)\left(\frac{x^2 - 40x + 144}{135}\right)$$
$$+ \left(10 + \frac{1}{25}(x-25)\right)\left(\frac{x^2 - 10x + 9}{384}\right)\left(\frac{x^2 - 20x + 64}{189}\right),$$

and

$$p_{\sigma 5}(x) = \left(2 + \frac{51}{10}((\sqrt{x}+1)^2 - 4)\right)\left(\frac{x^2 - 34x + 225}{192}\right)\left(\frac{x^2 - 52x + 576}{525}\right)$$
$$+ \left(4 + \frac{308}{135}((\sqrt{x}+1)^2 - 16)\right)\left(\frac{x^2 - 26x + 25}{128}\right)\left(\frac{x^2 - 40x + 144}{135}\right)$$
$$+ \left(10 + \frac{11}{325}((\sqrt{x}+1)^2 - 36)\right)\left(\frac{x^2 - 10x + 9}{384}\right)\left(\frac{x^2 - 20x + 64}{189}\right).$$

The example above shows that the Hermite and σ-Hermite interpolation polynomials can be different and, moreover, the σ-Hermite interpolation polynomial may not be a polynomial in the classical sense. On the other hand, the difference between these polynomials is not very large, as can be seen from Figure 1.6.

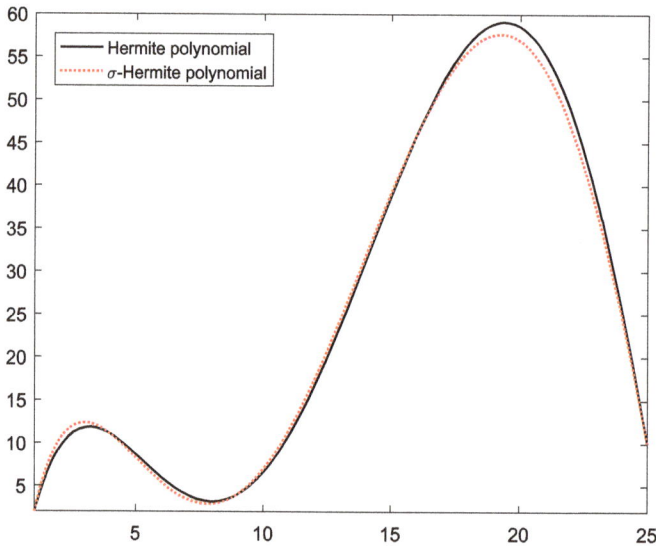

Figure 1.6: The graphs of the Hermite and σ-Hermite polynomials.

1.5 Delta differentiation

We conclude this chapter with a theoretical discussion of the error in the delta derivative of a function when it is approximated by the delta derivative of the related interpolation polynomial.

Theorem 1.49. *Suppose that $n \geq 0$, $a, b \in \mathbb{T}$, $a < b$, $x_j \in [a,b]$, $j \in \{0,1,\ldots,n\}$, are distinct, $f : [a,b] \to \mathbb{R}$, $p_n : [a,b] \to \mathbb{R}$ is a polynomial of degree n interpolating f at the points $x_j \in [a,b]$, $j \in \{0,1,\ldots,n\}$ and $f^{\Delta^k}(x)$ exist for any $x \in [a,b]$ and for any $k \in \{1,\ldots,n+1\}$. Then for any $x \in [a,b]$ there exist $\xi = \xi(x) \in (a,b)$ and distinct points $\eta_j, j \in \{1,\ldots,n\}$, in (a,b) such that*

$$f^{\Delta}(x) - p_n^{\Delta}(x) = \frac{f^{\Delta^{n+1}}(\xi)}{\pi_n^{*\Delta^{n+1}}(\xi)} \pi_n^*(x), \quad x \in [a,b],$$

or

$$H_{\min,n+1}(\xi) \leq \frac{f^{\Delta}(x) - p_n^{\Delta}(x)}{\pi_n^*(x)} \leq H_{\max,n+1}(\xi), \quad x \in [a,b],$$

where

$$H_{\max,n+1}(\xi) = \max\left\{ \frac{f^{\Delta^{n+1}}(\xi)}{\pi_n^{*\Delta^{n+1}}(\xi)}, \frac{f^{\Delta^{n+1}}(\rho(\xi))}{\pi_n^{*\Delta^{n+1}}(\rho(\xi))} \right\},$$

$$H_{\min,n+1}(\xi) = \min\left\{ \frac{f^{\Delta^{n+1}}(\xi)}{\pi_n^{*\Delta^{n+1}}(\xi)}, \frac{f^{\Delta^{n+1}}(\rho(\xi))}{\pi_n^{*\Delta^{n+1}}(\rho(\xi))} \right\},$$

$$\pi_n^*(x) = (x - \eta_1) \cdots (x - \eta_n), \quad x \in [a,b].$$

Proof. Let p_n be the Lagrange interpolation polynomial for the function f with interpolation points x_j, $j \in \{0,1,\ldots,n\}$. Then the function $f - p_n$ has at least $n+1$ GZs in $[a,b]$. Hence, by Rolle's theorem, it follows that there exist $\eta_j, j \in \{1,\ldots,n\}$, in (a,b) which are GZs of the function $f^{\Delta} - p_n^{\Delta}$. Define the function

$$\chi(t) = f^{\Delta}(t) - p_n^{\Delta}(t) - \frac{f^{\Delta}(x) - p_n^{\Delta}(x)}{\pi_n^*(x)} \pi_n^*(t), \quad t \in [a,b].$$

Then

$$\chi(\eta_j) = f^{\Delta}(\eta_j) - p_n^{\Delta}(\eta_j) - \frac{f^{\Delta}(x) - p_n^{\Delta}(x)}{\pi_n^*(x)} \pi_n^*(\eta_j)$$

$$= f^{\Delta}(\eta_j) - p_n^{\Delta}(\eta_j)$$

$$= 0, \quad x \in [a,b], \quad j \in \{0,1,\ldots,n\},$$

and $\chi(x) = 0$, $x \in [a, b]$. Thus, $\chi : [a, b] \to \mathbb{R}$ has at least $n + 1$ GZs. Hence, by Rolle's theorem, it follows that χ^{Δ^n} has at least one GZ on (a, b). Therefore, there exists $\xi = \xi(x) \in (a, b)$ such that

$$\chi^{\Delta^n}(\xi) = 0 \quad \text{or} \quad \chi^{\Delta^n}(\rho(\xi))\chi^{\Delta^n}(\xi) < 0.$$

Note that

$$\chi^{\Delta^n}(t) = f^{\Delta^{n+1}}(t) - \frac{f^{\Delta}(x) - p_n^{\Delta}(x)}{\pi_n^*(x)}\pi_n^{*\Delta^n}(t), \quad t \in [a, b].$$

We now consider the two cases separately.
1. Let $\chi^{\Delta^n}(\xi) = 0$. Then

$$f^{\Delta^{n+1}}(\xi) = \frac{f^{\Delta}(x) - p_n^{\Delta}(x)}{\pi_n^*(x)}\pi_n^{*\Delta^n}(\xi), \quad x \in [a, b],$$

or, equivalently,

$$f^{\Delta}(x) - p_n^{\Delta}(x) = \frac{f^{\Delta^{n+1}}(\xi)}{\pi_n^{*\Delta^n}(\xi)}\pi_n^*(x) = \frac{f^{\Delta^{n+1}}(\xi)}{\pi_n^{*\Delta^n}(\xi)}\pi_n^*(x), \quad x \in [a, b].$$

2. Let

$$\chi^{\Delta^n}(\rho(\xi))\chi^{\Delta^n}(\xi) < 0.$$

Then

$$\chi^{\Delta^n}(\rho(\xi)) = f^{\Delta^{n+1}}(\rho(\xi)) - \frac{f^{\Delta}(x) - p_n^{\Delta}(x)}{\pi_n^*(x)}\pi_n^{*\Delta^n}(\rho(\xi)), \quad x \in [a, b],$$

and

$$\chi^{\Delta^n}(\xi) = f^{\Delta^{n+1}}(\xi) - \frac{f^{\Delta}(x) - p_n^{\Delta}(x)}{\pi_n^*(x)}\pi_n^{*\Delta^n}(\xi), \quad x \in [a, b].$$

Hence,

$$0 > \chi^{\Delta^n}(\rho(\xi))\chi^{\Delta^n}(\xi)$$

$$= \left(f^{\Delta^{n+1}}(\rho(\xi)) - \frac{f^{\Delta}(x) - p_n^{\Delta}(x)}{\pi_n^*(x)}\pi_n^{*\Delta^n}(\rho(\xi))\right)$$

$$\times \left(f^{\Delta^{n+1}}(\xi) - \frac{f^{\Delta}(x) - p_n^{\Delta}(x)}{\pi_n^*(x)}\pi_n^{*\Delta^n}(\xi)\right)$$

$$= \left(\frac{f^{\Delta}(x) - p_n^{\Delta}(x)}{\pi_n^*(x)}\right)^2 \pi_n^{*\Delta^n}(\rho(\xi))\pi_n^{*\Delta^n}(\xi)$$

$$-\frac{f^{\Delta}(x)-p_n^{\Delta}(x)}{\pi_n^*(x)}(\pi_n^{*\Delta^n}(\rho(\xi))f^{\Delta^{n+1}}(\xi)+\pi_n^{*\Delta^n}(\xi)f^{\Delta^{n+1}}(\rho(\xi)))$$

$$+f^{\Delta^{n+1}}(\rho(\xi))f^{\Delta^{n+1}}(\xi),\quad x\in[a,b].$$

Therefore,

$$H_{\min,n+1}(\xi)\le\frac{f^{\Delta}(x)-p_n^{\Delta}(x)}{\pi_n^*(x)}\le H_{\max,n+1}(\xi),\quad x\in[a,b].$$

This completes the proof. □

Remark 1.50. Suppose that all the conditions of Theorem 1.49 hold. If

$$\lim_{n\to\infty}\max_{x\in[a,b]}\left(\frac{f^{\Delta^{n+1}}(\xi)}{\pi_n^{*\Delta^n}(\xi)}\pi_n^*(x)\right)=0$$

and

$$\lim_{n\to\infty}\max_{x\in[a,b]}\left(\frac{f^{\Delta^{n+1}}(\rho(\xi))}{\pi_n^{*\Delta^n}(\rho(\xi))}\pi_n^*(x)\right)=0,$$

then

$$\lim_{n\to\infty}\max_{x\in[a,b]}|f^{\Delta}(x)-p_n^{\Delta}(x)|=0.$$

1.6 Advanced practical problems

Problem 1.51. Let $\mathbb{T}=(\frac{1}{3})^{\mathbb{N}_0}$. Construct the Lagrange interpolation polynomial for the set

$$\left\{\left(\frac{1}{27},1\right),\left(\frac{1}{9},0\right),\left(\frac{1}{3},-1\right),(1,2)\right\}.$$

Problem 1.52. Let $\mathbb{T}=4^{\mathbb{N}_0}$, and suppose $f:\mathbb{T}\to\mathbb{R}$ is defined by

$$f(x)=e_1(x,1)+\sin_2(x,4)+x^2+x,\quad x\in\mathbb{T}.$$

Find the Lagrange interpolation polynomial for the function f with interpolation points

$$a=x_0=1,\quad x_1=4,\quad x_2=16,\quad x_3=64,\quad x_4=256=b.$$

Problem 1.53. Let $\mathbb{T} = 3^{\mathbb{N}_0}$,

$$x_0 = 1, \quad x_1 = 3, \quad x_2 = 9, \quad b = 27.$$

Find $\pi_3(x)$, $\Pi_3^2(x)$, $x \in \mathbb{T}$.

Problem 1.54. Let $\mathbb{T} = 3\mathbb{N}_0$. Check if the points

$$x_0 = 0, \quad x_1 = 3, \quad x_2 = 27, \quad x_3 = 180$$

are σ-distinct points.

Problem 1.55. Let $\mathbb{T} = 4^{\mathbb{N}_0}$. Check if the points

$$x_0 = 1, \quad x_1 = 16, \quad x_2 = 256$$

are σ-distinct points.

Problem 1.56. Let $\mathbb{T} = (\frac{1}{3})^{\mathbb{N}_0}$,

$$a = x_0 = \frac{1}{81}, \quad x_1 = \frac{1}{9}, \quad x_2 = 1 = b,$$

and suppose $f : \mathbb{T} \to \mathbb{R}$ is defined by

$$f(x) = \frac{x^4 + x^3 + 1}{x^5 + x^2 + x + 1}, \quad x \in \mathbb{T}.$$

Find the σ-Lagrange interpolation polynomial for the function f with σ-interpolation points x_0, x_1, x_2.

Problem 1.57. Let $\mathbb{T} = \{0, \frac{1}{4}, \frac{1}{3}, \frac{1}{2}, \frac{5}{6}, 1, 4, 5, 12\}$,

$$a = 0, \quad x_0 = 0, \quad x_1 = \frac{1}{3}, \quad x_2 = 4, \quad b = 12.$$

Find the σ-Lagrange and Lagrange interpolation polynomials for the function $f :$ $\mathbb{T} \to \mathbb{R}$ defined by

$$f(x) = \frac{x + 1}{x^2 - 2x + 7} + 4x + 1, \quad x \in \mathbb{T},$$

with σ-interpolation and interpolation points x_0, x_1, x_2.

Problem 1.58. Let $\mathbb{T} = \{0, \frac{1}{2}, 3, 11, 19, 108\}$,

$$a = 0, \quad x_0 = 0, \quad x_1 = \frac{1}{2}, \quad x_2 = 11, \quad b = 108.$$

Find $\Pi_{\sigma3}^3(19)$.

Problem 1.59. Let $\mathbb{T} = 4\mathbb{Z}$. Find the Hermite interpolation polynomial for the set

$$x_0 = -24, \quad x_1 = -4, \quad x_2 = 16,$$
$$y_0 = 1, \qquad y_1 = -3, \quad y_2 = -1,$$
$$z_0 = -1, \qquad z_1 = 1, \qquad z_2 = -1.$$

Problem 1.60. Let $\mathbb{T} = 2\mathbb{Z}$ and $f : \mathbb{T} \to \mathbb{R}$ be defined by

$$f(x) = \frac{x+1}{x^2+1} + x^3, \quad x \in \mathbb{T}.$$

Let also,

$$x_0 = -4, \quad x_1 = 0, \quad x_2 = 8, \quad b = 10.$$

Find the Hermite interpolation polynomial for the function f.

Problem 1.61. Let $\mathbb{T} = \{-3, -1, 0, 7, 8, 19, 29\}$,

$$a = -3, \quad x_0 = -3, \quad x_1 = 0, \quad x_2 = 8, \quad b = 29,$$

and $f : \mathbb{T} \to \mathbb{R}$ be defined by

$$f(x) = e_1(x, 0) + x^2 + x + 1, \quad x \in \mathbb{T}.$$

Find the σ-Hermite interpolation polynomial for the function f.

2 Numerical integration

Numerical integration is as a significant subject as the integration itself. Almost all real life problems represented by mathematical models require computation of certain definite integrals. In many cases, these integrals cannot be evaluated exactly. For this reason, development and use of efficient and reliable numerical methods which provide a good approximation to a given integral are among the main tasks in numerical analysis and methods.

The main purpose of this chapter is to approximately evaluate the Cauchy time scale integral. One natural approach is to apply the results of the previous chapter on polynomial interpolation to derive formulae for numerical integration. Therefore, we will give the detailed theoretical base and the derivation of some quadrature rules, as well as explain how one can estimate the associated approximation error. We will also give illustrative numerical examples.

Throughout this chapter, we assume that $\mathbb{T}$ is a time scale with forward jump operator σ and delta differentiation operator Δ. We also assume that $[a, b] \subset \mathbb{T}$ for some finite $a, b \in \mathbb{T}$.

2.1 Newton–Cotes formulae

First, we give the derivation of the so-called Newton–Cotes integration formulae, as they are called in classical numerical analysis [6, 7].

Suppose that $f : \mathbb{T} \to \mathbb{R}$ is rd-continuous. We assume that the integral $\int_a^b f(x)\Delta x$ may not be evaluated exactly and wish to find its approximate value.

Let $n \in \mathbb{N}_0$, $x_j \in [a, b]$, $j \in \{0, 1, \ldots, n\}$, be distinct. With $p_n \in \mathcal{P}_n$ we will denote the Lagrange interpolation polynomial for the function f. Then

$$p_n(x) = \sum_{k=0}^{n} L_k(x) f(x_k), \quad x \in [a, b],$$

and

$$\int_a^b f(x)\Delta x \approx \int_a^b p_n(x)\Delta x = \sum_{k=0}^{n} \left(\int_a^b L_k(x)\Delta x \right) f(x_k).$$

Set

$$w_k = \int_a^b L_k(x)\Delta x, \quad k \in \{0, 1, \ldots, n\}.$$

Therefore,

https://doi.org/10.1515/9783110787320-002

$$\int_a^b f(x)\Delta x \approx \sum_{k=0}^{n} w_k f(x_k). \tag{2.1}$$

Definition 2.1. The values w_k, $k \in \{0,1,\dots,n\}$, are called the quadrature weights, while the interpolation points x_k, $k \in \{0,1,\dots,n\}$, are called the quadrature points or nodes.

Definition 2.2. The numerical quadrature rule (2.1) is said to be the Newton–Cotes formula.

Now, for $t, s \in \mathbb{T}$, we recall the definition of the time scale monomials, which was given as follows [1, 2]:

$$h_0(t,s) = 1, \quad h_{k+1}(t,s) = \int_s^t h_k(\tau,s)\Delta\tau, \quad k \in \mathbb{N}_0.$$

The two basic Newton–Cotes formulas are known as trapezoid and Simpson rules. We will first derive these rules for a Cauchy integral on an arbitrary time scale. Consider the case $n = 1$, so that we take

$$a = x_0, \quad b = x_1.$$

Then

$$\begin{aligned}
p_1(x) &= L_0(x)f(a) + L_1(x)f(b) \\
&= \frac{x-b}{a-b}f(a) + \frac{x-a}{b-a}f(b) \\
&= \frac{1}{b-a}((b-x)f(a) + (x-a)f(b)), \quad x \in [a,b].
\end{aligned}$$

Hence,

$$\begin{aligned}
\int_a^b p_1(x)\Delta x &= \frac{1}{b-a}\left(f(a)\int_a^b (b-x)\Delta x + f(b)\int_a^b (x-a)\Delta x\right) \\
&= \frac{1}{b-a}\left(f(a)\int_b^a (x-b)\Delta x + f(b)\int_a^b (x-a)\Delta x\right) \\
&= \frac{1}{b-a}(f(a)h_2(a,b) + f(b)h_2(b,a))
\end{aligned}$$

and

$$\int_a^b f(x)\Delta x \approx \frac{1}{b-a}(f(a)h_2(a,b) + f(b)h_2(b,a)). \tag{2.2}$$

Definition 2.3. The numerical integration formula (2.2) will be called the trapezoid rule.

Below we apply the trapezoid rule to particular examples.

Example 2.4. Let $\mathbb{T} = 2^{\mathbb{N}_0}$, $a = 1$, $b = 8$, and $f : \mathbb{T} \to \mathbb{R}$ be defined as

$$f(t) = \frac{t-1}{t+1}, \quad t \in \mathbb{T}.$$

We will evaluate $\int_1^8 f(t)\Delta t$. Here $\sigma(t) = 2t$, $t \in \mathbb{T}$. Let

$$f_1(t) = \frac{1}{3}t^2 - t, \quad f_2(t) = \frac{1}{3}t^2 - 8t, \quad t \in \mathbb{T}.$$

Then

$$f_1^\Delta(t) = \frac{1}{3}(\sigma(t) + t) - 1 = \frac{1}{3}(2t + t) - 1 = t - 1,$$

$$f_2^\Delta(t) = \frac{1}{3}(\sigma(t) + t) - 8 = \frac{1}{3}(2t + t) - 8 = t - 8, \quad t \in [1, 8],$$

$$h_2(a, b) = \int_8^1 (t - 8)\Delta t = \int_8^1 f_2^\Delta(t)\Delta t = f_2(1) - f_2(8)$$

$$= \left(\frac{1}{3} - 8\right) - \left(\frac{64}{3} - 64\right) = \frac{1}{3} - 8 - \frac{64}{3} + 64 = 35,$$

$$h_2(b, a) = \int_1^8 (t - 1)\Delta t = \int_1^8 f_1^\Delta(t)\Delta t = f_1(8) - f_1(1)$$

$$= \left(\frac{64}{3} - 8\right) - \left(\frac{1}{3} - 1\right) = \frac{64}{3} - 8 - \frac{1}{3} + 1 = 14,$$

$$f(1) = 0, \quad f(8) = \frac{7}{9}.$$

Thus,

$$\int_1^8 \frac{t-1}{t+1}\Delta t \approx \frac{1}{7}\left(\frac{7}{9} \cdot 14\right) = \frac{14}{9}.$$

Example 2.5. Let $\mathbb{T} = 3^{\mathbb{N}_0}$ and $f : \mathbb{T} \to \mathbb{R}$ be defined by

$$f(t) = \frac{t^3 - 1}{t^3 + t + 1} + e_1(t, 1), \quad t \in \mathbb{T}.$$

We will evaluate the integral $\int_1^9 f(t)\Delta t$. Here

$$\sigma(t) = 3t, \quad t \in \mathbb{T}, \quad a = 1, \quad b = 9.$$

Let

$$f_1(t) = \frac{1}{4}t^2 - t, \quad f_2(t) = \frac{1}{4}t^2 - 9t, \quad t \in \mathbb{T}.$$

Then

$$f_1^\Delta(t) = \frac{1}{4}(\sigma(t) + t) - 1 = \frac{1}{4}(3t + t) - 1 = t - 1,$$

$$f_2^\Delta(t) = \frac{1}{4}(\sigma(t) + t) - 9 = \frac{1}{4}(3t + t) - 9 = t - 9, \quad t \in \mathbb{T},$$

$$h_2(9,1) = \int_1^9 (t-1)\Delta t = \int_1^9 f_1^\Delta(t)\Delta t = f_1(9) - f_1(1)$$

$$= \frac{81}{4} - 9 - \frac{1}{4} + 1 = 12,$$

$$h_2(1,9) = \int_9^1 (t-9)\Delta t = \int_9^1 f_2^\Delta(t)\Delta t = f_2(1) - f_2(9)$$

$$= \left(\frac{1}{4} - 9\right) - \left(\frac{81}{4} - 81\right) = \frac{1}{4} - \frac{81}{4} - 9 + 81 = 52,$$

$$f(1) = 1,$$

$$f(9) = \frac{729 - 1}{729 + 9 + 1} + e_1(9,1) = \frac{728}{739} + e_1(9,1).$$

Now, using the trapezoid rule, we get

$$\int_1^9 f(t)\Delta t \approx \frac{1}{8}\left(52 + 12 \cdot \left(\frac{728}{739} + e_1(9,1)\right)\right) = \frac{1}{2}\left(13 + 3 \cdot \left(\frac{728}{739} + e_1(9,1)\right)\right).$$

Exercise 2.6. Let $\mathbb{T} = 2\mathbb{Z}$. Using the trapezoid rule, evaluate the integral

$$\int_1^8 \left(\frac{t+1}{t^2 + 3t + 4} + e_2(t,1)\right)\Delta t.$$

Next, we will derive the Simpson rule. In its derivation we need the following definition and lemma.

Definition 2.7. For $t, s \in \mathbb{T}$, define the monomials

$$H_0(t,s) = 1, \quad H_{k+1}(t,s) = \int_s^t h_k(\sigma(\tau), s)\Delta\tau, \quad k \in \mathbb{N}_0.$$

Lemma 2.8. *Let $t, s \in \mathbb{T}$ and suppose $t_1, t_2 \in \mathbb{T}$ are between s and t. Then*

$$\int_s^t (y - t_1)(y - t_2)\Delta y = h_2(t, t_2)h_1(t, t_1) - h_2(s, t_1)h_1(s, t_2)$$

$$+ H_3(s, t_1) - H_3(t_2, t_1) - H_3(t, t_2).$$

Proof. We have

$$\int_s^t (y - t_1)(y - t_2)\Delta y$$

$$= \int_s^{t_1} (y - t_1)(y - t_2)\Delta y + \int_{t_1}^{t_2} (y - t_1)(y - t_2)\Delta y + \int_{t_2}^{t} (y - t_1)(y - t_2)\Delta y$$

$$= -\int_{t_1}^{s} (y - t_1)(y - t_2)\Delta y + \int_{t_1}^{t_2} h_2^{\Delta}(y, t_1)(y - t_2)\Delta y + \int_{t_2}^{t} h_2^{\Delta}(y, t_2)(y - t_1)\Delta y$$

$$= -\int_{t_1}^{s} h_2^{\Delta}(y, t_1)(y - t_2)\Delta y + h_2(y, t_1)(y - t_2)|_{y=t_1}^{y=t_2}$$

$$- \int_{t_1}^{t_2} h_2(\sigma(y), t_1)\Delta y + h_2(y, t_2)(y - t_1)|_{y=t_2}^{y=t} - \int_{t_2}^{t} h_2(\sigma(y), t_2)\Delta y$$

$$= -h_2(y, t_1)(y - t_2)|_{y=t_1}^{y=s} + \int_{t_1}^{s} h_2(\sigma(y), t_1)\Delta y$$

$$- H_3(t_2, t_1) + h_2(t, t_2)(t - t_1) - H_3(t, t_2)$$

$$= -h_2(s, t_1)h_1(s, t_2) + H_3(s, t_1) - H_3(t_2, t_1) + h_2(t, t_2)h_1(t, t_1) - H_3(t, t_2).$$

This completes the proof. □

Now, take $n = 2$ in the Newton–Cotes formulae (2.1), that is, we take three points

$$a = x_0 < x_1 < x_2 = b.$$

For $x \in [a, b]$, using Lemma 2.8, we get

$$\int_a^b L_0(x)\Delta x = \frac{1}{(x_0 - x_1)(x_0 - x_2)} \int_{x_0}^{x_2} (x - x_1)(x - x_2)\Delta x$$

$$= \frac{1}{(x_0 - x_1)(x_0 - x_2)} (h_2(x_2, x_2)h_1(x_2, x_1) - h_2(x_0, x_1)h_1(x_0, x_2))$$

$$+ H_3(x_0, x_1) - H_3(x_2, x_1) - H_3(x_2, x_2))$$

$$= \frac{1}{(x_0 - x_1)(x_0 - x_2)}(-h_2(x_0, x_1)h_1(x_0, x_2) + H_3(x_0, x_1) - H_3(x_2, x_1)),$$

$$\int_a^b L_1(x)\Delta x = \frac{1}{(x_1 - x_0)(x_1 - x_2)} \int_{x_0}^{x_2} (x - x_0)(x - x_2)\Delta x$$

$$= \frac{1}{(x_1 - x_0)(x_1 - x_2)}(h_2(x_2, x_2)h_1(x_2, x_0) - h_2(x_0, x_0)h_1(x_0, x_2)$$

$$+ H_3(x_0, x_0) - H_3(x_2, x_0) - H_3(x_2, x_2))$$

$$= -\frac{H_3(x_2, x_0)}{(x_1 - x_0)(x_1 - x_2)},$$

$$\int_a^b L_2(x)\Delta x = \frac{1}{(x_2 - x_0)(x_2 - x_1)} \int_{x_0}^{x_2} (x - x_0)(x - x_1)\Delta x$$

$$= \frac{1}{(x_2 - x_0)(x_2 - x_1)}(h_2(x_2, x_1)h_1(x_2, x_0) - h_2(x_0, x_0)h_1(x_0, x_1)$$

$$+ H_3(x_0, x_0) - H_3(x_1, x_0) - H_3(x_2, x_1))$$

$$= \frac{1}{(x_2 - x_0)(x_2 - x_1)}(h_2(x_2, x_1)h_1(x_2, x_0) - H_3(x_1, x_0) - H_3(x_2, x_1)).$$

Hence, using (2.1), we find

$$\int_a^b f(x)\Delta x \approx \frac{f(x_0)}{(x_0 - x_1)(x_0 - x_2)}(-h_2(x_0, x_1)h_1(x_0, x_2) + H_3(x_0, x_1) - H_3(x_2, x_1))$$

$$- f(x_1)\frac{H_3(x_2, x_0)}{(x_1 - x_0)(x_1 - x_2)}$$

$$+ \frac{f(x_2)}{(x_2 - x_0)(x_2 - x_1)}(h_2(x_2, x_1)h_1(x_2, x_0) - H_3(x_1, x_0) - H_3(x_2, x_1)). \quad (2.3)$$

Definition 2.9. The numerical integration formula (2.3) is said to be the Simpson rule.

We apply the Simpson rule to specific examples.

Example 2.10. Let $\mathbb{T} = 2^{\mathbb{N}_0}$, $a = 1$, $b = 8$, and $f : \mathbb{T} \to \mathbb{R}$ be defined as

$$f(t) = \frac{t - 1}{t + 1}, \quad t \in \mathbb{T}.$$

We will evaluate $\int_1^8 f(t)\Delta t$ using the Simpson rule. For this problem, we have $\sigma(t) = 2t$, $\mu(t) = 2t - t = t, t \in \mathbb{T}$. Let

$$g_1(t) = \frac{t^2}{3}, \quad g_2(t) = \frac{t^3}{7}, \quad t \in \mathbb{T}.$$

We have

$$g_1^\Delta(t) = \frac{\sigma(t) + t}{3} = t, \quad g_2^\Delta(t) = \frac{(\sigma(t))^2 + t\sigma(t) + t^2}{7} = t^2, \quad t \in \mathbb{T}.$$

Then we compute

$$h_0(t,s) = 1, \quad h_1(t,s) = \int_s^t \Delta\tau = t - s,$$

$$h_2(t,s) = \int_s^t (\tau - s)\Delta\tau$$

$$= \int_s^t g_1^\Delta \Delta\tau - s(t - s)$$

$$= g_1(\tau)|_{\tau=s}^{\tau=t} - st + s^2 = \frac{\tau^2}{3}\Big|_{\tau=s}^{\tau=t} - st + s^2 = \frac{t^2}{3} - st + \frac{2s^2}{3},$$

$$H_3(t,s) = \int_s^t h_2(\sigma(\tau), s)\Delta\tau$$

$$= \int_s^t \left(\frac{(2\tau)^2}{3} - s(2\tau) + \frac{2s^2}{3} \right)\Delta\tau$$

$$= \int_s^t \left(\frac{4}{3}g_2^\Delta - 2sg_1^\Delta + \frac{2s^2}{3} \right)\Delta\tau$$

$$= \left(\frac{4}{3}g_2(\tau) - 2sg_1(\tau) + \frac{2s^2}{3}\tau \right)\Big|_{\tau=s}^{\tau=t}$$

$$= \frac{4}{21}(t^3 - s^3) - \frac{2s}{3}(t^2 - s^2) + \frac{2s^2}{3}(t - s)$$

$$= \frac{4}{21}t^3 - \frac{2}{3}st^2 + \frac{2}{3}s^2t - \frac{4}{21}s^3, \quad t, s \in \mathbb{T}.$$

We will apply the Simpson rule with $a = x_0 = 1$, $x_1 = 4$, $x_2 = b = 8$. Using the values $f(1) = 0$, $f(4) = \frac{3}{5}$, and $f(8) = \frac{7}{9}$, we compute

$$\int_1^8 f(t)\Delta t \approx \frac{f(1)}{(1-4)(1-8)}(-h_2(1,4)h_1(1,8) + H_3(1,4) - H_3(8,4))$$

$$- \frac{f(4)}{(4-1)(4-8)}H_3(8,1)$$

$$+ \frac{f(8)}{(8-1)(8-4)}(h_2(8,4)h_1(8,1) - H_3(4,1) - H_3(8,4))$$

$$= 0 - \frac{\frac{3}{5}}{-12}60 + \frac{\frac{7}{9}}{28}(-4) = \frac{26}{9} \approx 2.88889.$$

Recall that the approximate value of the same integral computed by using the trape-zoid rule was obtained in Example 2.4 as $\frac{14}{9} \approx 1.55556$. In fact, it is possible to compute the exact value of the given integral as

$$\int\limits_1^8 f(t)\Delta t = \overset{\sigma(1)}{\int\limits_1} f(t)\Delta t + \overset{\sigma(2)}{\int\limits_2} f(t)\Delta t + \overset{\sigma(4)}{\int\limits_4} f(t)\Delta t$$

$$= f(1)\mu(1) + f(2)\mu(2) + f(4)\mu(4)$$

$$= 0 + 2\frac{1}{3} + 4\frac{3}{5} = \frac{2}{3} + \frac{12}{5} = \frac{46}{15} \approx 3.06667.$$

This example demonstrates that the Simpson rule is more accurate than the trape-zoid rule since we approximate the function to be integrated by a polynomial of one higher degree.

Exercise 2.11. Let $\mathbb{T} = 3\mathbb{Z}$. Using the Simpson rule, evaluate the integral

$$\int\limits_{-6}^9 \frac{\Delta t}{t^2 + t + 1}.$$

2.2 σ-Newton–Cotes formulae

Recalling the alternative interpolation defined in Chapter 1, called σ interpolation, we consider an alternative to the Newton–Cotes formulae which we will call σ-New-ton–Cotes formulae. In this section, we describe this approach and evaluate approxi-mately a Cauchy time scale integral using the σ-Lagrange interpolation polynomial.

Suppose that $f : \mathbb{T} \to \mathbb{R}$ is rd-continuous, $n \in \mathbb{N}_0$, $a, b \in \mathbb{T}$, $\sigma(a) < b$, $x_j \in [a, b] \subset \mathbb{T}$, $j \in \{0, 1, \ldots, n\}$, are σ-distinct points. Then

$$p_{\sigma n}(x) = \sum_{k=0}^n L_{\sigma k}(x) f(x_k), \quad x \in [a, b],$$

and

$$\int\limits_a^b p_{\sigma n}(x)\Delta x = \sum_{k=0}^n \left(\int\limits_a^b L_{\sigma k}(x)\Delta x \right) f(x_k).$$

Hence,

$$\int\limits_a^b f(x)\Delta x \approx \int\limits_a^b p_{\sigma n}(x)\Delta x = \sum_{k=0}^n \left(\int\limits_a^b L_{\sigma k}(x)\Delta x \right) f(x_k).$$

Denote

$$w_{\sigma k} = \int_a^b L_{\sigma k}(x)\Delta x, \quad k \in \{0,1,\dots,n\}.$$

Then

$$\int_a^b f(x)\Delta x \approx \sum_{k=0}^n w_{\sigma k} f(x_k). \tag{2.4}$$

Definition 2.12. The values $w_{\sigma k}$, $k \in \{0,1,\dots,n\}$, will be called the σ-quadrature weights and the σ-interpolation points x_j, $j \in \{0,1,\dots,n\}$, will be called the σ-quadrature points (nodes).

Definition 2.13. The numerical quadrature formula (2.4) is said to be σ-Newton–Cotes formula.

Next, we recall the definition of the polynomials [1, 2, 8]

$$g_0(t,s) = 1, \quad g_{k+1}(t,s) = \int_s^t g_k(\sigma(\tau),s)\Delta\tau, \quad k \in \mathbb{N}_0,$$

for $t,s \in \mathbb{T}$. First, we take $n = 1$ in the σ-Newton–Cotes formula given in (2.4), that is, we take

$$a = x_0 \le \sigma(x_0) < x_1 \le \sigma(x_1) = b.$$

Then

$$\begin{aligned}
p_{\sigma 1}(x) &= L_{\sigma 0}(x)f(x_0) + L_{\sigma 1}(x)f(x_1)\\
&= \frac{\sigma(x) - \sigma(x_1)}{\sigma(x_0) - \sigma(x_1)}f(x_0) + \frac{\sigma(x) - \sigma(x_0)}{\sigma(x_1) - \sigma(x_0)}f(x_1)\\
&= \frac{\sigma(x) - b}{\sigma(a) - b}f(a) + \frac{\sigma(x) - \sigma(a)}{b - \sigma(a)}f(x_1)\\
&= \frac{1}{b - \sigma(a)}(-(\sigma(x) - b)f(a) + (\sigma(x) - \sigma(a))f(x_1)), \quad x \in [a,b],
\end{aligned}$$

and

$$\begin{aligned}
\int_a^b p_{\sigma 1}(x)\Delta x &= \frac{1}{b - \sigma(a)}\left(-f(a)\int_a^b(\sigma(x) - b)\Delta x + f(x_1)\int_a^b(\sigma(x) - \sigma(a))\Delta x\right)\\
&= \frac{1}{b - \sigma(a)}\left(f(a)\int_b^a(\sigma(x) - b)\Delta x + f(x_1)\int_a^{\sigma(a)}(\sigma(x) - \sigma(a))\Delta x\right)
\end{aligned}$$

$$+ f(x_1) \int_{\sigma(a)}^{b} (\sigma(x) - \sigma(a))\Delta x \bigg)$$

$$= \frac{1}{b - \sigma(a)} (f(a)g_2(a, b) + f(x_1)\mu(a)(\sigma(a) - \sigma(a)) + f(x_1)g_2(b, \sigma(a)))$$

$$= \frac{1}{b - \sigma(a)} (f(a)g_2(a, b) + f(x_1)g_2(b, \sigma(a))).$$

Consequently, we get

$$\int_{a}^{b} f(x)\Delta x \approx \frac{1}{b - \sigma(a)} (f(a)g_2(a, b) + f(x_1)g_2(b, \sigma(a))). \tag{2.5}$$

Definition 2.14. The numerical integration formula (2.5) will be called the σ-trapezoid rule.

Below, we apply the σ-trapezoid rule to an example.

Example 2.15. Let $\mathbb{T} = \{\frac{1}{4}, \frac{1}{3}, \frac{1}{2}, 1\}$,

$$a = \frac{1}{4} = x_0, \quad x_1 = \frac{1}{2}, \quad b = \sigma\left(\frac{1}{2}\right) = 1,$$

and suppose that the function $f : \mathbb{T} \rightarrow \mathbb{R}$ is defined by

$$f(t) = \frac{t - 1}{t + 1}, \quad t \in \mathbb{T}.$$

We will evaluate

$$\int_{\frac{1}{4}}^{1} f(t)\Delta t.$$

We have

$$\sigma(x_0) = \sigma\left(\frac{1}{4}\right) = \frac{1}{3} < b, \quad \mu\left(\frac{1}{4}\right) = \sigma\left(\frac{1}{4}\right) - \frac{1}{4} = \frac{1}{3} - \frac{1}{4} = \frac{1}{12},$$

$$\sigma\left(\frac{1}{3}\right) = \frac{1}{2}, \quad \mu\left(\frac{1}{3}\right) = \sigma\left(\frac{1}{3}\right) - \frac{1}{3} = \frac{1}{2} - \frac{1}{3} = \frac{1}{6},$$

$$\sigma(x_1) = \sigma\left(\frac{1}{2}\right) = 1, \quad \mu\left(\frac{1}{2}\right) = \sigma\left(\frac{1}{2}\right) - \frac{1}{2} = 1 - \frac{1}{2} = \frac{1}{2}.$$

Furthermore,

$$g_2(a, b) = \int_1^{\frac{1}{4}} (\sigma(t) - 1)\Delta t$$

$$= -\int_{\frac{1}{4}}^1 (\sigma(t) - 1)\Delta t$$

$$= -\mu\left(\frac{1}{4}\right)\left(\sigma\left(\frac{1}{4}\right) - 1\right) - \mu\left(\frac{1}{3}\right)\left(\sigma\left(\frac{1}{3}\right) - 1\right)$$
$$\quad - \mu\left(\frac{1}{2}\right)\left(\sigma\left(\frac{1}{2}\right) - 1\right)$$

$$= -\frac{1}{12}\left(\frac{1}{3} - 1\right) - \frac{1}{6}\left(\frac{1}{2} - 1\right) - \frac{1}{2}(1 - 1)$$

$$= -\frac{1}{12}\left(-\frac{2}{3}\right) - \frac{1}{6}\left(-\frac{1}{2}\right) = \frac{5}{36},$$

$$g_2(b, \sigma(a)) = \int_{\frac{1}{3}}^1 \left(\sigma(t) - \frac{1}{3}\right)\Delta t$$

$$= \int_{\frac{1}{3}}^{\frac{1}{2}} \left(\sigma(t) - \frac{1}{3}\right)\Delta t + \int_{\frac{1}{2}}^1 \left(\sigma(t) - \frac{1}{3}\right)\Delta t$$

$$= \mu\left(\frac{1}{3}\right)\left(\sigma\left(\frac{1}{3}\right) - \frac{1}{3}\right) + \mu\left(\frac{1}{2}\right)\left(\sigma\left(\frac{1}{2}\right) - \frac{1}{3}\right)$$

$$= \frac{1}{6}\left(\frac{1}{2} - \frac{1}{3}\right) + \frac{1}{2}\left(1 - \frac{1}{3}\right) = \frac{13}{36},$$

$$f(a) = f\left(\frac{1}{4}\right) = \frac{\frac{1}{4} - 1}{\frac{1}{4} + 1} = -\frac{3}{5}, \quad f(x_1) = f\left(\frac{1}{2}\right) = \frac{\frac{1}{2} - 1}{\frac{1}{2} + 1} = -\frac{1}{3}.$$

Hence,

$$\int_{\frac{1}{4}}^1 \frac{t - 1}{t + 1}\Delta t \approx \frac{1}{1 - \sigma(\frac{1}{4})}\left(f\left(\frac{1}{4}\right)g_2\left(\frac{1}{4}, 1\right) + f\left(\frac{1}{2}\right)g_2\left(1, \frac{1}{3}\right)\right)$$

$$= \frac{1}{1 - \frac{1}{3}}\left(-\frac{3}{5} \cdot \frac{5}{36} - \frac{1}{3} \cdot \frac{13}{36}\right)$$

$$= \frac{3}{2}\left(-\frac{1}{12} - \frac{13}{108}\right) = -\frac{11}{36}.$$

Exercise 2.16. Let

$$\mathbb{T} = \left\{-2, -\frac{3}{2}, -1, -\frac{5}{6}, -\frac{1}{2}, -\frac{1}{3}, 0, \frac{1}{8}, \frac{1}{4}, \frac{3}{2}, 2\right\},$$

$$a = -2 = x_0, \quad x_1 = -\frac{5}{6}, \quad b = 2,$$

and suppose that $f : \mathbb{T} \to \mathbb{R}$ is defined by

$$f(t) = t^4 + t^3 + \frac{t+1}{t^2+1}, \quad t \in \mathbb{T}.$$

Using the σ-trapezoid rule, evaluate the integral $\int_{-2}^{2} f(t)\Delta t$.

We proceed with the following definition and lemma which are needed to derive the σ-Simpson rule.

Definition 2.17. For $s, t \in \mathbb{T}$, define the polynomials

$$G_0(t,s) = 1, \quad G_{k+1}(t,s) = \int_s^t g_k(\tau, s)\Delta\tau, \quad k \in \mathbb{N}_0.$$

Lemma 2.18. *Let $t, s \in \mathbb{T}$ and $t_1, t_2, \sigma(t_1), \sigma(t_2)$ be between s and t. Then*

$$\int_s^t (\sigma(y) - \sigma(t_1))(\sigma(y) - \sigma(t_2))\Delta y$$

$$= -g_2(s, \sigma(t_1))g_1(s, \sigma(t_2)) + g_1(t, \sigma(t_1))g_2(t, \sigma(t_2))$$
$$+ G_3(s, \sigma(t_1)) - G_3(\sigma(t_2), \sigma(t_1)) - G_3(t, \sigma(t_2)).$$

Proof. We have

$$\int_s^t (\sigma(y) - \sigma(t_1))(\sigma(y) - \sigma(t_2))\Delta y$$

$$= \int_s^{\sigma(t_1)} (\sigma(y) - \sigma(t_1))(\sigma(y) - \sigma(t_2))\Delta y$$

$$+ \int_{\sigma(t_1)}^{\sigma(t_2)} (\sigma(y) - \sigma(t_1))(\sigma(y) - \sigma(t_2))\Delta y + \int_{\sigma(t_2)}^{t} (\sigma(y) - \sigma(t_1))(\sigma(y) - \sigma(t_2))\Delta y$$

$$= -\int_{\sigma(t_1)}^{s} g_2^\Delta(y, \sigma(t_1))(\sigma(y) - \sigma(t_2))\Delta y$$

$$+ \int_{\sigma(t_1)}^{\sigma(t_2)} g_2^\Delta(y, \sigma(t_1))(\sigma(y) - \sigma(t_2))\Delta y + \int_{\sigma(t_2)}^{t} (\sigma(y) - \sigma(t_1))g_2^\Delta(y, \sigma(t_2))\Delta y$$

$$= -g_2(y, \sigma(t_1))(y - \sigma(t_2))|_{y=\sigma(t_1)}^{y=s} + \int_{\sigma(t_1)}^{s} g_2(y, \sigma(t_1))\Delta y$$

$$+ g_2(y, \sigma(t_1))(y - \sigma(t_2))|_{y=\sigma(t_1)}^{y=\sigma(t_2)} - \int_{\sigma(t_1)}^{\sigma(t_2)} g_2(y, \sigma(t_1))\Delta y$$

$$+ (y - \sigma(t_1))g_2(y, \sigma(t_2))|_{y=\sigma(t_2)}^{y=t} - \int_{\sigma(t_2)}^{t} g_2(y, \sigma(t_2))\Delta y$$

$$= -g_2(s, \sigma(t_1))(s - \sigma(t_2)) + g_2(\sigma(t_1), \sigma(t_1))(\sigma(t_1) - \sigma(t_2)) + G_3(s, \sigma(t_1))$$
$$+ g_2(\sigma(t_2), \sigma(t_1))(\sigma(t_2) - \sigma(t_2))$$
$$- g_2(\sigma(t_1), \sigma(t_1))(\sigma(t_1) - \sigma(t_2)) - G_3(\sigma(t_2), \sigma(t_1))$$
$$+ (t - \sigma(t_1))g_2(t, \sigma(t_2)) - (\sigma(t_2) - \sigma(t_1))g_2(\sigma(t_2), \sigma(t_2)) - G_3(t, \sigma(t_2))$$
$$= -g_2(s, \sigma(t_1))g_1(s, \sigma(t_2)) + g_1(t, \sigma(t_1))g_2(t, \sigma(t_2))$$
$$+ G_3(s, \sigma(t_1)) - G_3(\sigma(t_2), \sigma(t_1)) - G_3(t, \sigma(t_2)).$$

This completes the proof. □

Let now

$$a = x_0 < x_1 \le \sigma(x_1) < x_2 \le \sigma(x_2) = b.$$

Then, using Lemma 2.18, we get

$$\int_{x_0}^{\sigma(x_2)} L_{\sigma 0}(y)\Delta y = \frac{1}{(\sigma(x_0) - \sigma(x_1))(\sigma(x_0) - \sigma(x_2))}(-g_2(x_0, \sigma(x_1)), g_1(x_0, \sigma(x_2))$$

$$+ g_1(\sigma(x_2), \sigma(x_1))g_2(\sigma(x_2), \sigma(x_2))$$
$$+ G_3(x_0, \sigma(x_1)) - G_3(\sigma(x_2), \sigma(x_1)) - G_3(\sigma(x_2), \sigma(x_2)))$$

$$= \frac{1}{(\sigma(x_0) - \sigma(x_1))(\sigma(x_0) - \sigma(x_2))}(-g_2(x_0, \sigma(x_1))g_1(x_0, \sigma(x_2))$$
$$+ G_3(x_0, \sigma(x_1)) - G_3(\sigma(x_2), \sigma(x_1))),$$

$$\int_{x_0}^{\sigma(x_2)} L_{\sigma 1}(y)\Delta y = \frac{1}{(\sigma(x_1) - \sigma(x_0))(\sigma(x_1) - \sigma(x_2))}(-g_2(x_0, \sigma(x_0))g_1(x_0, \sigma(x_2))$$

$$+ g_1(\sigma(x_2), \sigma(x_0))g_2(\sigma(x_2), \sigma(x_2))$$
$$+ G_3(x_0, \sigma(x_0)) - G_3(\sigma(x_2), \sigma(x_0)) - G_3(\sigma(x_2), \sigma(x_2)))$$

$$= -\frac{1}{(\sigma(x_1) - \sigma(x_0))(\sigma(x_1) - \sigma(x_2))} G_3(\sigma(x_2), \sigma(x_0)),$$

$$\int_{x_0}^{\sigma(x_2)} L_{\sigma 2}(y)\Delta y = \frac{1}{(\sigma(x_2) - \sigma(x_0))(\sigma(x_2) - \sigma(x_1))}(-g_2(x_0, \sigma(x_0))g_1(x_0, \sigma(x_1))$$

$$+ g_1(\sigma(x_2), \sigma(x_0))g_2(\sigma(x_2), \sigma(x_1))$$
$$+ G_3(x_0, \sigma(x_0)) - G_3(\sigma(x_1), \sigma(x_0)) - G_3(\sigma(x_2), \sigma(x_1)))$$

$$= \frac{1}{(\sigma(x_2) - \sigma(x_0))(\sigma(x_2) - \sigma(x_1))}$$
$$\times (g_1(\sigma(x_2), \sigma(x_0))g_2(\sigma(x_2), \sigma(x_1))$$
$$- G_3(\sigma(x_1), \sigma(x_0)) - G_3(\sigma(x_2), \sigma(x_1))).$$

Hence and (2.4), we get

$$\int_a^b f(x)\Delta x \approx \frac{f(x_0)}{(\sigma(x_0) - \sigma(x_1))(\sigma(x_0) - \sigma(x_2))}(-g_2(x_0, \sigma(x_1))g_1(x_0, \sigma(x_2))$$

$$+ G_3(x_0, \sigma(x_1)) - G_3(\sigma(x_2), \sigma(x_1)))$$

$$- \frac{f(x_1)}{(\sigma(x_1) - \sigma(x_0))(\sigma(x_1) - \sigma(x_2))}G_3(\sigma(x_2), \sigma(x_0))$$

$$+ \frac{f(x_2)}{(\sigma(x_2) - \sigma(x_0))(\sigma(x_2) - \sigma(x_1))}(g_1(\sigma(x_2), \sigma(x_0))g_2(\sigma(x_2), \sigma(x_1))$$

$$- G_3(\sigma(x_1), \sigma(x_0)) - G_3(\sigma(x_2), \sigma(x_1))). \tag{2.6}$$

Definition 2.19. The numerical integration formula (2.6) will be called the σ-Simpson rule.

Example 2.20. Consider the time scale $\mathbb{T} = 2^{\mathbb{N}_0}$ and let $[a, b] = [1, 32]$. We have $\sigma(t) = 2t$, $\mu(t) = t$, $t \in \mathbb{T}$. On this time scale, we compute

$$g_0(t, s) = 1,$$

$$g_1(t, s) = \int_s^t g_0((\sigma(\tau), s)\Delta\tau = t - s,$$

$$g_2(t, s) = \int_s^t g_1((\sigma(\tau), s)\Delta\tau$$

$$= \int_s^t (2\tau - s)\Delta\tau = \left(\frac{2\tau^2}{3} - s\tau\right)\Big|_{\tau=s}^{\tau=t}$$

$$= \frac{2t^2}{3} - st + \frac{s^2}{3},$$

$$G_3(t,s) = \int_s^t g_2(\tau, s)\Delta\tau$$

$$= \int_s^t \left(\frac{2t^2}{3} - st + \frac{s^2}{3} \right)\Delta\tau$$

$$= \left(\frac{2}{21}\tau^3 - \frac{1}{3}s\tau^2 + \frac{s^2}{3}\tau \right)\Big|_{\tau=s}^{\tau=t}$$

$$= \frac{2}{21}t^3 - \frac{1}{3}st^2 + \frac{1}{3}s^2t - \frac{2}{21}s^3, \quad t, s \in \mathbb{T}.$$

Let $a = x_0 = 1$, $x_1 = 16$, $\sigma(x_1) = b = 32$. Applying the σ-trapezoid rule for any function $f : \mathbb{T} \to \mathbb{R}$ gives

$$\int_1^{32} f(t)\Delta t \approx \frac{1}{32 - 2}(f(1)g_2(1, 32) + f(16)g_2(32, 2)).$$

Let $a = x_0 = 1$, $x_1 = 4$, $x_2 = 16$, $\sigma(x_2) = b = 32$. The σ-Simpson rule for any function $f : \mathbb{T} \to \mathbb{R}$ yields

$$\int_1^{32} f(t)\Delta t \approx \frac{f(1)}{(2 - 8)(2 - 32)}(-g_2(1, 8)g_1(1, 32) + G_3(1, 8) - G_3(32, 8))$$

$$- \frac{f(4)}{(8 - 2)(8 - 32)}G_3(32, 2)$$

$$+ \frac{f(16)}{(32 - 2)(32 - 8)}(g_2(32, 8)g_1(32, 2) - G_3(8, 2) - G_3(32, 8)).$$

It is easy to compute the exact value of the integral as

$$\int_1^{32} f(t)\Delta t = f(1)\mu(1) + f(2)\mu(2) + f(4)\mu(4) + f(8)\mu(8) + f(16)\mu(16)$$

$$= f(1) + 2f(2) + 4f(4) + 8f(8) + 16f(16).$$

The σ-trapezoid rule will give the exact value of the integral if the integrand function is a linear polynomial. Taking $f(t) = 2t + 5$, $t \in \mathbb{T}$, we compute the integral with the σ-trapezoid rule as

$$\int_1^{32}(2t + 5)\Delta t \approx \frac{1}{30}(7g_2(1, 32) + 37g_2(32, 2)) = 837,$$

and with the σ-Simpson rule as

$$\int_1^{32}(2t+5)\Delta t \approx \frac{7}{180}(-g_2(1,8)g_1(1,32) + G_3(1,8) - G_3(32,8))$$

$$+ \frac{13}{144}G_3(32,2) + \frac{37}{720}(g_2(32,8)g_1(32,2) - G_3(8,2) - G_3(32,8))$$

$$= 837.$$

The exact value of the integral is

$$\int_1^{32}(2t+5)\Delta t = 7 + 18 + 52 + 168 + 592 = 837.$$

The σ-Simpson rule will give the exact value of the integral if the integrand function is a second degree polynomial. Taking $f(t) = t^2 + t + 2$, $t \in \mathbb{T}$, we compute the integral with the σ-trapezoid rule as

$$\int_1^{32}(t^2+t+2)\Delta t \approx \frac{1}{30}(4g_2(1,32) + 274g_2(32,2)) = 5704,$$

and with the σ-Simpson rule as

$$\int_1^{32}(t^2+t+2)\Delta t \approx \frac{4}{180}(-g_2(1,8)g_1(1,32) + G_3(1,8) - G_3(32,8))$$

$$+ \frac{22}{144}G_3(32,2) + \frac{274}{720}(g_2(32,8)g_1(32,2) - G_3(8,2) - G_3(32,8))$$

$$= 5084.$$

The exact value of the integral is

$$\int_1^{32}(t^2+t+2)\Delta t = 4 + 16 + 88 + 592 + 4384 = 5084.$$

Finally, we choose $f(t) = \frac{t+2}{t^2+7}$, $t \in \mathbb{T}$. Then we compute the integral with the σ-trapezoid rule as

$$\int_1^{32}\frac{t+2}{t^2+7}\Delta t \approx \frac{1}{30}\left(\frac{3}{8}g_2(1,32) + \frac{18}{263}g_2(32,2)\right) \approx 5.2894,$$

and with the σ-Simpson rule as

$$\int_1^{32} \frac{t+2}{t^2+7} \Delta t \approx \frac{\frac{3}{8}}{180}(-g_2(1,8)g_1(1,32) + G_3(1,8) - G_3(32,8))$$

$$+ \frac{\frac{6}{23}}{144} G_3(32,2) + \frac{\frac{18}{263}}{720}(g_2(32,8)g_1(32,2) - G_3(8,2) - G_3(32,8))$$

$$\approx 4.3798.$$

The exact value of the integral is

$$\int_1^{32} \frac{t+2}{t^2+7} \Delta t = \frac{3}{8} + \frac{8}{11} + \frac{24}{23} + \frac{80}{71} + \frac{288}{263} \approx 4.3676.$$

Exercise 2.21. Let

$$\mathbb{T} = \left\{2, \frac{5}{2}, \frac{8}{3}, 3, \frac{13}{4}, \frac{15}{4}, 4\right\},$$

$$a = 2 = x_0, \quad x_1 = \frac{8}{3}, \quad x_2 = \frac{15}{4}, \quad b = 4,$$

and suppose that $f : \mathbb{T} \to \mathbb{R}$ is defined by

$$f(t) = t^2 + \frac{1+t+t^2}{1+t^4+t^6}, \quad t \in \mathbb{T}.$$

Using the σ-Simpson rule, evaluate the integral $\int_a^b f(t)\Delta t$.

2.3 Error estimates

Our aim in this section is to evaluate the size of the error in the numerical integration formula (2.1). The error in (2.1) is defined by

$$E_n(f) = \int_a^b f(x)\Delta x - \sum_{k=0}^n w_k f(x_k).$$

Theorem 2.22. *Let $n \in \mathbb{N}$, $a, b \in \mathbb{T}$, $a < b$. Suppose that $f : [a,b] \to \mathbb{R}, f \in C^{n+1}([a,b])$. Then*

$$E_n(f) = \int_a^b \frac{f^{\Delta^{n+1}}(\xi)}{\Pi_{n+1}^{n+1}(\xi)} \pi_{n+1}(x)\Delta x,$$

or

$$\int_a^b G_{min,n+1}(\xi,x)\Delta x \le E_n(f) \le \int_a^b G_{max,n+1}(\xi,x)\Delta x,$$

where

$$G_{max,n+1}(\xi,x) = \begin{cases} F_{max,n+1}(\xi)\pi_{n+1}(x) & \text{if } \pi_{n+1}(x) > 0, \\ F_{min,n+1}(\xi)\pi_{n+1}(x) & \text{if } \pi_{n+1}(x) \le 0, \end{cases}$$

$$G_{min,n+1}(\xi,x) = \begin{cases} F_{min,n+1}(\xi)\pi_{n+1}(x) & \text{if } \pi_{n+1}(x) > 0, \\ F_{max,n+1}(\xi)\pi_{n+1}(x) & \text{if } \pi_{n+1}(x) \le 0. \end{cases}$$

Proof. We have that

$$E_n(f) = \int_a^b (f(x) - p_n(x))\Delta x,$$

where $p_n \in \mathcal{P}_n$ is the Lagrange interpolation polynomial for the function f. Hence, by Theorem 1.10, we get the desired result. This completes the proof. $\qquad\square$

2.4 σ-Error estimates

Now, we will evaluate the size of the error in the numerical integration formula (2.4). We define the σ-error in (2.4) as follows:

$$E_{\sigma n}(f) = \int_a^b f(x)\Delta x - \sum_{k=0}^n w_{\sigma k} f(x_k).$$

Theorem 2.23. *Let* $n \in \mathbb{N}$, $a, b \in \mathbb{T}$, $a < b$. *Suppose that* $f : [a,b] \to \mathbb{R}$, $f \in C^{n+1}([a,b])$. *Then*

$$E_{\sigma n}(f) = \int_a^b \frac{f^{\Delta^{n+1}}(\xi)}{\Pi_{\sigma n+1}^{n+1}(\xi)} \pi_{\sigma n+1}(x)\Delta x,$$

or

$$\int_a^b G_{\sigma\,min,n+1}(\xi,x)\Delta x \le E_{\sigma n}(f) \le \int_a^b G_{\sigma\,max,n+1}(\xi,x)\Delta x,$$

where

$$G_{\sigma\,max,n+1}(\xi,x) = \begin{cases} F_{\sigma\,max,n+1}(\xi)\pi_{\sigma n+1}(x) & \text{if } \pi_{\sigma n+1}(x) > 0, \\ F_{\sigma\,min,n+1}(\xi)\pi_{\sigma n+1}(x) & \text{if } \pi_{\sigma n+1}(x) \le 0, \end{cases}$$

$$G_{\sigma\min,n+1}(\xi, x) = \begin{cases} F_{\sigma\min,n+1}(\xi)\pi_{\sigma n+1}(x) & \text{if } \pi_{\sigma n+1}(x) > 0, \\ F_{\sigma\max,n+1}(\xi)\pi_{\sigma n+1}(x) & \text{if } \pi_{\sigma n+1}(x) \leq 0. \end{cases}$$

Proof. Observe that

$$E_{\sigma n}(f) = \int_a^b (f(x) - p_{\sigma n}(x))\Delta x,$$

where $p_{\sigma n} \in \mathcal{P}_n^\sigma$ is the σ-Lagrange interpolation polynomial for the function f. Hence, by Theorem 1.28, we get the expected σ-error estimate. This completes the proof. □

2.5 Composite quadrature rules

In this section, we will derive the composite trapezoid and Simpson rules. The idea of composite quadrature rules is to increase the number of nodes in the interval of integration and thus decompose it into a union of disjoint subintervals. Then the related quadrature rule is applied on each of the subintervals. Naturally, this approach will reduce the error in the computation of the approximate value of the integral.

Suppose that $a, b \in \mathbb{T}$, $a < b$, and $f : [a, b] \to \mathbb{R}$ is a given continuous function. Let

$$a = x_0 < x_1 < \cdots < x_n = b.$$

Then

$$\int_a^b f(x)\Delta x = \sum_{j=0}^{n-1} \int_{x_j}^{x_{j-1}} f(x)\Delta x.$$

Now, applying the trapezoid rule, we find

$$\int_{x_j}^{x_{j+1}} f(x)\Delta x \approx \frac{1}{x_{j+1} - x_j}(f(x_j)h_2(x_j, x_{j+1}) + f(x_{j+1})h_2(x_{j+1}, x_j)), \quad j \in \{0, 1, \ldots, n\}.$$

Hence,

$$\int_a^b f(x)\Delta x \approx \sum_{j=0}^{n-1} \frac{1}{x_{j+1} - x_j}(f(x_j)h_2(x_j, x_{j+1}) + f(x_{j+1})h_2(x_{j+1}, x_j)). \tag{2.7}$$

Definition 2.24. The numerical integration formula (2.7) is said to be the composite trapezoid rule.

Example 2.25. Let $\mathbb{T} = 2^{\mathbb{N}_0}$ and $f : \mathbb{T} \to \mathbb{R}$ be defined by

$$f(t) = \frac{t+1}{t^2+t+1}, \quad t \in \mathbb{T}.$$

Let also

$$1 = x_0 < 4 = x_1 < 16 = x_2 = b.$$

Here $\sigma(t) = 2t$, $t \in \mathbb{T}$, and $h_1(t,s) = t - s$, $t,s \in \mathbb{T}$. For a fixed $s \in \mathbb{T}$, denote

$$f_1(t) = \frac{1}{3}t^2 - st, \quad t \in \mathbb{T}.$$

Then

$$f_1^\Delta(t) = \frac{1}{3}(\sigma(t) + t) - s = \frac{1}{3}(2t + t) - s = t - s = h_1(t,s), \quad t \in \mathbb{T},$$

and

$$h_2(t,s) = \int_s^t h_1(\tau, s)\Delta\tau = \int_s^t (\tau - s)\Delta\tau = \int_s^t f_1^\Delta(\tau)\Delta\tau = f_1(\tau)|_{\tau=s}^{\tau=t}$$

$$= \frac{1}{3}t^2 - st - \frac{1}{3}s^2 + s^2 = \frac{t^2 + 2s^2 - 3st}{3}, \quad t \in \mathbb{T}.$$

Hence,

$$h_2(x_1, x_0) = h_2(4, 1) = \frac{16 + 2 - 12}{3} = 2,$$

$$h_2(x_0, x_1) = h_2(1, 4) = \frac{1 + 32 - 12}{3} = 7,$$

$$h_2(x_2, x_1) = h_2(16, 4) = \frac{256 + 32 - 192}{3} = \frac{96}{3} = 32,$$

$$h_2(x_1, x_2) = h_2(4, 16) = \frac{16 + 512 - 192}{3} = \frac{336}{3} = 112,$$

$$f(1) = \frac{2}{3}, \quad f(4) = \frac{5}{21}, \quad f(16) = \frac{17}{273}.$$

Thus,

$$\int_1^{16} \frac{t+1}{t^2+t+1}\Delta t \approx \frac{1}{x_1 - x_0}\left(f(x_0)h_2(x_0, x_1) + f(x_1)h_2(x_1, x_0)\right)$$

$$+ \frac{1}{x_2 - x_1}\left(f(x_1)h_2(x_1, x_2) + f(x_2)h_2(x_2, x_1)\right)$$

$$= \frac{1}{3}\left(\frac{2}{3} \cdot 7 + \frac{5}{21} \cdot 2\right) + \frac{1}{12}\left(\frac{5}{21} \cdot 112 + \frac{17}{273} \cdot 32\right)$$

$$= \frac{14}{9} + \frac{10}{63} + \frac{280}{126} + \frac{136}{819}$$

$$= \frac{248}{63} + \frac{136}{819} = \frac{3360}{819} = \frac{160}{39}.$$

Exercise 2.26. Let $\mathbb{T} = \{-1, -\frac{1}{3}, -\frac{1}{7}, 0, \frac{1}{2}, \frac{1}{3}, \frac{1}{4}\}$. Using the composite trapezoid rule, evaluate the integral

$$\int_{-1}^{\frac{1}{4}} \left(\frac{t^2 + 1}{t^4 + 1} + \sin_1(t, 0) \right) \Delta t.$$

Next, we derive the composite Simpson rule. Note that since the Simpson rule requires 3 points, that is, $a = x_0 < x_1 < x_2 = b$, we need to take an odd number of points in the interval $[a, b]$ including a and b. Suppose that

$$a = x_0 < x_1 < x_2 < \cdots < x_{2m} = b.$$

Then

$$\int_a^b f(x) \Delta x = \sum_{j=1}^m \int_{x_{2j-2}}^{x_{2j}} f(x) \Delta x.$$

Hence, using the Simpson rule, we find

$$\int_a^b f(x) \Delta x \approx \sum_{j=1}^m \Bigg(\frac{f(x_{2j-2})}{(x_{2j-2} - x_{2j-1})(x_{2j-2} - x_{2j})} (-h_2(x_{2j-2}, x_{2j-1}) h_1(x_{2j-2}, x_{2j})$$

$$+ H_3(x_{2j-2}, x_{2j-1}) - H_3(x_{2j}, x_{2j-1}))$$

$$- f(x_{2j-1}) \frac{H_3(x_{2j}, x_{2j-2})}{(x_{2j-1} - x_{2j-2})(x_{2j-1} - x_{2j})}$$

$$+ \frac{f(x_{2j})}{(x_{2j} - x_{2j-2})(x_{2j} - x_{2j-1})} (h_2(x_{2j}, x_{2j-1}) h_1(x_{2j}, x_{2j-2})$$

$$- H_3(x_{2j-1}, x_{2j-2}) - H_3(x_{2j}, x_{2j-1})) \Bigg). \tag{2.8}$$

Definition 2.27. The numerical integration formula (2.8) is said to be the composite Simpson rule.

Example 2.28. Let $\mathbb{T} = 2\mathbb{Z}$ and $[a, b] = [-10, 16]$. Consider the integral $\int_{-10}^{16} f(t) \Delta t$. We have $\sigma(t) = t + 2$ and $\mu(t) = t + 2 - t = 2$, $t \in \mathbb{T}$. Let

$$f_1(t) = \frac{t^2}{2} - t,$$

$$f_2(t) = \frac{t^3}{3} - t^2 + \frac{2t}{3}, \quad t \in \mathbb{T}.$$

We compute

$$f_1^\Delta(t) = \frac{\sigma(t) + t}{2} - 1 = \frac{2t + 2}{2} - 1 = t,$$

$$f_2^\Delta(t) = \frac{(\sigma(t))^2 + t\sigma(t) + t^2}{3} - (\sigma(t) + t) + \frac{2}{3}$$

$$= \frac{(t + 2)^2 + t(t + 2) + t^2}{3} - (t + 2 + t) + \frac{2}{3}$$

$$= \frac{3t^2 + 6t + 4}{3} - 2t - 2 + \frac{2}{3}$$

$$= t^2, \quad t \in \mathbb{T}.$$

On this time scale, we have

$$h_0(t,s) = 1,$$

$$h_1(t,s) = t - s,$$

$$h_2(t,s) = \int_s^t (\tau - s)\Delta\tau = f_1(\tau) - s\tau|_{\tau=s}^{\tau=t}$$

$$= \left(\frac{t^2}{2} - t - st\right) - \left(\frac{s^2}{2} - s - s^2\right)$$

$$= \frac{t^2}{2} - (1 + s)t + s + \frac{s^2}{2}, \quad t,s \in \mathbb{T},$$

and

$$H_3(t,s) = \int_s^t h_2(\sigma(\tau),s)\Delta\tau$$

$$= \int_s^t \left(\frac{(\tau + 2)^2}{2} - (1 + s)(\tau + 2) + s + \frac{s^2}{2}\right)\Delta\tau$$

$$= \int_s^t \left(\frac{\tau^2}{2} + (1 - s)\tau - s + \frac{s^2}{2}\right)\Delta\tau$$

$$= \frac{f_2(\tau)}{2} + (1 - s)f_1(\tau) + \left(\frac{s^2}{2} - s\right)\tau|_{\tau=s}^{\tau=t}$$

$$= \frac{t^3}{6} - \frac{st^2}{2} + \left(\frac{s^2}{2} - \frac{2}{3}\right)t + \frac{2s}{3} - \frac{s^3}{6}, \quad t,s \in \mathbb{T}.$$

Then for $n = 2$ and $x_0 = -10$, $x_1 = 2$, $x_2 = 16$, the Simpson rule gives

$$\int_{-10}^{16} f(t)\Delta t \approx S_2(f)$$

$$= \frac{f(-10)}{(-10-2)(-10-16)}(-h_2(-10,2)h_1(-10,16) + H_3(-10,2) - H_3(16,2))$$

$$- \frac{f(2)}{(2+10)(2-16)}H_3(16,-10)$$

$$+ \frac{f(16)}{(16+10)(16-2)}(h_2(16,2)h_1(16,-10) - H_3(2,-10) - H_3(16,2))$$

For $n = 4$ and choosing $x_0 = -10$, $x_1 = -4$, $x_2 = 0$, $x_3 = 8$, $x_4 = 16$, the composite Simpson rule yields

$$\int_{-10}^{16} f(t)\Delta t \approx S_4(f)$$

$$= \frac{f(-10)}{(-10+4)(-10-0)}(-h_2(-10,-4)h_1(-10,0) + H_3(-10,-4) - H_3(0,-4))$$

$$- \frac{f(-4)}{(-4+10)(-4-0)}H_3(0,-10)$$

$$+ \frac{f(0)}{(0+10)(0+4)}(h_2(0,-4)h_1(0,-10) - H_3(-4,-10) - H_3(0,-4))$$

$$+ \frac{f(0)}{(0-8)(0-16)}(-h_2(0,8)h_1(0,16) + H_3(0,8) - H_3(16,8))$$

$$- \frac{f(8)}{(8-0)(8-16)}H_3(16,0)$$

$$+ \frac{f(16)}{(16-0)(16-8)}(h_2(16,8)h_1(16,0) - H_3(8,0) - H_3(16,8)).$$

Finally, for $n = 6$, choosing $x_0 = -10$, $x_1 = -4$, $x_2 = 0$, $x_3 = 4$, $x_4 = 8$, $x_5 = 12$, and $x_6 = 16$, the composite Simpson rule gives

$$\int_{-10}^{16} f(t)\Delta t \approx S_6(f)$$

$$= \frac{f(-10)}{(-10+4)(-10-0)}(-h_2(-10,-4)h_1(-10,0) + H_3(-10,-4) - H_3(0,-4))$$

$$- \frac{f(-4)}{(-4+10)(-4-0)}H_3(0,-10)$$

$$+ \frac{f(0)}{(0+10)(0+4)}(h_2(0,-4)h_1(0,-10) - H_3(-4,-10) - H_3(0,-4))$$

$$+ \frac{f(0)}{(0-4)(0-8)}(-h_2(0,4)h_1(0,8) + H_3(0,4) - H_3(8,4))$$

$$- \frac{f(4)}{(4-0)(4-8)}H_3(8,0)$$

$$+ \frac{f(8)}{(8-0)(8-4)}(h_2(8,4)h_1(8,0) - H_3(4,0) - H_3(8,4))$$

$$+ \frac{f(8)}{(8-12)(8-16)}(-h_2(8,12)h_1(8,16) + H_3(8,12) - H_3(16,12))$$

$$- \frac{f(12)}{(12-8)(12-16)}H_3(16,8)$$

$$+ \frac{f(16)}{(16-8)(16-12)}(h_2(16,12)h_1(16,8) - H_3(12,8) - H_3(16,12)).$$

Let $f(t) = \sqrt{t^2 + 1}$, $t \in \mathbb{T}$. Then we compute the values

$$S_2(\sqrt{t^2+1}) = 149.7828, \quad S_4(\sqrt{t^2+1}) = 177.4453, \quad S_6(\sqrt{t^2+1}) = 176.5128.$$

On the other hand, the exact value of the integral is computed as

$$I = \int_{-10}^{16} \sqrt{t^2+1}\, 1\Delta t = \sum_{i=-5}^{7} \mu(2i)f(2i) = 2\sum_{i=-5}^{7} \sqrt{(2i)^2+1} = 176.3708.$$

Exercise 2.29. Let $\mathbb{T} = 3^{\mathbb{N}_0}$,

$$a = 1 = x_0, \quad x_1 = 9, \quad x_2 = b = 81,$$

and suppose that $f : \mathbb{T} \to \mathbb{R}$ is defined by

$$f(t) = 1 + t - \frac{3t}{1+t^4} + e_2(t,1), \quad t \in \mathbb{T}.$$

Using the composite Simpson rule, evaluate the integral $\int_1^{81} f(t)\Delta t$.

2.6 σ-Composite quadrature rules

In a similar way, we can define the σ-composite trapezoid and Simpson rules. First, we derive the σ-composite trapezoid rule. Suppose that $a, b \in \mathbb{T}$, $a < b$, $f : [a,b] \to \mathbb{R}$ is rd-continuous, $n \in \mathbb{N}$ and that $x_j \in [a,b] \subset \mathbb{T}, j \in \{0,1,\ldots,n\}$ are σ-distinct, i. e.,

$$a = x_0 \le \sigma(x_0) < x_1 \le \sigma(x_1) < \cdots < x_n \le \sigma(x_n) = b.$$

Then

$$\int_a^b f(x)\Delta x = \int_{x_0}^{\sigma(x_1)} f(x)\Delta x + \int_{\sigma(x_1)}^{\sigma(x_2)} f(x)\Delta x + \cdots + \int_{\sigma(x_{n-1})}^{\sigma(x_n)} f(x)\Delta x$$

$$= \int_{x_0}^{\sigma(x_1)} f(x)\Delta x + \left(\int_{x_1}^{\sigma(x_2)} f(x)\Delta x - \int_{x_1}^{\sigma(x_1)} f(x)\Delta x \right)$$

$$+ \cdots + \left(\int_{x_{n-1}}^{\sigma(x_n)} f(x)\Delta x - \int_{x_{n-1}}^{\sigma(x_{n-1})} f(x)\Delta x \right)$$

$$= \sum_{j=0}^{n-1} \int_{x_j}^{\sigma(x_{j+1})} f(x)\Delta x - \sum_{j=0}^{n-1} \int_{x_j}^{\sigma(x_j)} f(x)\Delta x$$

$$= \sum_{j=0}^{n-1} \int_{x_j}^{\sigma(x_{j+1})} f(x)\Delta x - \sum_{j=0}^{n-1} \mu(x_j)f(x_j).$$

We apply the σ-trapezoid rule and obtain

$$\int_a^b f(x)\Delta x \approx \sum_{j=0}^{n-1} \frac{1}{\sigma(x_{j+1}) - \sigma(x_j)} \left(f(x_j)g_2(x_j, \sigma(x_{j+1})) + f(x_{j+1})g_2(\sigma(x_{j+1}), \sigma(x_j)) \right)$$

$$- \sum_{j=0}^{n-1} \mu(x_j)f(x_j). \tag{2.9}$$

Definition 2.30. The numerical integral formula (2.9) is said to be the σ-composite trapezoid rule.

Example 2.31. Let

$$\mathbb{T} = \left\{ 0, \frac{1}{8}, \frac{1}{6}, \frac{1}{4}, \frac{1}{2}, 1 \right\},$$

$$a = x_0 = 0, \quad x_1 = \frac{1}{6}, \quad x_2 = \frac{1}{2}, \quad b = 1.$$

We have

$$x_0 = 0 < \sigma(x_0) = \frac{1}{8} < x_1 = \frac{1}{6} < \sigma(x_1) = \frac{1}{4} < x_2 = \frac{1}{2} < \sigma(x_2) = 1 = b.$$

Let $f(t) = t$, $t \in \mathbb{T}$. We will evaluate the integral $\int_a^b f(t)\Delta t$ using the σ-composite trapezoid rule. We have

$$g_2(x_0, \sigma(x_1)) = \int_{\sigma(x_1)}^{x_0} (\sigma(t) - \sigma(x_1))\Delta t$$

$$= \int_{\frac{1}{4}}^{\frac{1}{8}} \left(\sigma(t) - \frac{1}{4}\right)\Delta t$$

$$= -\int_{\frac{1}{8}}^{\frac{1}{4}} \left(\sigma(t) - \frac{1}{4}\right)\Delta t$$

$$= -\mu\left(\frac{1}{8}\right)\left(\sigma\left(\frac{1}{8}\right) - \frac{1}{4}\right) - \mu\left(\frac{1}{6}\right)\left(\sigma\left(\frac{1}{6}\right) - \frac{1}{4}\right)$$

$$= -\left(\frac{1}{6} - \frac{1}{8}\right)\left(\frac{1}{6} - \frac{1}{4}\right) - \left(\frac{1}{4} - \frac{1}{6}\right)\left(\frac{1}{4} - \frac{1}{4}\right)$$

$$= -\frac{1}{24}\left(-\frac{1}{12}\right) = \frac{1}{288},$$

$$g_2(\sigma(x_1), \sigma(x_0)) = \int_{\sigma(x_0)}^{\sigma(x_1)} (\sigma(t) - \sigma(x_0))\Delta t$$

$$= \int_{\frac{1}{8}}^{\frac{1}{4}} \left(\sigma(t) - \frac{1}{8}\right)\Delta t$$

$$= \mu\left(\frac{1}{8}\right)\left(\sigma\left(\frac{1}{8}\right) - \frac{1}{8}\right) + \mu\left(\frac{1}{6}\right)\left(\sigma\left(\frac{1}{6}\right) - \frac{1}{8}\right)$$

$$= \left(\frac{1}{6} - \frac{1}{8}\right)\left(\frac{1}{6} - \frac{1}{8}\right) + \left(\frac{1}{4} - \frac{1}{6}\right)\left(\frac{1}{4} - \frac{1}{8}\right)$$

$$= \frac{1}{576} + \frac{1}{12}\left(\frac{1}{8}\right) = \frac{1}{576} + \frac{1}{96} = \frac{7}{576},$$

$$g_2(x_1, \sigma(x_2)) = \int_{\sigma(x_2)}^{x_1} (\sigma(t) - \sigma(x_2))\Delta t$$

$$= \int_{1}^{\frac{1}{6}} (\sigma(t) - 1)\Delta t$$

$$= -\int_{\frac{1}{6}}^{1} (\sigma(t) - 1)\Delta t$$

$$= -\mu\left(\frac{1}{6}\right)\left(\sigma\left(\frac{1}{6}\right) - 1\right) - \mu\left(\frac{1}{4}\right)\left(\sigma\left(\frac{1}{4}\right) - 1\right)$$

$$- \mu\left(\frac{1}{2}\right)\left(\sigma\left(\frac{1}{2}\right) - 1\right)$$

$$= -\left(\frac{1}{4} - \frac{1}{6}\right)\left(\frac{1}{4} - 1\right) - \left(\frac{1}{2} - \frac{1}{4}\right)\left(\frac{1}{2} - 1\right) - \left(1 - \frac{1}{2}\right)(1 - 1)$$

$$= -\frac{1}{12}\left(-\frac{3}{4}\right) - \frac{1}{4}\left(-\frac{1}{2}\right) = \frac{1}{16} + \frac{1}{8} = \frac{3}{16},$$

$$g_2(\sigma(x_2), \sigma(x_1)) = \int_{\sigma(x_1)}^{\sigma(x_2)} (\sigma(t) - \sigma(x_1))\Delta t$$

$$= \int_{\frac{1}{4}}^{1}\left(\sigma(t) - \frac{1}{4}\right)\Delta t$$

$$= \mu\left(\frac{1}{4}\right)\left(\sigma\left(\frac{1}{4}\right) - \frac{1}{4}\right) + \mu\left(\frac{1}{2}\right)\left(\sigma\left(\frac{1}{2}\right) - \frac{1}{4}\right)$$

$$= \left(\frac{1}{2} - \frac{1}{4}\right)\left(\frac{1}{2} - \frac{1}{4}\right) + \left(1 - \frac{1}{2}\right)\left(1 - \frac{1}{4}\right)$$

$$= \frac{1}{16} + \frac{3}{8} = \frac{7}{16},$$

$$f(0) = 0, \quad f\left(\frac{1}{6}\right) = \frac{1}{6}, \quad f\left(\frac{1}{2}\right) = \frac{1}{2}.$$

and the σ-composite trapezoid rule, we get

$$\int_0^1 t\Delta t \approx \frac{1}{\sigma(x_1) - \sigma(x_0)}(f(x_0)g_2(x_0, \sigma(x_1)) + f(x_1)g_2(\sigma(x_1), \sigma(x_0)))$$

$$+ \frac{1}{\sigma(x_2) - \sigma(x_1)}(f(x_1)g_2(x_1, \sigma(x_2)) + f(x_2)g_2(\sigma(x_2), \sigma(x_1)))$$

$$- \mu(x_1)f(x_1)$$

$$= \frac{1}{\frac{1}{4}}\left(\frac{1}{6} \cdot \frac{7}{576}\right) + \frac{1}{1 - \frac{1}{4}}\left(\frac{1}{6} \cdot \frac{3}{16} + \frac{1}{2} \cdot \frac{7}{16}\right) - \frac{1}{4} \cdot \frac{1}{6}$$

$$= \frac{14}{1728} + \frac{4}{3}\left(\frac{3}{96} + \frac{7}{32}\right) - \frac{1}{24}$$

$$= \frac{14}{1728} + \frac{4}{3}\left(\frac{1}{4}\right) - \frac{1}{24} = \frac{14}{1728} + \frac{7}{24} = \frac{259}{864}.$$

Exercise 2.32. Let

$$\mathbb{T} = \left\{-2, -\frac{3}{2}, -1, -\frac{5}{6}, -\frac{2}{3}, -\frac{1}{2}, -\frac{1}{3}, -\frac{1}{8}, 0\right\},$$

$$a = -2 = x_0, \quad x_1 = -1, \quad x_2 = -\frac{1}{8}, \quad b = 0,$$

and suppose that $f : \mathbb{T} \to \mathbb{R}$ is defined by

$$f(t) = \frac{t^2 - t + 1}{t^4 + t^2 + 1} + t, \quad t \in \mathbb{T}.$$

Using the σ-composite trapezoid rule, evaluate the integral $\int_a^b f(t)\Delta t$.

Next, we will obtain the σ-composite Simpson rule. Let $m \in \mathbb{N}$, $m \geq 2$, and

$$a = x_0 \leq \sigma(x_0) < x_1 \leq \sigma(x_1) < \cdots < x_{2m} \leq \sigma(x_{2m}) = b.$$

Then

$$\int_a^b f(x)\Delta x = \int_{x_0}^{\sigma(x_2)} f(x)\Delta x + \int_{\sigma(x_2)}^{\sigma(x_4)} f(x)\Delta x + \cdots + \int_{\sigma(x_{2m-2})}^{\sigma(x_{2m})} f(x)\Delta x$$

$$= \int_{x_0}^{\sigma(x_2)} f(x)\Delta x + \left(\int_{x_2}^{\sigma(x_4)} f(x)\Delta x - \int_{x_2}^{\sigma(x_2)} f(x)\Delta x \right)$$

$$+ \cdots + \left(\int_{x_{2m-2}}^{\sigma(x_{2m})} f(x)\Delta x - \int_{x_{2m-2}}^{\sigma(x_{2m-2})} f(x)\Delta x \right)$$

$$= \sum_{j=1}^{m} \int_{x_{2j-2}}^{\sigma(x_{2j})} f(x)\Delta x - \sum_{j=1}^{m} \int_{x_{2j-2}}^{\sigma(x_{2j-2})} f(x)\Delta x$$

$$= \sum_{j=1}^{m} \int_{x_{2j-2}}^{\sigma(x_{2j})} f(x)\Delta x - \sum_{j=1}^{m} \mu(x_{2j-2}) f(x_{2j-2}).$$

Now, applying the σ-Simpson rule to each integral in the first sum, we arrive at

$$\int_a^b f(x)\Delta x \approx \sum_{j=1}^{m} \frac{f(x_{2j-2})}{(\sigma(x_{2j-2}) - \sigma(x_{2j-1}))(\sigma(x_{2j-2}) - \sigma(x_{2j}))}$$

$$\times (-g_2(x_{2j-2}, \sigma(x_{2j-1}))g_1(x_{2j-2}, \sigma(x_{2j}))$$

$$+ G_3(x_{2j-2}, \sigma(x_{2j-1})) - G_3(\sigma(x_{2j}), \sigma(x_{2j-1})))$$

$$- \frac{f(x_{2j-1})}{(\sigma(x_{2j-1}) - \sigma(x_{2j-2}))(\sigma(x_{2j-1}) - \sigma(x_{2j}))} G_3(\sigma(x_{2j}), \sigma(x_{2j-2}))$$

$$+ \frac{f(x_{2j})}{(\sigma(x_{2j}) - \sigma(x_{2j-2}))(\sigma(x_{2j}) - \sigma(x_{2j-1}))}$$

$$\times (g_1(\sigma(x_{2j}), \sigma(x_{2j-2}))g_2(\sigma(x_{2j}), \sigma(x_{2j-1}))$$

$$- G_3\big(\sigma(x_{2j-1}), \sigma(x_{2j-2})\big) - G_3\big(\sigma(x_{2j}), \sigma(x_{2j-1})\big)\big)$$

$$- \sum_{j=1}^{m} \mu(x_{2j-2}) f(x_{2j-2}). \qquad (2.10)$$

Definition 2.33. The numerical integration formula (2.10) will be called the σ-composite Simpson rule.

In the following example, we will apply the σ-composite Simpson rule to approximately find the integral in Example 2.28.

Example 2.34. Let $\mathbb{T} = 2\mathbb{Z}$ and $[a, b] = [-10, 16]$. We consider the integral $\int_{-10}^{16} f(t)\Delta t$, where f is as in Example 2.28. We have $\sigma(t) = t + 2$ and $\mu(t) = t + 2 - t = 2$, $t \in \mathbb{T}$. For the functions

$$f_1(t) = \frac{t^2}{2} - t, \quad f_2(t) = \frac{t^3}{3} - t^2 + \frac{2t}{3}, \quad t \in \mathbb{T},$$

we have computed

$$f_1^\Delta(t) = t, \quad f_2^\Delta(t) = t^2, \quad t \in \mathbb{T},$$

in Example 2.28. We also compute

$$g_0(t, s) = 1,$$
$$g_1(t, s) = t - s,$$
$$g_2(t, s) = \int_s^t (\sigma(\tau) - s)\Delta\tau$$
$$= \int_s^t (\tau + 2 - s)\Delta\tau$$
$$= \left(\frac{t^2}{2} - t\right) + (2 - s)\tau\big|_s^t$$
$$= \left(\frac{t^2}{2} - t - st\right) - \left(\frac{s^2}{2} - s - s^2\right)$$
$$= \frac{t^2}{2} + (1 - s)t - s + \frac{s^2}{2}, \quad t, s \in \mathbb{T},$$

and

$$G_3(t, s) = \int_s^t g_2(\tau, s)\Delta\tau$$
$$= \int_s^t \left(\frac{\tau^2}{2} + (1 - s)\tau + s + \frac{s^2}{2} - s\right)\Delta\tau$$

$$= \frac{\tau^3}{6} - \frac{\tau^2}{2} + \frac{\tau}{3} + (1-s)\left(\frac{\tau^2}{2} - \tau\right) + \left(\frac{s^2}{2} - s\right)\tau\big|_{\tau=s}^{\tau=t}$$

$$= \frac{t^3}{6} - \frac{st^2}{2} + \left(\frac{s^2}{2} - \frac{2}{3}\right)t + \frac{2s}{3} - \frac{s^3}{6}, \quad t, s \in \mathbb{T}.$$

Then for $m = 1$, taking $a = x_0 = -10$, $\sigma(x_0) = -8$, $x_1 = 2$, $\sigma(x_1) = 4$, $x_2 = 14$, and $\sigma(x_2) = 16 = b$, the σ-Simpson rule gives

$$\int_{-10}^{16} f(t)\Delta t \approx \sigma S_2(f)$$

$$= \frac{f(-10)}{(-8-4)(-8-16)}\left(-g_2(-10,4)g_1(-10,16) + G_3(-10,4) - G_3(16,4)\right)$$

$$- \frac{f(2)}{(4+8)(4-16)}G_3(16,-8)$$

$$+ \frac{f(14)}{(16+8)(16-4)}\left(g_2(16,4)g_1(16,-8) - G_3(4,-8) - G_3(16,4)\right).$$

For $m = 2$, choosing

$$x_0 = -10, \quad \sigma(x_0) = -8, \quad x_1 = -6, \quad \sigma(x_1) = -4, \quad x_2 = 0, \quad \sigma(x_2) = 2,$$

$$x_3 = 8, \quad \sigma(x_3) = 10, \quad x_4 = 14, \quad \sigma(x_4) = 16,$$

the σ-composite Simpson rule becomes

$$\int_{-10}^{16} f(t)\Delta t \approx \sigma S_4(f)$$

$$= \frac{f(-10)}{(-8+4)(-8-2)}\left(-g_2(-10,-4)g_1(-10,2) + G_3(-10,-4) - G_3(2,-4)\right)$$

$$- \frac{f(-6)}{(-4+8)(-4-2)}G_3(2,-8)$$

$$+ \frac{f(0)}{(2+8)(2+4)}\left(g_2(2,-4)g_1(2,-8) - G_3(-4,-8) - G_3(2,-4)\right)$$

$$+ \frac{f(0)}{(2-10)(2-16)}\left(-g_2(0,10)g_1(0,16) + G_3(0,10) - G_3(16,10)\right)$$

$$- \frac{f(8)}{(10-2)(10-16)}G_3(16,2)$$

$$+ \frac{f(14)}{(16-2)(16-10)}\left(g_2(16,10)g_1(16,2) - G_3(10,2) - G_3(16,10)\right) - 2f(0).$$

Finally, for $m = 3$, we choose

$$x_0 = -10, \quad \sigma(x_0) = -8, \quad x_1 = -6, \quad \sigma(x_1) = -4,$$
$$x_2 = -2, \quad \sigma(x_2) = 0, \quad x_3 = 2, \quad \sigma(x_3) = 4,$$
$$x_4 = 6, \quad \sigma(x_4) = 8, \quad x_5 = 10, \quad \sigma(x_5) = 12,$$
$$x_6 = 14, \quad \sigma(x_6) = 16,$$

and then the σ-composite Simpson rule is written as

$$\int_{-10}^{16} f(t)\Delta t \approx \sigma S_6(f)$$

$$= \frac{f(-10)}{(-8+4)(-8-0)}\left(-g_2(-10,-4)g_1(-10,0) + G_3(-10,-4) - G_3(0,-4)\right)$$

$$- \frac{f(-6)}{(-4+8)(-4-0)}G_3(0,-8)$$

$$+ \frac{f(-2)}{(0+8)(0+4)}\left(g_2(0,-4)g_1(0,-8) - G_3(-4,-8) - G_3(2,-4)\right)$$

$$+ \frac{f(-2)}{(0-4)(0-8)}\left(-g_2(-2,4)g_1(-2,8) + G_3(-2,4) - G_3(8,4)\right)$$

$$- \frac{f(2)}{(4-0)(4-8)}G_3(8,0)$$

$$+ \frac{f(6)}{(8-0)(8-4)}\left(g_2(8,4)g_1(8,0) - G_3(4,0) - G_3(8,4)\right)$$

$$+ \frac{f(6)}{(8-12)(8-16)}\left(-g_2(6,12)g_1(6,16) + G_3(6,12) - G_3(16,12)\right)$$

$$- \frac{f(10)}{(12-8)(12-16)}G_3(16,8)$$

$$+ \frac{f(14)}{(16-8)(16-12)}\left(g_2(16,12)g_1(16,8) - G_3(12,8) - G_3(16,12)\right)$$

$$- 2(f(-2) + f(6)).$$

As in Example 2.28 we take $f(t) = \sqrt{t^2 + 1}$, $t \in \mathbb{T}$. Then the approximate values of $\int_{-10}^{16} \sqrt{t^2 + 1}\Delta t$ are obtained as

$$\sigma S_2(\sqrt{t^2 + 1}) = 157.2944,$$
$$\sigma S_4(\sqrt{t^2 + 1}) = 178.6610,$$
$$\sigma S_6(\sqrt{t^2 + 1}) = 177.0139.$$

Recall that the exact value of the integral is

$$I = \int_{-10}^{16} \sqrt{1 + t^2}\Delta t = 176.3708.$$

Exercise 2.35. Let $\mathbb{T} = 2\mathbb{Z}$,

$$a = x_0 = 0, \quad x_1 = 4, \quad x_2 = 8, \quad x_3 = 12, \quad x_4 = 16 = b,$$

and suppose that $f : \mathbb{T} \to \mathbb{R}$ is defined by

$$f(t) = 1 + t + t^2 + e_2(t, 0), \quad t \in \mathbb{T}.$$

Using the σ-composite Simpson rule, evaluate the integral $\int_a^b f(t)\Delta t$.

2.7 The Euler–Maclauren expansion

The Euler–Maclauren expansion is a formula which represents the difference between a definite integral and its approximation by a composite trapezoid rule. It is employed in the derivation of the so-called Romberg integration. In this section we prove the Euler– Maclauren expansion for an arbitrary time scale.

Theorem 2.36. *Let $a, b \in \mathbb{T}, a < b, k, m \in \mathbb{N}, x_j \in \mathbb{T}, j \in \{0, 1, \ldots, m\}$,*

$$a = x_0 < x_1 < \cdots < x_m = b,$$

$f \in C_{rd}^{2k}([a, b])$ and let $T(m)$ be the result of the approximation of the integral $I = \int_a^b f(x)\Delta x$ by the composite trapezoid rule with m subintervals $[x_{j-1}, x_j], j \in \{1, \ldots, m\}$. Then

$$I - T(m) = \sum_{j=1}^m \left(\frac{1}{h_1(x_j, x_{j-1})} \sum_{l=1}^{2k-1} \left((-1)^l (H_{l+1}(x_j, x_{j-1}) f^{\Delta^l}(x_j) + H_{l+1}(x_{j-1}, x_j) f^{\Delta^l}(x_{j-1}) \right) \right.$$

$$\left. + \int_{x_{j-1}}^{x_j} (H_{2k+1}(\sigma(\tau), x_{j-1}) - H_{2k+1}(\sigma(\tau), x_j)) f^{\Delta^{2k}}(\tau)\Delta\tau \right). \tag{2.11}$$

Definition 2.37. The formula (2.11) is said to be the Euler–Maclauren expansion.

Proof. Fix $j \in \{1, \ldots, m\}$. Then, using the trapezoid rule, we find

$$\int_{x_{j-1}}^{x_j} f(x)\Delta x - \frac{1}{h_1(x_j, x_{j-1})} (f(x_{j-1})h_2(x_{j-1}, x_j) + f(x_j)h_2(x_j, x_{j-1}))$$

$$= \frac{1}{h_1(x_j, x_{j-1})} \left(\int_{x_{j-1}}^{x_j} h_1(\tau, x_{j-1}) f(\tau)\Delta\tau - \int_{x_{j-1}}^{x_j} h_1(\tau, x_j) f(\tau)\Delta\tau \right)$$

$$- \frac{1}{h_1(x_j, x_{j-1})} (f(x_{j-1})h_2(x_{j-1}, x_j) + f(x_j)h_2(x_j, x_{j-1}))$$

$$= \frac{1}{h_1(x_j, x_{j-1})} \left(\int_{x_{j-1}}^{x_j} h_2^\Delta(\tau, x_{j-1}) f(\tau)\Delta\tau - \int_{x_{j-1}}^{x_j} h_2^\Delta(\tau, x_j) f(\tau)\Delta\tau \right)$$

$$- \frac{1}{h_1(x_j, x_{j-1})} (f(x_{j-1})h_2(x_{j-1}, x_j) + f(x_j)h_2(x_j, x_{j-1}))$$

$$= \frac{1}{h_1(x_j, x_{j-1})} (f(x_{j-1})h_2(x_{j-1}, x_j) + f(x_j)h_2(x_j, x_{j-1}))$$

$$- \frac{1}{h_1(x_j, x_{j-1})} \int_{x_{j-1}}^{x_j} (h_2(\sigma(\tau), x_{j-1}) - h_2(\sigma(\tau), x_j)) f^\Delta(\tau)\Delta\tau$$

$$- \frac{1}{h_1(x_j, x_{j-1})} (f(x_{j-1})h_2(x_{j-1}, x_j) + f(x_j)h_2(x_j, x_{j-1}))$$

$$= -\frac{1}{h_1(x_j, x_{j-1})} \int_{x_{j-1}}^{x_j} (H_3^\Delta(\tau, x_{j-1}) - H_3^\Delta(\tau, x_j)) f^\Delta(\tau)\Delta\tau$$

$$= -\frac{1}{h_1(x_j, x_{j-1})} (H_3(x_j, x_{j-1}) f^\Delta(x_j) + H_3(x_{j-1}, x_j) f^\Delta(x_{j-1}))$$

$$+ \frac{1}{h_1(x_j, x_{j-1})} \int_{x_{j-1}}^{x_j} (H_4^\Delta(\tau, x_{j-1}) - H_4^\Delta(\tau, x_j)) f^{\Delta^2}(\tau)\Delta\tau$$

$$= -\frac{1}{h_1(x_j, x_{j-1})} (H_3(x_j, x_{j-1}) f^\Delta(x_j) + H_3(x_{j-1}, x_j) f^\Delta(x_{j-1}))$$

$$+ \frac{1}{h_1(x_j, x_{j-1})} (H_4(x_j, x_{j-1}) f^{\Delta^2}(x_j) + H_4(x_{j-1}, x_j) f^{\Delta^2}(x_{j-1}))$$

$$- \frac{1}{h_1(x_j, x_{j-1})} \int_{x_{j-1}}^{x_j} (H_5^\Delta(\tau, x_{j-1}) - H_5^\Delta(\tau, x_j)) f^{\Delta^3}(\tau)\Delta\tau$$

$$= \cdots$$

$$= \frac{1}{h_1(x_j, x_{j-1})} \sum_{l=1}^{2k-1} \left((-1)^l (H_{l+1}(x_j, x_{j-1}) f^{\Delta^l}(x_j) + H_{l+1}(x_{j-1}, x_j) f^{\Delta^l}(x_{j-1})) \right.$$

$$\left. + \int_{x_{j-1}}^{x_j} (H_{2k+1}(\sigma(\tau), x_{j-1}) - H_{2k+1}(\sigma(\tau), x_j)) f^{\Delta^{2k}}(\tau)\Delta\tau \right).$$

Summing over all the subintervals $[x_{j-1}, x_j]$, $j \in \{1, \ldots, m\}$, gives the required result. This completes the proof. $\qquad\qquad\square$

2.8 The σ-Euler–Maclauren Expansion

The Euler–Maclauren expansion can be also given for the σ-composite trapezoid rule. In this section, we give the σ-Euler–Maclauren expansion on an arbitrary time scale and discuss its proof.

Theorem 2.38. *Let $a, b \in \mathbb{T}, a < b, k, m \in \mathbb{N}, x_j \in \mathbb{T}, j \in \{0, 1, \ldots, m\}$,*

$$a = x_0 \leq \sigma(x_0) < x_1 \leq \sigma(x_1) < \cdots < x_m \leq \sigma(x_m) = b,$$

$f \in C_{rd}^{2k}([a, b])$, and let $T_\sigma(m)$ be the result of the approximation of the integral $I = \int_a^b f(x)\Delta x$ by the σ-composite trapezoid rule with m subintervals $[x_{j-1}, x_j], j \in \{1, \ldots, m\}$.
Then

$$I - T_\sigma(m) = \sum_{j=1}^m \frac{\mu(x_j)}{\sigma(x_j) - \sigma(x_{j-1})} f^\Delta(x_j)$$

$$+ \sum_{j=1}^m \left(\frac{1}{g_1(\sigma(x_j), \sigma(x_{j-1}))} \sum_{l=1}^{2k-1} \left((-1)^l \big(g_{l+1}(\sigma(x_j), \sigma(x_{j-1}))f^{\Delta^l}(\sigma(x_j)) \right. \right.$$

$$+ g_{l+1}(x_{j-1}, \sigma(x_j))f^{\Delta^l}(x_{j-1}))$$

$$+ \int_{x_{j-1}}^{\sigma(x_j)} \left(g_{2k+1}(\sigma(\tau), \sigma(x_{j-1})) - g_{2k+1}(\sigma(\tau), \sigma(x_j)) \right) f^{\Delta^{2k}}(\tau)\Delta\tau \Bigg) \Bigg). \qquad (2.12)$$

Definition 2.39. The formula (2.12) is said to be the σ-Euler–Maclaurin expansion.

Proof. Take $j \in \{1, \ldots, m\}$ arbitrarily. Applying the σ-trapezoid rule, we obtain

$$\int_{x_{j-1}}^{\sigma(x_j)} f(x)\Delta x - \frac{1}{g_1(\sigma(x_j), \sigma(x_{j-1}))} \left(f(x_{j-1})g_2(x_{j-1}, \sigma(x_j)) + f(x_j)g_2(\sigma(x_j), \sigma(x_{j-1})) \right)$$

$$= \frac{1}{g_1(\sigma(x_j), \sigma(x_{j-1}))}$$

$$\times \left(\int_{x_{j-1}}^{\sigma(x_j)} g_1(\sigma(\tau), \sigma(x_{j-1}))f(\tau)\Delta\tau - \int_{x_{j-1}}^{\sigma(x_j)} g_1(\sigma(\tau), \sigma(x_j))f(\tau)\Delta\tau \right)$$

$$- \frac{1}{g_1(\sigma(x_j), \sigma(x_{j-1}))} \left(f(x_{j-1})g_2(x_{j-1}, \sigma(x_j)) + f(x_j)g_2(\sigma(x_j), \sigma(x_{j-1})) \right)$$

$$= \frac{1}{g_1(\sigma(x_j), \sigma(x_{j-1}))}$$

$$\times \left(\int_{x_{j-1}}^{\sigma(x_j)} g_2^\Delta(\tau, \sigma(x_{j-1}))f(\tau)\Delta\tau - \int_{x_{j-1}}^{\sigma(x_j)} g_2^\Delta(\tau, \sigma(x_j))f(\tau)\Delta\tau \right)$$

$$- \frac{1}{g_1(\sigma(x_j), \sigma(x_{j-1}))} \left(f(x_{j-1})g_2(x_{j-1}, \sigma(x_j)) + f(x_j)g_2(\sigma(x_j), \sigma(x_{j-1})) \right)$$

$$= \frac{1}{g_1(\sigma(x_j), \sigma(x_{j-1}))} \left(f(x_{j-1})g_2(x_{j-1}, \sigma(x_j)) + f(\sigma(x_j))g_2(\sigma(x_j), \sigma(x_{j-1})) \right)$$

$$- \frac{1}{g_1(\sigma(x_j), \sigma(x_{j-1}))} \int_{x_{j-1}}^{\sigma(x_j)} \left(g_2(\sigma(\tau), \sigma(x_{j-1})) - g_2(\sigma(\tau), \sigma(x_j)) \right) f^\Delta(\tau)\Delta\tau$$

$$- \frac{1}{g_1(\sigma(x_j), \sigma(x_{j-1}))} \left(f(x_{j-1}) g_2(x_{j-1}, \sigma(x_j)) + f(x_j) g_2(\sigma(x_j), \sigma(x_{j-1})) \right)$$

$$= \frac{\mu(x_j)}{\sigma(x_j) - \sigma(x_{j-1})} f^\Delta(x_j)$$

$$- \frac{1}{g_1(\sigma(x_j), \sigma(x_{j-1}))} \int_{x_{j-1}}^{\sigma(x_j)} \left(g_3^\Delta(\tau, \sigma(x_{j-1})) - g_3^\Delta(\tau, \sigma(x_j)) \right) f^\Delta(\tau) \Delta\tau$$

$$= \frac{\mu(x_j)}{\sigma(x_j) - \sigma(x_{j-1})} f^\Delta(x_j)$$

$$- \frac{1}{g_1(\sigma(x_j), \sigma(x_{j-1}))} \left(g_3(\sigma(x_j), \sigma(x_{j-1})) f^\Delta(\sigma(x_j)) + g_3(x_{j-1}, \sigma(x_j)) f^\Delta(x_{j-1}) \right)$$

$$+ \frac{1}{g_1(\sigma(x_j), \sigma(x_{j-1}))} \int_{x_{j-1}}^{\sigma(x_j)} \left(g_4^\Delta(\tau, \sigma(x_{j-1})) - g_4^\Delta(\tau, \sigma(x_j)) \right) f^{\Delta^2}(\tau) \Delta\tau$$

$$= \frac{\mu(x_j)}{\sigma(x_j) - \sigma(x_{j-1})} f^\Delta(x_j)$$

$$- \frac{1}{g_1(\sigma(x_j), \sigma(x_{j-1}))} \left(g_3(\sigma(x_j), \sigma(x_{j-1})) f^\Delta(\sigma(x_j)) + g_3(x_{j-1}, \sigma(x_j)) f^\Delta(x_{j-1}) \right)$$

$$+ \frac{1}{g_1(\sigma(x_j), \sigma(x_{j-1}))} \left(g_4(\sigma(x_j), \sigma(x_{j-1})) f^{\Delta^2}(\sigma(x_j)) + g_4(x_{j-1}, \sigma(x_j)) f^{\Delta^2}(x_{j-1}) \right)$$

$$- \frac{1}{g_1(\sigma(x_j), \sigma(x_{j-1}))} \int_{x_{j-1}}^{\sigma(x_j)} \left(g_5^\Delta(\tau, \sigma(x_{j-1})) - g_5^\Delta(\tau, \sigma(x_j)) \right) f^{\Delta^3}(\tau) \Delta\tau.$$

Continuing in the same way, we get

$$\int_{x_{j-1}}^{\sigma(x_j)} f(x) \Delta x - \frac{1}{g_1(\sigma(x_j), \sigma(x_{j-1}))} \left(f(x_{j-1}) g_2(x_{j-1}, \sigma(x_j)) + f(x_j) g_2(\sigma(x_j), \sigma(x_{j-1})) \right)$$

$$= \cdots$$

$$= \frac{\mu(x_j)}{\sigma(x_j) - \sigma(x_{j-1})} f^\Delta(x_j) + \frac{1}{g_1(\sigma(x_j), \sigma(x_{j-1}))}$$

$$\times \sum_{l=1}^{2k-1} \left((-1)^l \left(g_{l+1}(\sigma(x_j), \sigma(x_{j-1})) f^{\Delta^l}(x_j) + g_{l+1}(x_{j-1}, \sigma(x_j)) f^{\Delta^l}(x_{j-1}) \right) \right.$$

$$+ \left. \int_{x_{j-1}}^{\sigma(x_j)} \left(g_{2k+1}(\sigma(\tau), \sigma(x_{j-1})) - g_{2k+1}(\sigma(\tau), \sigma(x_j)) \right) f^{\Delta^{2k}}(\tau) \Delta\tau \right).$$

Summing over all the subintervals $[x_{j-1}, x_j], j \in \{1, \ldots, m\}$, gives the desired result. This completes the proof. □

2.9 Construction of Gauss quadrature rules

From the classical numerical analysis it is known that the Gauss quadrature rules can be defined so that the numerical integration provides an exact result whenever the integrand function is a polynomial of highest possible degree. In this sense, both the nodes and weights are determined from this requirement. Below, we derive the Gauss quadrature rule for an arbitrary time scale.

Let $a, b \in \mathbb{T}$, $a < b$, $f : [a, b] \to \mathbb{R}$, $w : [a, b] \to [0, \infty)$, $f \in C^1_{rd}([a, b])$, $w \in C_{rd}([a, b])$. We wish to construct a quadrature formula for the approximate evaluation of the integral

$$\int_a^b w(x) f(x) \Delta x.$$

Suppose that $n \in \mathbb{N}$, $x_j \in [a, b] \subset \mathbb{T}$, $j \in \{0, 1, \ldots, n\}$, and

$$a = x_0 \leq \sigma(x_0) < x_1 \leq \sigma(x_1) < \cdots < x_n \leq \sigma(x_n) = b.$$

Define the polynomials

$$P_j(x) = \left(1 - \frac{M_j^\Delta(x_j) L_j(x_j) + M_j(\sigma(x_j)) L_j^\Delta(x_j)}{M_j(\sigma(x_j)) L_j(\sigma(x_j))} (x - x_j)\right) M_j(x) L_j(x),$$

$$K_j(x) = \frac{x - x_j}{M_j(\sigma(x_j)) L_j(\sigma(x_j))} M_j(x) L_j(x), \quad j \in \{0, 1, \ldots, n\}, \quad x \in [a, b].$$

where the polynomials M_j, L_j, $j \in \{0, 1, \ldots, n\}$, are defined as in Chapter 1. Then the Hermite interpolation polynomial of degree $2n + 1$ for the function f is given by the expression

$$p_{2n+1}(x) = \sum_{j=0}^{n} (P_j(x) f(x_j) + K_j(x) f^\Delta(x_j)), \quad x \in [a, b].$$

Thus,

$$\int_a^b w(x) f(x) \Delta x \approx \int_a^b w(x) p_{2n+1}(x) \Delta x$$

$$= \int_a^b w(x) \left(\sum_{j=0}^{n} (P_j(x) f(x_j) + K_j(x) f^\Delta(x_j))\right) \Delta x$$

$$= \sum_{j=0}^{n} \left(\int_a^b w(x) P_j(x) \Delta x\right) f(x_j) + \sum_{j=0}^{n} \left(\int_a^b w(x) K_j(x) \Delta x\right) f^\Delta(x_j).$$

Let

$$W_j = \int_a^b w(x)P_j(x)\Delta x,$$

$$V_j = \int_a^b w(x)K_j(x)\Delta x, \quad j \in \{0,1,\dots,n\}.$$

Therefore,

$$\int_a^b w(x)f(x)\Delta x \approx \sum_{j=0}^n (W_j f(x_j) + V_j f^\Delta(x_j)). \tag{2.13}$$

Now we take the function w so that $V_j \equiv 0$, $j \in \{0,1,\dots,n\}$. Such a function and time scale exist. We demonstrate the Gauss quadrature rule in the following example in which we take a general time scale.

Example 2.40. Let

$$\mathbb{T} = \left\{0, \frac{1}{8}, \frac{1}{6}, \frac{1}{5}, \frac{1}{4}, \frac{1}{3}, \frac{1}{2}, 1\right\},$$

$$a = x_0 = 0, \quad x_1 = \frac{1}{6}, \quad x_2 = \frac{1}{4}, \quad x_3 = \frac{1}{2}, \quad b = \sigma(x_3) = 1,$$

$$w(x) = \left(x - \frac{1}{8}\right)\left(x - \frac{1}{5}\right)\left(x - \frac{1}{3}\right)\text{sign}\left(\left(x - \frac{1}{8}\right)\left(x - \frac{1}{5}\right)\left(x - \frac{1}{3}\right)\right), \quad x \in [0,1].$$

Note that

$$K_j(x_k) = 0, \quad j,k \in \{0,1,2,3\},$$

$$w\left(\frac{1}{8}\right) = w\left(\frac{1}{5}\right) = w\left(\frac{1}{3}\right) = 0.$$

Thus,

$$\int_0^1 w(x)K_j(x)\Delta x = \mu(0)w(0)K_j(0) + \mu\left(\frac{1}{8}\right)w\left(\frac{1}{8}\right)K_j\left(\frac{1}{8}\right)$$

$$+ \mu\left(\frac{1}{6}\right)w\left(\frac{1}{6}\right)K_j\left(\frac{1}{6}\right) + \mu\left(\frac{1}{4}\right)w\left(\frac{1}{4}\right)K_j\left(\frac{1}{4}\right)$$

$$+ \mu\left(\frac{1}{3}\right)w\left(\frac{1}{3}\right)K_j\left(\frac{1}{3}\right) + \mu\left(\frac{1}{2}\right)w\left(\frac{1}{2}\right)K_j\left(\frac{1}{2}\right)$$

$$= 0, \quad j \in \{0,1,2,3\}.$$

Then, the formula (2.13) takes the form

$$\int_a^b w(x)f(x)\Delta x \approx \sum_{j=0}^n W_j f(x_j). \tag{2.14}$$

Definition 2.41. The numerical integration formula (2.14) is said to be the Gauss quadrature rule, the quantities $W_j, j \in \{0, 1, \ldots, n\}$, are said to be quadrature weights.

Example 2.42. Let $\mathbb{T} = \{0, \frac{1}{4}, \frac{1}{3}, 1\}$ and

$$a = x_0 = 0, \quad x_1 = \frac{1}{3}, \quad \sigma(x_1) = 1 = b, \quad w(x) = 1 - x, \quad f(x) = x, \quad x \in [0,1].$$

We will evaluate

$$\int_a^b w(x)f(x)\Delta x,$$

using the Gauss quadrature rule. We have

$$L_0(x) = \frac{x - x_1}{x_0 - x_1} = -3x + 1, \qquad L_0(x_0) = 1,$$

$$L_0^\Delta(x) = -3, \qquad L_0(\sigma(x_0)) = L_0\left(\frac{1}{4}\right) = -\frac{3}{4} + 1 = \frac{1}{4},$$

$$L_1(x) = \frac{x - x_0}{x_1 - x_0} = 3x, \qquad L_1(x_1) = 1,$$

$$L_1^\Delta(x) = 3, \qquad L_1(\sigma(x_1)) = L_1(1) = 3,$$

$$M_0(x) = \frac{x - \sigma(x_1)}{x_0 - \sigma(x_1)} = -x + 1, \qquad M_0^\Delta(x) = -1,$$

$$M_0(\sigma(x_0)) = -\frac{1}{4} + 1 = \frac{3}{4}, \qquad M_1(x) = \frac{x - \sigma(x_0)}{x_1 - \sigma(x_0)} = \frac{x - \frac{1}{4}}{\frac{1}{3} - \frac{1}{4}} = 12x - 3,$$

$$M_1^\Delta(x) = 12, \qquad M_1(\sigma(x_1)) = 12 - 3 = 9,$$

Moreover,

$$P_0(x) = \left(1 - \frac{M_0^\Delta(x_0)L_0(x_0) + M_0(\sigma(x_0))L_0^\Delta(x_0)}{M_0(\sigma(x_0))L_0(\sigma(x_0))}(x - x_0)\right)M_0(x)L_0(x)$$

$$= \left(1 - \frac{-1 + \frac{3}{4}(-3)}{\frac{3}{4} \cdot \frac{1}{4}}x\right)(1 - x)(1 - 3x)$$

$$= \left(1 - \frac{-\frac{13}{4}}{\frac{3}{16}}x\right)(1 - x)(1 - 3x)$$

$$= \left(1 - \frac{52}{3}x\right)(1 - x)(1 - 3x),$$

$$P_1(x) = \left(1 - \frac{M_1^\Delta(x_1)L_1(x_1) + M_1(\sigma(x_1))L_1^\Delta(x_1)}{M_1(\sigma(x_1))L_1(\sigma(x_1))}(x - x_1)\right)M_1(x)L_1(x)$$

$$= \left(1 - \frac{12 + 9 \cdot 3}{9 \cdot 3}\left(x - \frac{1}{3}\right)\right)(12x - 3)(3x)$$

$$= \left(1 - \frac{13}{9}\left(x - \frac{1}{3}\right)\right)(12x - 3)(3x)$$

$$= 3x(12x - 3) - 13x\left(x - \frac{1}{3}\right)(4x - 1), \quad x \in [0,1],$$

$$P_1(0) = 0, \quad P_1\left(\frac{1}{4}\right) = 0, \quad P\left(\frac{1}{3}\right) = 1,$$

$$w(0) = 1, \quad w\left(\frac{1}{4}\right) = \frac{3}{4}, \quad w\left(\frac{1}{3}\right) = \frac{2}{3},$$

$$\mu\left(\frac{1}{4}\right) = \sigma\left(\frac{1}{4}\right) - \frac{1}{4} = \frac{1}{3} - \frac{1}{4} = \frac{1}{12}, \quad \mu\left(\frac{1}{3}\right) = \sigma\left(\frac{1}{3}\right) - \frac{1}{3} = 1 - \frac{1}{3} = \frac{2}{3}.$$

Hence,

$$\int_0^1 (1 - x)x\Delta x \approx f(x_0)\int_0^1 w(x)P_0(x)\Delta x + f(x_1)\int_0^1 w(x)P_1(x)\Delta x$$

$$= \frac{1}{3}\int_0^1 w(x)P_1(x)\Delta x$$

$$= \frac{1}{3}\mu(0)w(0)P_1(0) + \frac{1}{3}\mu\left(\frac{1}{4}\right)w\left(\frac{1}{4}\right)P_1\left(\frac{1}{4}\right)$$

$$+ \frac{1}{3}\mu\left(\frac{1}{3}\right)w\left(\frac{1}{3}\right)P_1\left(\frac{1}{3}\right)$$

$$= \frac{1}{3} \cdot \frac{2}{3} \cdot \frac{2}{3} = \frac{4}{27}.$$

Exercise 2.43. Let

$$\mathbb{T} = \left\{-1, -\frac{2}{3}, -\frac{1}{2}, 0, \frac{1}{3}, \frac{2}{3}\right\},$$

$$a = x_0 = -1, \quad x_1 = -\frac{1}{2}, \quad x_2 = \frac{1}{3}, \quad b = \frac{2}{3},$$

$$w(x) = \left|x^3 - \frac{4}{9}x\right|, \quad f(x) = \frac{1 + x}{1 + x + x^2}, \quad x \in \mathbb{T}.$$

Using the Gauss quadrature rule, evaluate

$$\int_a^b w(x)f(x)\Delta x, \quad \int_a^b (w(x))^2 f(x)\Delta x, \quad \int_a^b (w(x))^3 f(x)\Delta x.$$

2.10 Error estimation for Gauss quadrature rules

Now, we give the error estimate for the approximation of an integral by a Gauss quadrature rule. This estimate employs the polynomials used in the error estimates in the polynomial interpolation discussed in Chapter 1.

Let $a, b, f, w, x_j, j \in \{0, 1, \ldots, n\}$, be as in Section 2.9. From Section 1.3, we have

$$G_{\min,2n+2}(\xi) \le \frac{f(x) - p_{2n+1}(x)}{\pi_{n+1}(x)\zeta_{n+1}(x)} \le G_{\max,2n+2}(\xi)$$

for some $\xi = \xi(x) \in (a, b)$, $x \in [a, b]$. Observe that,

$$\operatorname{sign}(\pi_{n+1}(x)) = \operatorname{sign}(\zeta_{n+1}(x)), \quad x \in [a, b].$$

Then, for $x \in [a, b]$,

$$G_{\min,2n+2}(\xi)\pi_{n+1}(x)\zeta_{n+1}(x) \le f(x) - p_{2n+1}(x) \le G_{\max,2n+2}(\xi)\pi_{n+1}(x)\zeta_{n+1}(x). \tag{2.15}$$

Next,

$$\int_a^b w(x)f(x)\Delta x - \sum_{j=0}^n W_j f(x_j) = \int_a^b w(x)(f(x) - p_{2n+1}(x))\Delta x.$$

Now, applying (2.15), we arrive at

$$\int_a^b w(x)G_{\min,2n+2}(\xi)\pi_{n+1}(x)\zeta_{n+1}(x)\Delta x \le \int_a^b w(x)f(x)\Delta x - \sum_{j=0}^n W_j f(x_j)$$

$$\le \int_a^b w(x)G_{\max,2n+2}(\xi)\pi_{n+1}(x)\zeta_{n+1}(x)\Delta x.$$

2.11 σ-Gauss quadrature rules

While working on time scales, one can use the σ-Gauss quadrature rule as an alternative to the Gauss quadrature rule. We define the σ-Gauss quadrature rule in this section.

Throughout this section, suppose that σ is delta differentiable. Let $a, b \in \mathbb{T}$, $a < b$, $f : [a, b] \to \mathbb{R}$, $w : [a, b] \to [0, \infty)$, $f \in C_{rd}^1([a, b])$, $w \in C_{rd}([a, b])$. We wish to construct a quadrature formula for the approximate evaluation of the integral

$$\int_a^b w(x)f(x)\Delta x.$$

Suppose that $n \in \mathbb{N}$, $x_j \in \mathbb{T}$, $j \in \{0, 1, \ldots, n\}$, and

$$a = x_0 \leq \sigma(x_0) < x_1 \leq \sigma(x_1) < \cdots < x_n \leq \sigma(x_n) = b.$$

Define the polynomials

$$P_{\sigma j}(x) = \left(1 - \frac{M_j^{\Delta}(x_j) + M_j(\sigma(x_j))L_j^{\Delta}(x_j)}{\sigma^{\Delta}(x_j)M_j(\sigma(x_j))L_j(\sigma(x_j))}(\sigma(x) - \sigma(x_j))\right) M_j(x)L_j(x),$$

$$K_{\sigma j}(x) = \frac{\sigma(x) - \sigma(x_j)}{\sigma^{\Delta}(x_j)M_j(\sigma(x_j))L_j(\sigma(x_j))} M_j(x)L_j(x), \quad x \in [a, b], \; j \in \{0, 1, \ldots, n\},$$

where the polynomials L_j and M_j are defined as in Chapter 1. Then the σ-Hermite interpolation polynomial of degree $2n + 1$ for the function f is given in the following way.

$$P_{\sigma 2n+1}(x) = \sum_{j=0}^{n}(P_{\sigma j}(x)f(x_j) + K_{\sigma j}(x)f^{\Delta}(x_j)), \quad x \in [a, b].$$

Thus,

$$\int_a^b w(x)f(x)\Delta x \approx \int_a^b w(x)p_{\sigma 2n+1}(x)\Delta x$$

$$= \int_a^b w(x)\left(\sum_{j=0}^{n}(P_{\sigma j}(x)f(x_j) + K_{\sigma j}(x)f^{\Delta}(x_j))\right)\Delta x$$

$$= \sum_{j=0}^{n}\left(\int_a^b w(x)P_{\sigma j}(x)\Delta x\right)f(x_j) + \sum_{j=0}^{n}\left(\int_a^b w(x)K_{\sigma j}(x)\Delta x\right)f^{\Delta}(x_j).$$

Set

$$W_{\sigma j} = \int_a^b w(x)P_{\sigma j}(x)\Delta x,$$

$$V_{\sigma j} = \int_a^b w(x)K_{\sigma j}(x)\Delta x, \quad j \in \{0, 1, \ldots, n\}.$$

Therefore

$$\int_a^b w(x)f(x)\Delta x \approx \sum_{j=0}^{n}(W_{\sigma j}f(x_j) + V_{\sigma j}f^{\Delta}(x_j)). \tag{2.16}$$

Now we take the function w so that $V_{\sigma j} \equiv 0$, $j \in \{0, 1, \ldots, n\}$. Thus, (2.16) takes the form

$$\int_a^b w(x)f(x)\Delta x \approx \sum_{j=0}^n W_{\sigma j}f(x_j). \tag{2.17}$$

Definition 2.44. The numerical integration formula (2.17) is said to be the σ-Gauss quadrature rule and the quantities $W_{\sigma j}, j \in \{0, 1, \dots, n\}$, are said to be the σ-quadrature weights.

Exercise 2.45. Let $\mathbb{T} = \{0, \frac{1}{3}, \frac{5}{6}, 1, 3, 7, 11, 12\}$,

$$a = x_0 = 0, \quad x_1 = \frac{5}{6}, \quad x_2 = 3, \quad x_3 = 11, \quad b = 12,$$

and consider

$$f(t) = t + t^2, \quad w(t) = \left(t - \frac{1}{3}\right)^2 (t-1)^2 (t-7)^2 (t-12)^2, \quad t \in \mathbb{T}.$$

Using the σ-Gauss quadrature rule, evaluate the integrals

$$\int_a^b w(t)f(t)\Delta t, \quad \int_a^b (w(t))^2 f(t)\Delta t.$$

2.12 Error estimation for σ-Gauss quadrature rules

We conclude this chapter with the error estimate in the approximation of a given integral by a σ-Gauss quadrature rule. Let $\sigma, a, b, f, w, x_j, j \in \{0, 1, \dots, n\}$, be as in the previous section. From Section 1.4, we have

$$G_{\min, 2n+2}(\xi) \le \frac{f(x) - p_{\sigma 2n+1}(x)}{\pi_{n+1}(x)\zeta_{n+1}(x)} \le G_{\max, 2n+2}(\xi),$$

for some $\xi = \xi(x) \in (a, b), x \in [a, b]$. Since

$$\operatorname{sign}(\pi_{n+1}(x)) = \operatorname{sign}(\zeta_{n+1}(x)), \quad x \in [a, b],$$

we find, for $x \in [a, b]$,

$$G_{\min, 2n+2}(\xi)\pi_{n+1}(x)\zeta_{n+1}(x) \le f(x) - p_{\sigma 2n+1}(x) \le G_{\max, 2n+2}(\xi)\pi_{n+1}(x)\zeta_{n+1}(x),$$

and

$$\int_a^b w(x)f(x)\Delta x - \sum_{j=0}^n W_j f(\sigma x_j) = \int_a^b w(x)(f(x) - p_{\sigma 2n+1}(x))\Delta x,$$

whereupon we arrive at

$$\int_a^b w(x)G_{\min,2n+2}(\xi)\pi_{n+1}(x)\zeta_{n+1}(x)\Delta x \le \int_a^b w(x)f(x)\Delta x - \sum_{j=0}^n W_jf(x_j)$$

$$\le \int_a^b w(x)G_{\max,2n+2}(\xi)\pi_{n+1}(x)\zeta_{n+1}(x)\Delta x.$$

2.13 Advanced practical problems

Problem 2.46. Let $\mathbb{T} = (\frac{1}{2})^{\mathbb{N}_0}$. Using the trapezoid rule, evaluate the integral

$$\int_{\frac{1}{256}}^1 \frac{t^2 + t + 1}{t^4 - 3t^3 + 7t^2 + 4t + 1}\Delta t.$$

Problem 2.47. Let $\mathbb{T} = 3^{\mathbb{N}_0}$. Using the Simpson rule, evaluate the integral

$$\int_1^{27} \frac{t^5 + t + 1}{t^2 + 3t + 10}\Delta t.$$

Problem 2.48. Let $\mathbb{T} = 3^{\mathbb{N}_0}$. Using the composite trapezoid rule, evaluate the integral

$$\int_1^{81} \left(\frac{t - 1}{t^4 + t + 1} + \cos_1(t, 1)\right)\Delta t.$$

Problem 2.49. Let

$$\mathbb{T} = \left\{-1, -\frac{1}{8}, \frac{1}{2}, \frac{5}{6}, \frac{4}{3}, \frac{15}{7}, 3\right\}, \quad a = -1 = x_0, \quad x_1 = \frac{1}{2}, \quad b = 3,$$

and $f : \mathbb{T} \to \mathbb{R}$ be defined by

$$f(t) = \cos_1(t, -1) + \sin_2(t, -1) + t^2, \quad t \in \mathbb{T}.$$

Using the σ-trapezoid rule, evaluate the integral $\int_{-1}^3 f(t)\Delta t$.

Problem 2.50. Let

$$\mathbb{T} = \left\{-3, 0, \frac{1}{8}, \frac{1}{3}, \frac{1}{2}, \frac{5}{6}, 1, \frac{4}{3}, 2\right\}, \quad a = -3 = x_0, \quad x_1 = \frac{1}{2}, \quad x_2 = \frac{4}{3}, \quad b = 2,$$

and $f : \mathbb{T} \to \mathbb{R}$ be defined by

$$f(t) = \sin_2(t, 0) + e_1(t, 0) - \frac{t}{1 + t^2}, \quad t \in \mathbb{T}.$$

Using the σ-Simpson rule, evaluate the integral $\int_a^b f(t)\Delta t$.

Problem 2.51. Let $\mathbb{T} = 2^{\mathbb{N}_0}$,

$$a = 1 = x_0, \quad x_1 = 8, \quad x_2 = b = 64,$$

and $f : \mathbb{T} \to \mathbb{R}$ be defined by

$$f(t) = \frac{1 + t + t^2}{1 + t + t^2 + t^3 + t^4 + t^5 + t^6} + \cosh_1(t, 3), \quad t \in \mathbb{T}.$$

Using the composite trapezoid rule, evaluate the integral $\int_a^b f(t)\Delta t$.

Problem 2.52. Let

$$\mathbb{T} = \left\{1, \frac{5}{4}, \frac{4}{3}, \frac{3}{2}, \frac{5}{3}, 2, \frac{7}{3}, \frac{31}{12}, 3, 4\right\},$$

$$a = x_0 = 1, \quad x_1 = \frac{4}{3}, \quad x_2 = \frac{5}{3}, \quad x_3 = \frac{7}{3}, \quad x_4 = 4, \quad b = 4,$$

and $f : \mathbb{T} \to \mathbb{R}$ be defined by

$$f(t) = t^4 + t^2 + \frac{1}{1 + t} + e_1(t, 1), \quad t \in \mathbb{T}.$$

Using the σ-composite trapezoid rule, evaluate the integral $\int_a^b f(t)\Delta t$.

Problem 2.53. Let $\mathbb{T} = 3^{\mathbb{N}_0}$,

$$a = 1 = x_0, \quad x_1 = 9, \quad x_2 = 81, \quad x_3 = 729, \quad x_4 = b = 6561,$$

and suppose $f : \mathbb{T} \to \mathbb{R}$ is defined by

$$f(t) = \frac{1 + t}{1 + t^6} + \sin_1(t, 1) + \cosh_2(t, 1), \quad t \in \mathbb{T}.$$

Using the σ-composite Simpson rule, evaluate the integral $\int_a^b f(t)\Delta t$.

Problem 2.54. Let

$$\mathbb{T} = \left\{0, \frac{1}{10}, \frac{1}{9}, \frac{1}{8}, \frac{1}{7}, \frac{1}{6}, \frac{1}{5}, \frac{1}{4}, \frac{1}{3}, \frac{1}{2}, \frac{2}{3}, \frac{5}{6}, 1\right\},$$

$$a = x_0 = 0, \quad x_1 = \frac{1}{9}, \quad x_2 = \frac{1}{7}, \quad x_3 = \frac{1}{5}, \quad x_4 = \frac{1}{3}, \quad x_5 = \frac{5}{6}, \quad b = 1,$$

$$w(x) = \left(x - \frac{1}{10}\right)^2 \left(x - \frac{1}{8}\right)^2 \left(x - \frac{1}{6}\right)^2 \left(x - \frac{1}{4}\right)^2 \left(x - \frac{1}{10}\right)^2 \left(x - \frac{2}{3}\right)^2 (x - 1)^2,$$

$$f(x) = 1 + x + e_1(x, 0), \quad x \in [0, 1].$$

Using the Gauss quadrature rule, evaluate

$$\int_a^b w(x)f(x)\Delta x, \quad \int_a^b (w(x))^2 f(x)\Delta x, \quad \int_a^b (w(x))^3 f(x)\Delta x.$$

Problem 2.55. Let

$$\mathbb{T} = \left\{ -1, -\frac{1}{4}, 0, \frac{1}{8}, 1, 7 \right\},$$

$$a = x_0 = -1, \quad x_1 = 0, \quad x_2 = 1, \quad b = 7,$$

$$f(t) = 1 + t + t^2, \quad w(t) = \left(t + \frac{1}{4} \right)^2 \left(t - \frac{1}{8} \right)^2 (t - 7)^2, \quad t \in \mathbb{T}.$$

Using the σ-Gauss quadrature rule, evaluate the integral

$$\int_a^b (w(t))^2 (f(t) + (f(t))^3) \Delta t.$$

3 Piecewise polynomial approximation

The focus of our discussions in the previous chapters has been the question of approximation of a given rd-continuous function f, defined on an interval $[a, b]$, by a polynomial on that interval either through Lagrange, σ-Lagrange, Hermite, or σ-Hermite interpolation polynomials. Each of these constructions was global in nature, in the sense that the approximation was defined by the same analytical expression on the whole interval $[a, b]$. In this chapter, we will present an alternative and more flexible way for approximation of a function f, and this approach is based on dividing the interval $[a, b]$ into a number of subintervals and looking for a piecewise approximation by polynomials of low degree. Such piecewise-polynomial approximations are called splines, and the endpoints of the subintervals are known as the knots.

Let $\mathbb{T}$ be a time scale with forward jump operator σ and delta differentiation operator Δ.

3.1 Linear interpolating splines

We first discuss the piecewise approximation by the lowest degree polynomials, that is, by linear functions, called linear splines. Let $a, b \in \mathbb{T}$, $a < b$, and $m \in \mathbb{N}$, $m \geq 2$.

Definition 3.1. Suppose that $f \in C_{rd}([a, b])$ and $K = \{x_0, x_1, \ldots, x_m\}$ is a subset of $[a, b]$ such that

$$a = x_0 < x_1 < \cdots < x_m = b.$$

The linear spline s_L, interpolating f at the points x_j, $j \in \{0, 1, \ldots, m\}$, is defined by

$$s_L(x) = \frac{x_j - x}{x_j - x_{j-1}} f(x_{j-1}) + \frac{x - x_{j-1}}{x_j - x_{j-1}} f(x_j), \quad x \in [x_{j-1}, x_j], \quad j \in \{1, \ldots, m\}.$$

The points x_j, $j \in \{0, 1, \ldots, m\}$, are the knots of the spline, and K is said to be the set of knots.

By Definition 3.1, it follows that

$$s_L(x_{j-1}) = f(x_{j-1}), \quad s_L(x_j) = f(x_j), \quad j \in \{1, \ldots, m\}.$$

We will now give an error estimate for the maximal error for the linear spline interpolation of a given function. Below denote

$$r_{j+1} = x_{j+1} - x_j, \quad j \in \{0, 1, \ldots, m - 1\}, \quad r = \max_{j \in \{1, \ldots, m\}} r_j.$$

https://doi.org/10.1515/9783110787320-003

Theorem 3.2. *Let $f \in C_{rd}^1([a,b])$. Then*

$$\max_{x \in [a,b]} |s_L(x) - f(x)| \le 2r \max_{x \in [a,b]} |f^{\Delta}(x)|.$$

Proof. Let $j \in \{1, \ldots, m\}$ be arbitrarily chosen and fixed. We have

$$|s_L(x) - f(x)| = \left| \frac{x_j - x}{x_j - x_{j-1}} f(x_{j-1}) + \frac{x - x_{j-1}}{x_j - x_{j-1}} f(x_j) - f(x) \right|$$

$$= \left| \frac{x_j - x}{x_j - x_{j-1}} (f(x_{j-1}) - f(x)) + \frac{x - x_{j-1}}{x_j - x_{j-1}} (f(x_j) - f(x)) \right|$$

$$\le \frac{x_j - x}{x_j - x_{j-1}} |f(x_{j-1}) - f(x)| + \frac{x - x_{j-1}}{x_j - x_{j-1}} |f(x_j) - f(x)|, \quad x \in [a,b].$$

By the mean value theorem, it follows that

$$|f(x_{j-1}) - f(x)| \le \max_{x \in [x_{j-1}, x_j]} |f^{\Delta}(x)| (x - x_{j-1})$$

$$\le \max_{x \in [a,b]} |f^{\Delta}(x)| (x - x_{j-1}), \quad x \in [x_{j-1}, x_j],$$

$$|f(x_j) - f(x)| \le \max_{x \in [x_{j-1}, x_j]} |f^{\Delta}(x)| (x_j - x)$$

$$\le \max_{x \in [a,b]} |f^{\Delta}(x)| (x_j - x), \quad x \in [x_{j-1}, x_j].$$

Therefore,

$$|s_L(x) - f(x)| \le \frac{(x_j - x)^2}{r_j} \max_{x \in [a,b]} |f^{\Delta}(x)| + \frac{(x - x_{j-1})^2}{r_j} \max_{x \in [a,b]} |f^{\Delta}(x)|$$

$$\le 2r_j \max_{x \in [a,b]} |f^{\Delta}(x)|$$

$$\le 2r \max_{x \in [a,b]} |f^{\Delta}(x)|, \quad x \in [x_{j-1}, x_j].$$

Since $j \in \{1, \ldots, m\}$ was arbitrarily chosen, we obtain

$$\sup_{x \in [a,b]} |s_L(x) - f(x)| \le 2r \max_{x \in [a,b]} |f^{\Delta}(x)|.$$

This completes the proof. □

We illustrate the computations of this error bound in the following example.

Example 3.3. Let

$$\mathbb{T} = \left\{ 0, \frac{1}{8}, \frac{1}{7}, \frac{1}{6}, \frac{1}{5}, \frac{1}{4}, \frac{1}{3}, \frac{1}{2}, 1, 2 \right\},$$

$$a = x_0 = 0, \quad x_1 = \frac{1}{8}, \quad x_2 = \frac{1}{7}, \quad x_3 = \frac{1}{4}, \quad x_4 = b = 1,$$

$$f(t) = t^2 + t + 1, \quad t \in \mathbb{T}.$$

We have

$$r_1 = x_1 - x_0 = \frac{1}{8}, \quad r_2 = x_2 - x_1 = \frac{1}{56},$$
$$r_3 = x_3 - x_2 = \frac{3}{28}, \quad r_4 = x_4 - x_3 = \frac{3}{4}.$$

Hence, $r = \frac{3}{4}$. Since all points of $\mathbb{T}$ are right-scattered, we have

$$f^\Delta(t) = \sigma(t) + t + 1, \quad t \in \mathbb{T}.$$

From here,

$$\max_{t \in [0,1]} f^\Delta(t) = \sigma(1) + 1 + 1 = 2 + 2 = 4.$$

Thus,

$$\sup_{x \in [0,1]} \left| s_L(x) - f(x) \right| \le 2 \cdot \frac{3}{4} \cdot 4 = 6.$$

Exercise 3.4. Let $\mathbb{T} = \mathbb{Z}$,

$$a = x_0 = 0, \quad x_1 = 2, \quad x_2 = 4, \quad x_3 = 6, \quad x_4 = 8, \quad x_5 = 10 = b,$$
$$f(t) = t^3 + t, \quad t \in \mathbb{T}.$$

Prove that

$$\sup_{x \in [0,10]} \left| s_L(x) - f(x) \right| \le 1328.$$

Note that, as stated in the following remark, for the local error on each interval we can use the error estimates given in Chapter 1.

Remark 3.5. Fix $j \in \{1, \ldots, m\}$ arbitrarily. Suppose that $f \in C_{rd}^2([a, b])$. Then, by Theorem 1.10, we have

$$f(x) - s_L(x) = \frac{f^{\Delta^2}(\xi)}{\Pi_2^2(\xi)} \pi_2(x), \quad x \in [x_{j-1}, x_j],$$

or

$$F_{\min,2}(\xi) \le \frac{f(x) - s_L(x)}{\pi_2(x)} \le F_{\max,2}(\xi), \quad x \in [x_{j-1}, x_j].$$

Now, suppose that s_L is a linear spline with knots x_j, $j \in \{0, 1, \ldots, m\}$. We can express s_L as a linear combination of suitable "basis functions" ϕ_j, $j \in \{0, 1, \ldots, m\}$, as follows:

$$s_L(x) = \sum_{j=0}^{m} \phi_j(x)f(x_j), \quad x \in [a, b].$$

We require that each function $\phi_j, j \in \{0, 1, \ldots, m\}$, is itself a linear spline that vanishes at every knot except x_j and $\phi_j(x_j) = 1$.

Definition 3.6. The functions $\phi_j, j \in \{0, 1, \ldots, m\}$, are said to be linear basis splines or hat functions.

Definition 3.7. The formal definition of $\phi_j, j \in \{0, 1, \ldots, m\}$, is given as follows:

$$\phi_j(x) = \begin{cases} 0 & \text{if } x \leq x_{j-1}, \\ \frac{x - x_{j-1}}{x_j - x_{j-1}} & \text{if } x_{j-1} \leq x \leq x_j, \\ \frac{x_{j+1} - x}{x_{j+1} - x_j} & \text{if } x_j \leq x \leq x_{j+1}, \\ 0 & \text{if } x_{j+1} \leq x, \end{cases}$$

for $j \in \{1, \ldots, m-1\}$, and

$$\phi_0(x) = \begin{cases} \frac{x_1 - x}{x_1 - x_0} & \text{if } a = x_0 \leq x \leq x_1, \\ 0 & \text{if } x_1 \leq x, \end{cases}$$

and

$$\phi_m(x) = \begin{cases} 0 & \text{if } x \leq x_{m-1}, \\ \frac{x - x_{m-1}}{x_m - x_{m-1}} & \text{if } x_{m-1} \leq x \leq x_m = b. \end{cases}$$

Example 3.8. Let $\mathbb{T} = \mathbb{Z}$ and let

$$a = x_0 = -3, \quad x_1 = -2, \quad x_2 = 0, \quad x_3 = 1 = b,$$
$$y_0 = 1, \quad y_1 = 2, \quad y_2 = 1, \quad y_3 = 0.$$

We compute the linear basis spline functions as

$$\phi_0(x) = \begin{cases} -2 - x & \text{if } -3 \leq x < -2, \\ 0 & \text{if } -2 \leq x \leq 1, \end{cases}$$

$$\phi_1(x) = \begin{cases} 0 & \text{if } x < -3, \\ x + 3 & \text{if } -3 \leq x < -2, \\ -\frac{x}{2} & \text{if } -2 \leq x \leq 0, \\ 0 & \text{if } 0 < x, \end{cases}$$

$$\phi_2(x) = \begin{cases} 0 & \text{if } x < -2, \\ \frac{x+2}{2} & \text{if } -2 \le x < 0, \\ 1-x & \text{if } 0 \le x \le 1, \\ 0 & \text{if } 1 < x, \end{cases}$$

$$\phi_3(x) = \begin{cases} 0 & \text{if } x < 0, \\ x & \text{if } 0 \le x \le 1. \end{cases}$$

The graph of the linear spline s_L defined as

$$s_L(x) = \sum_{j=0}^{4} \phi_j(x) f(x_j), \quad x \in [-3, 1],$$

is given in Figure 3.1. It is clear that, as was expected, the linear spline s_L is a union of straight line segments joining the points (x_i, y_i) and (x_{i+1}, y_{i+1}) for $i = 0, 1, 2$.

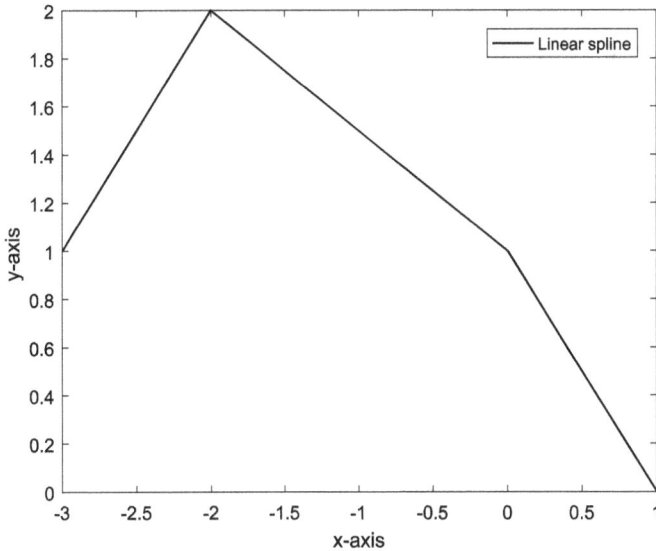

Figure 3.1: The graph of the linear spline s_L.

3.2 Linear interpolating σ-splines

Following the σ-interpolation concept introduced in Chapter 1, we present below an alternative spline interpolation, called the linear σ-spline interpolation. Let $a, b \in \mathbb{T}$, $a < b$, and $m \in \mathbb{N}$, $m \ge 2$.

Definition 3.9. Suppose that $f \in C_{rd}([a,b])$, $K = \{x_0, x_1, \ldots, x_m\}$ is a subset of $[a,b]$ such that

$$a = x_0 \leq \sigma(x_0) < x_1 \leq \sigma(x_1) < \cdots < x_m \leq \sigma(x_m) = b.$$

The linear σ-spline $s_{\sigma L}$, interpolating f at the points x_j, $j \in \{0, 1, \ldots, m\}$, is defined by

$$s_{\sigma L}(x) = \frac{\sigma(x_j) - \sigma(x)}{\sigma(x_j) - \sigma(x_{j-1})} f(x_{j-1}) + \frac{\sigma(x) - \sigma(x_{j-1})}{\sigma(x_j) - \sigma(x_{j-1})} f(x_j), \quad x \in [x_{j-1}, x_j],$$

for $j \in \{1, \ldots, m\}$. The points x_j, $j \in \{0, 1, \ldots, m\}$, are the knots of the σ-spline, and K is said to be the set of knots.

By Definition 3.9, it follows that

$$s_{\sigma L}(x_{j-1}) = f(x_{j-1}), \quad s_{\sigma L}(x_j) = f(x_j), \quad j \in \{1, \ldots, m\}.$$

In the following discussion, we obtain an estimate for the global error in the σ-linear spline interpolation. Below, denote

$$r_{j+1} = x_{j+1} - x_j, \quad j \in \{0, 1, \ldots, m-1\}, \quad r = \max_{j \in \{1, \ldots, m\}} r_j.$$

Theorem 3.10. *Let $f \in C_{rd}^1([a,b])$. Then*

$$\max_{x \in [a,b]} |s_{\sigma L}(x) - f(x)| \leq 2r \max_{x \in [a,b]} |f^{\Delta}(x)|.$$

Proof. Take $j \in \{1, \ldots, m\}$ arbitrarily. Then for $x \in [a,b]$,

$$|s_{\sigma L}(x) - f(x)| = \left| \frac{\sigma(x_j) - \sigma(x)}{\sigma(x_j) - \sigma(x_{j-1})} f(x_{j-1}) + \frac{\sigma(x) - \sigma(x_{j-1})}{\sigma(x_j) - \sigma(x_{j-1})} f(x_j) - f(x) \right|$$

$$= \left| \frac{\sigma(x_j) - \sigma(x)}{\sigma(x_j) - \sigma(x_{j-1})} (f(x_{j-1}) - f(x)) + \frac{\sigma(x) - \sigma(x_{j-1})}{\sigma(x_j) - \sigma(x_{j-1})} (f(x_j) - f(x)) \right|$$

$$\leq \frac{\sigma(x_j) - \sigma(x)}{\sigma(x_j) - \sigma(x_{j-1})} |f(x_{j-1}) - f(x)| + \frac{\sigma(x) - \sigma(x_{j-1})}{\sigma(x_j) - \sigma(x_{j-1})} |f(x_j) - f(x)|,$$

Now, applying the mean value theorem, we get

$$|f(x_{j-1}) - f(x)| \leq \max_{x \in [x_{j-1}, x_j]} |f^{\Delta}(x)| (x - x_{j-1})$$

$$\leq \max_{x \in [a,b]} |f^{\Delta}(x)| (x - x_{j-1}), \quad x \in [x_{j-1}, x_j],$$

and

$$|f(x_j) - f(x)| \le \max_{x \in [x_{j-1}, x_j]} |f^\Delta(x)|(x_j - x)$$

$$\le \max_{x \in [a,b]} |f^\Delta(x)|(x_j - x), \quad x \in [x_{j-1}, x_j].$$

Therefore,

$$|s_{\sigma L}(x) - f(x)| \le r_j \max_{x \in [a,b]} |f^\Delta(x)| + r_j \max_{x \in [a,b]} |f^\Delta(x)|$$

$$= 2r_j \max_{x \in [a,b]} |f^\Delta(x)|$$

$$\le 2r \max_{x \in [a,b]} |f^\Delta(x)|, \quad x \in [x_{j-1}, x_j].$$

Since $j \in \{1, \ldots, m\}$ was arbitrarily chosen, we arrive at

$$\sup_{x \in [a,b]} |s_{\sigma L}(x) - f(x)| \le 2r \max_{x \in [a,b]} |f^\Delta(x)|.$$

This completes the proof. □

We compute an estimate for the global error in the following example.

Example 3.11. Let

$$\mathbb{T} = \left\{ 0, \frac{1}{6}, \frac{5}{6}, 1, \frac{4}{3}, \frac{11}{6}, 2, \frac{5}{2}, \frac{8}{3}, 3 \right\},$$

$$a = x_0 = 0, \quad x_1 = \frac{5}{6}, \quad x_2 = \frac{4}{3}, \quad x_3 = 2, \quad x_4 = \frac{8}{3}, \quad b = 3,$$

$$f(t) = \frac{1}{t+1}, \quad t \in \mathbb{T}.$$

We have

$$x_0 < \sigma(x_0) < x_1 < \sigma(x_1) < x_2 < \sigma(x_2) < x_3 < \sigma(x_3) < x_4 < \sigma(x_4) < \sigma(x_4) = b.$$

Next, since all points of $\mathbb{T}$ are right-scattered, we have

$$f^\Delta(t) = -\frac{1}{(t+1)(\sigma(t)+1)},$$

$$|f^\Delta(t)| = \frac{1}{(t+1)(\sigma(t)+1)}, \quad t \in \mathbb{T}.$$

Thus,

$$\max_{t \in [a,b]} |f^\Delta(t)| = \frac{1}{\sigma(0)+1} = \frac{1}{\frac{1}{6}+1} = \frac{6}{7}.$$

Next,

$$r_1 = x_1 - x_0 = \frac{5}{6}, \qquad r_2 = x_2 - x_1 = \frac{4}{3} - \frac{5}{6} = \frac{1}{2},$$

$$r_3 = x_3 - x_2 = 2 - \frac{4}{3} = \frac{2}{3}, \quad r_4 = x_4 - x_3 = \frac{8}{3} - 2 = \frac{2}{3}.$$

Consequently, $r = \frac{5}{6}$ and

$$|s_{\sigma L}(x) - f(x)| \le 2 \cdot \frac{5}{6} \cdot \frac{6}{7} = \frac{10}{7}.$$

Exercise 3.12. Let

$$\mathbb{T} = \left\{ -3, -\frac{5}{2}, -\frac{7}{3}, -2, -\frac{7}{6}, -1, 0, \frac{1}{2}, 1, \frac{4}{3}, 2 \right\},$$

$$a = x_0 = 3, \quad x_1 = -\frac{7}{3}, \quad x_2 = -\frac{7}{6}, \quad x_3 = 0, \quad x_4 = 1, \quad b = \frac{4}{3},$$

$$f(t) = \frac{1}{(t+10)(t^2+t+1)}, \quad t \in \mathbb{T}.$$

Estimate

$$|s_{\sigma L}(x) - f(x)|, \quad x \in [a, b].$$

For the local error, we can give the following remark recalling the error in the σ-polynomial interpolation given in Chapter 1.

Remark 3.13. Fix $j \in \{1, \ldots, m\}$ arbitrarily. Suppose that $f \in C_{rd}^2([a, b])$. Then, by Theorem 1.28, we have

$$f(x) - s_{\sigma L}(x) = \frac{f^{\Delta^2}(\xi)}{\Pi_{\sigma 2}^2(\xi)} \pi_{\sigma 2}(x), \quad x \in [x_{j-1}, x_j],$$

or

$$F_{\sigma\min,2}(\xi) \le \frac{f(x) - s_{\sigma L}(x)}{\pi_{\sigma 2}(x)} \le F_{\sigma\max,2}(\xi), \quad x \in [x_{j-1}, x_j].$$

Now, we will formally define the σ-linear spline by using the relevant basis functions. Suppose that $s_{\sigma L}$ is a linear spline with knots $x_j, j \in \{0, 1, \ldots, m\}$. We can express $s_{\sigma L}$ as a linear combination of suitable "σ-basis functions" $\phi_{\sigma j}, j \in \{0, 1, \ldots, m\}$, as follows:

$$s_{\sigma L}(x) = \sum_{j=0}^{m} \phi_{\sigma j}(x) f(x_j), \quad x \in [a, b].$$

We require that each function $\phi_{\sigma j}, j \in \{0, 1, \ldots, m\}$, is itself a σ-linear spline that vanishes at every knot except x_j and $\phi_{\sigma j}(x_j) = 1$.

Definition 3.14. The functions $\phi_{\sigma j}, j \in \{0, 1, \ldots, m\}$, are said to be linear basis σ-splines, or σ-hat functions.

Definition 3.15. The formal definition of $\phi_{\sigma j}, j \in \{0, 1, \ldots, m\}$, is as follows:

$$\phi_{\sigma j}(x) = \begin{cases} 0 & \text{if } x \leq x_{j-1}, \\ \frac{\sigma(x) - \sigma(x_{j-1})}{\sigma(x_j) - \sigma(x_{j-1})} & \text{if } x_{j-1} \leq x \leq x_j, \\ \frac{\sigma(x_{j+1}) - \sigma(x)}{\sigma(x_{j+1}) - \sigma(x_j)} & \text{if } x_j \leq x \leq x_{j+1}, \\ 0 & \text{if } x_{j+1} \leq x, \end{cases}$$

for $j \in \{1, \ldots, m-1\}$, and

$$\phi_{\sigma 0}(x) = \begin{cases} \frac{\sigma(x_1) - \sigma(x)}{\sigma(x_1) - \sigma(x_0)} & \text{if } a = x_0 \leq x \leq x_1, \\ 0 & \text{if } x_1 \leq x, \end{cases}$$

and

$$\phi_{\sigma m}(x) = \begin{cases} 0 & \text{if } x \leq x_{m-1}, \\ \frac{\sigma(x) - \sigma(x_{m-1})}{\sigma(x_m) - \sigma(x_{m-1})} & \text{if } x_{m-1} \leq x \leq x_m = b. \end{cases}$$

In the next example, we compute the linear and σ-linear spline approximation for a given function. We consider a time scale with a nonlinear forward jump operator in order to observe the difference between the linear and σ-linear splines.

Example 3.16. Let $\mathbb{T} = \mathbb{N}_0^2 = \{0, 1, 4, 9, 16, \ldots\}$ and $f(x) = \frac{1}{\sqrt{x}+1}$. Take

$$a = x_0 = 1, \quad x_1 = 9, \quad x_2 = 25, \quad x_3 = 49 = b.$$

Then we have

$$y_0 = \frac{1}{2}, \quad y_1 = \frac{1}{4}, \quad y_2 = \frac{1}{6}, \quad y_3 = \frac{1}{8}.$$

Note that on this time scale $\sigma(t) = (\sqrt{t} + 1)^2$. We will find the linear basis spline and σ-spline functions for the given knots $x_j, j \in \{0, 1, 2, 3\}$. First, we compute the linear basis spline functions as

$$\phi_0(x) = \begin{cases} \frac{9-x}{8} & \text{if } 1 \leq x < 9, \\ 0 & \text{if } 9 < x, \end{cases}$$

$$\phi_1(x) = \begin{cases} 0 & \text{if } x < 1, \\ \frac{x-1}{8} & \text{if } 1 \leq x < 9, \\ \frac{25-x}{16} & \text{if } 9 \leq x \leq 25, \\ 0 & \text{if } 25 < x, \end{cases}$$

$$\phi_2(x) = \begin{cases} 0 & \text{if } x < 9, \\ \frac{x-9}{16} & \text{if } 9 \le x < 25, \\ \frac{49-x}{24} & \text{if } 25 \le x \le 49, \\ 0 & \text{if } 49 < x, \end{cases}$$

$$\phi_3(x) = \begin{cases} 0 & \text{if } x < 25, \\ \frac{x-25}{24} & \text{if } 25 \le x \le 49. \end{cases}$$

The graph of the linear spline $s_L(x)$ defined as

$$s_L(x) = \sum_{j=0}^{3} \phi_j(x) f(x_j), \quad x \in [1, 49],$$

is given in Figure 3.2 and is compared with the graph of $f(x) = \frac{1}{\sqrt{x}+1}$.

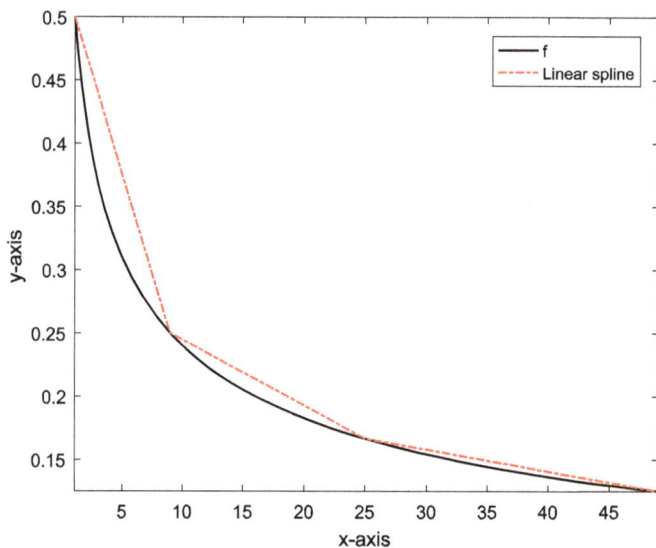

Figure 3.2: The graphs of the linear spline s_L and f.

Now we compute the linear basis σ-spline functions. First, note that

$$\sigma(x_0) = \sigma(1) = 4, \qquad \sigma(x_1) = \sigma(9) = 16,$$
$$\sigma(x_2) = \sigma(25) = 36, \qquad \sigma(x_3) = \sigma(49) = 64.$$

Then we have,

$$\phi_{\sigma 0}(x) = \begin{cases} \frac{16-(\sqrt{x}+1)^2}{12} & \text{if } 1 \leq x < 9, \\ 0 & \text{if } 9 < x, \end{cases}$$

$$\phi_{\sigma 1}(x) = \begin{cases} 0 & \text{if } x < 1, \\ \frac{(\sqrt{x}+1)^2-4}{12} & \text{if } 1 \leq x < 9, \\ \frac{36-(\sqrt{x}+1)^2}{20} & \text{if } 9 \leq x \leq 25 \\ 0 & \text{if } 25 < x, \end{cases}$$

$$\phi_{\sigma 2}(x) = \begin{cases} 0 & \text{if } x < 9, \\ \frac{(\sqrt{x}+1)^2-16}{20} & \text{if } 9 \leq x < 25, \\ \frac{64-(\sqrt{x}+1)^2}{28} & \text{if } 25 \leq x \leq 49, \\ 0 & \text{if } 49 < x, \end{cases}$$

$$\phi_{\sigma 3}(x) = \begin{cases} 0 & \text{if } x < 25, \\ \frac{(\sqrt{x}+1)^2-36}{28} & \text{if } 25 \leq x \leq 49. \end{cases}$$

Then, the linear σ-spline $s_{\sigma L}$ is defined as

$$s_{\sigma L}(x) = \sum_{j=0}^{3} \phi_{\sigma j}(x) f(x_j), \quad x \in [1, 49],$$

its graph is given in Figure 3.3 and compared with the graph of f. We compute the values of the linear spline s_L and the linear σ-spline $s_{\sigma L}$ and compare them with the values of the function f in Table 3.1. Note that at the knots both the linear spline and linear σ-spline coincide with f.

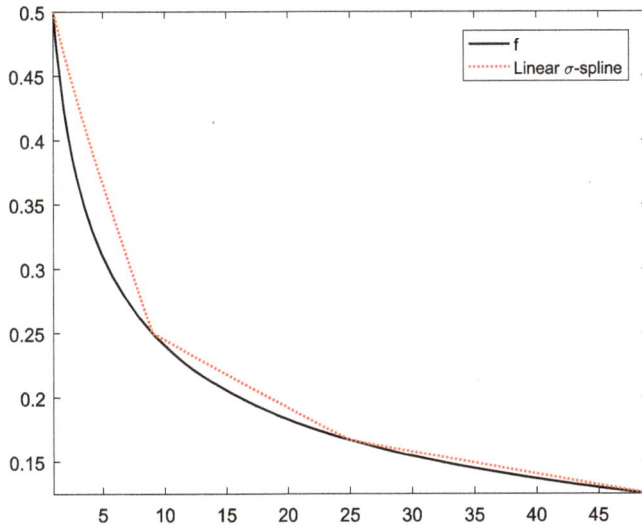

Figure 3.3: The graphs of the linear σ-spline $s_{\sigma L}$ and f.

Table 3.1: The values of f, s_L, and s_{oL} at the points of $\mathbb{T}$ in $[1, 49]$.

x	$f(x)$	$s_L(x)$	$s_{oL}(x)$
1	0.5000	0.5000	0.5000
4	0.3333	0.4063	0.3958
9	0.2500	0.2500	0.2500
16	0.2000	0.2135	0.2125
25	0.1667	0.1667	0.1667
36	0.1429	0.1475	0.1473
49	0.1250	0.1250	0.1250

3.3 Cubic splines

The most common piecewise polynomial interpolation is the interpolation by cubic polynomials, known as cubic spline interpolation. The disadvantage of linear spline is that usually at the endpoints of the subintervals the spline is not delta differentiable. To deal with this problem, cubic spline interpolation is more suitable than a quadratic one since it ensures the continuity of both the first and second order delta derivatives of the piecewise interpolation function at the knots.

In this section, we construct a natural cubic spline to interpolate a function on an arbitrary time scale. Suppose that $a, b \in \mathbb{T}$, $a < b$,

$$K = \{a = x_0 \leq \sigma(x_0) < x_1 \leq \sigma(x_1) < \cdots < x_m \leq \sigma(x_m) = b\},$$

$m \in \mathbb{N}$, $m \geq 3$, and $f \in C_{rd}([a, b])$.

Definition 3.17. The natural cubic spline interpolating a function at the set K is denoted by s_2 and it satisfies the end conditions

$$s_2^{\Delta^2}(x_0) = s_2^{\Delta^2}(x_m) = 0.$$

Now, we will construct the natural cubic spline. Let

$$a_j = s_2^{\Delta^2}(x_j), \quad j \in \{0, 1, \ldots, m\}.$$

We have $a_0 = a_m = 0$. In addition, suppose that

$$s_2^{\Delta}(x_j-) = s_2^{\Delta}(x_j+), \quad j \in \{1, \ldots, m-1\},$$

and

$$s_2^{\Delta^2}(x) = \frac{x_j - x}{x_j - x_{j-1}} a_{j-1} + \frac{x - x_{j-1}}{x_j - x_{j-1}} a_j, \quad x \in [x_{j-1}, x_j], \quad j \in \{1, \ldots, m\}. \tag{3.1}$$

Fix $j \in \{1, \ldots, m-1\}$. We integrate the equality (3.1) from x to x_j, $x \in [x_{j-1}, x_j]$, and find for $x \in [x_{j-1}, x_j]$,

$$\int_x^{x_j} s_2^{\Delta^2}(t)\Delta t = \frac{a_{j-1}}{x_j - x_{j-1}} \int_x^{x_j} (x_j - t)\Delta t + \frac{a_j}{x_j - x_{j-1}} \int_x^{x_j} (t - x_{j-1})\Delta t$$

$$= \frac{a_{j-1}}{x_j - x_{j-1}} \int_{x_j}^x (t - x_j)\Delta t + \frac{a_j}{x_j - x_{j-1}} \int_x^{x_j} (t - x_j)\Delta t$$

$$+ \frac{a_j}{x_j - x_{j-1}} \int_x^{x_j} (x_j - x_{j-1})\Delta t$$

$$= \frac{a_{j-1}}{h_1(x_j, x_{j-1})} h_2(x, x_j) - \frac{a_j}{h_1(x_j, x_{j-1})} h_2(x, x_j) + a_j h_1(x_j, x),$$

or

$$s_2^{\Delta}(x_j-) - s_2^{\Delta}(x) = \frac{a_{j-1}}{h_1(x_j, x_{j-1})} h_2(x, x_j) - \frac{a_j}{h_1(x_j, x_{j-1})} h_2(x, x_j) + a_j h_1(x_j, x), \quad x \in [x_{j-1}, x_j],$$

or

$$s_2^{\Delta}(x) = s_2^{\Delta}(x_j-) - \frac{a_{j-1}}{h_1(x_j, x_{j-1})} h_2(x, x_j) + \frac{a_j}{h_1(x_j, x_{j-1})} h_2(x, x_j) - a_j h_1(x_j, x), \quad x \in [x_{j-1}, x_j].$$

Now, we integrate the latter equality from x to x_j and find for $x \in [x_{j-1}, x_j]$,

$$\int_x^{x_j} s_2^{\Delta}(t)\Delta t = s_2^{\Delta}(x_j-) h_1(x_j, x) + \frac{a_{j-1}}{h_1(x_j, x_{j-1})} \int_{x_j}^x h_2(t, x_j)\Delta t$$

$$- \frac{a_j}{h_1(x_j, x_{j-1})} \int_{x_j}^x h_2(t, x_j)\Delta t - a_j \int_{x_j}^x (t - x_j)\Delta t$$

$$= s^{\Delta}(x_j-) h_1(x_j, x) + a_{j-1} \frac{h_3(x, x_j)}{h_1(x_j, x_{j-1})} - a_j \frac{h_3(x, x_j)}{h_1(x_j, x_{j-1})} - a_j h_2(x, x_j)$$

and

$$s_2(x_j) - s_2(x) = s_2^{\Delta}(x_j-) h_1(x_j, x) + a_{j-1} \frac{h_3(x, x_j)}{h_1(x_j, x_{j-1})}$$

$$- a_j \left(\frac{h_3(x, x_j)}{h_1(x_j, x_{j-1})} + h_2(x, x_j) \right), \quad x \in [x_{j-1}, x_j]. \tag{3.2}$$

In particular, for $x = x_{j-1}$, we have

$$f(x_j) - f(x_{j-1}) = s_2^{\Delta}(x_j-)h_1(x_j, x_{j-1}) + a_{j-1}\frac{h_3(x_{j-1}, x_j)}{h_1(x_j, x_{j-1})}$$
$$- a_j\left(\frac{h_3(x_{j-1}, x_j)}{h_1(x_j, x_{j-1})} + h_2(x_{j-1}, x_j)\right)$$

and

$$s_2^{\Delta}(x_j-) = \frac{f(x_j) - f(x_{j-1})}{h_1(x_j, x_{j-1})} - a_{j-1}\frac{h_3(x_{j-1}, x_j)}{(h_1(x_j, x_{j-1}))^2}$$
$$+ a_j\left(\frac{h_3(x_{j-1}, x_j)}{(h_1(x_j, x_{j-1}))^2} + \frac{h_2(x_{j-1}, x_j)}{h_1(x_j, x_{j-1})}\right). \tag{3.3}$$

Now, we consider the equality

$$s_2^{\Delta^2}(x) = \frac{x_{j+1} - x}{x_{j+1} - x_j}a_j + \frac{x - x_j}{x_{j+1} - x_j}a_{j+1}, \quad x \in [x_j, x_{j+1}],$$

which we integrate from x_j to x, $x \in [x_j, x_{j+1}]$, and find

$$s_2^{\Delta}(x) - s_2^{\Delta}(x_j+) = \frac{a_j}{x_{j+1} - x_j}\int_{x_j}^{x}(x_{j+1} - t)\Delta t + \frac{a_{j+1}}{x_{j+1} - x_j}\int_{x_j}^{x}(t - x_j)\Delta t$$

$$= \frac{a_j}{h_1(x_{j+1}, x_j)}\int_{x_j}^{x}(x_{j+1} - x_j)\Delta t - \frac{a_j}{h_1(x_{j+1}, x_j)}\int_{x_j}^{x}(t - x_j)\Delta t$$

$$+ \frac{a_{j+1}}{h_1(x_{j+1}, x_j)}h_2(x, x_j)$$

$$= a_j h_1(x, x_j) - \frac{a_j}{h_1(x_{j+1}, x_j)}h_2(x, x_j) + \frac{a_{j+1}}{h_1(x_{j+1}, x_j)}h_2(x, x_j),$$

as well as

$$s_2^{\Delta}(x) = s_2^{\Delta}(x_j+) + a_j h_1(x, x_j) - \frac{a_j}{h_1(x_{j+1}, x_j)}h_2(x, x_j) + \frac{a_{j+1}}{h_1(x_{j+1}, x_j)}h_2(x, x_j).$$

We integrate the latter equality from x_j to x and obtain

$$s_2(x) - s_2(x_j) = s_2^{\Delta}(x_j+)h_1(x, x_j) + a_j\int_{x_j}^{x}h_1(t, x_j)\Delta t$$

$$- \frac{a_j}{h_1(x_{j+1}, x_j)}\int_{x_j}^{x}h_2(t, x_j)\Delta t + \frac{a_{j+1}}{h_1(x_{j+1}, x_j)}\int_{x_j}^{x}h_2(t, x_j)\Delta t$$

$$= s_2^{\Delta}(x_j+)h_1(x, x_j) + a_j h_2(x, x_j) - a_j\frac{h_3(x, x_j)}{h_1(x_{j+1}, x_j)}$$

$$+ a_{j+1}\frac{h_3(x, x_j)}{h_1(x_{j+1}, x_j)}, \quad x \in [x_j, x_{j+1}].$$

In particular, for $x = x_{j+1}$, we have

$$f(x_{j+1}) - f(x_j) = s_2^{\Delta}(x_j+)h_1(x_{j+1}, x_j) + a_j\left(h_2(x_{j+1}, x_j) - \frac{h_3(x_{j+1}, x_j)}{h_1(x_{j+1}, x_j)} \right)$$

$$+ a_{j+1}\frac{h_3(x_{j+1}, x_j)}{h_1(x_{j+1}, x_j)}.$$

Thus,

$$s_2^{\Delta}(x_j+) = \frac{f(x_{j+1}) - f(x_j)}{h_1(x_{j+1}, x_j)} - a_j\left(\frac{h_2(x_{j+1}, x_j)}{h_1(x_{j+1}, x_j)} - \frac{h_3(x_{j+1}, x_j)}{(h_1(x_{j+1}, x_j))^2} \right)$$

$$- a_{j+1}\frac{h_3(x_{j+1}, x_j)}{(h_1(x_{j+1}, x_j))^2}.$$

Using the latter equality and (3.3), we arrive at

$$\frac{f(x_j) - f(x_{j-1})}{h_1(x_j, x_{j-1})} - a_{j-1}\frac{h_3(x_{j-1}, x_j)}{(h_1(x_j, x_{j-1}))^2} + a_j\left(\frac{h_3(x_{j-1}, x_j)}{(h_1(x_j, x_{j-1}))^2} + \frac{h_2(x_{j-1}, x_j)}{h_1(x_j, x_{j-1})} \right)$$

$$= \frac{f(x_{j+1}) - f(x_j)}{h_1(x_{j+1}, x_j)} - a_j\left(\frac{h_2(x_{j+1}, x_j)}{h_1(x_{j+1}, x_j)} - \frac{h_3(x_{j+1}, x_j)}{(h_1(x_{j+1}, x_j))^2} \right) - a_{j+1}\frac{h_3(x_{j+1}, x_j)}{(h_1(x_{j+1}, x_j))^2},$$

whereupon

$$- a_{j-1}\frac{h_3(x_{j-1}, x_j)}{(h_1(x_j, x_{j-1}))^2} + a_j\left(\frac{h_3(x_{j-1}, x_j)}{(h_1(x_j, x_{j-1}))^2} + \frac{h_2(x_{j-1}, x_j)}{h_1(x_j, x_{j-1})} + \frac{h_2(x_{j+1}, x_j)}{h_1(x_{j+1}, x_j)} \right.$$

$$\left. - \frac{h_3(x_{j+1}, x_j)}{(h_1(x_{j+1}, x_j))^2} \right) + a_{j+1}\frac{h_3(x_{j+1}, x_j)}{(h_1(x_{j+1}, x_j))^2}$$

$$= \frac{f(x_{j+1}) - f(x_j)}{h_1(x_{j+1}, x_j)} - \frac{f(x_j) - f(x_{j-1})}{h_1(x_j, x_{j-1})}, \quad j \in \{1, \ldots, m-1\}. \tag{3.4}$$

Note that

$$h_2(x, x_j) = \int_{x_j}^{x} (t - x_j)\Delta t$$

$$= \int_{x}^{x_j} (x_j - t)\Delta t \le (x_j - x)^2, \quad x \in [x_{j-1}, x_j],$$

$$h_3(x_{j-1}, x_j) = \int_{x_j}^{x_{j-1}} h_2(t, x_j)\Delta t$$

$$= -\int_{x_{j-1}}^{x_j} h_2(t, x_j)\Delta t$$

$$\geq - \int_{x_{j-1}}^{x_j} (x_j - t)^2 \Delta t$$

$$\geq -(x_j - x_{j-1}) \int_{x_{j-1}}^{x_j} (x_j - t) \Delta t$$

$$= -h_1(x_j, x_{j-1}) \int_{x_j}^{x_{j-1}} (t - x_j) \Delta t$$

$$= -h_1(x_j, x_{j-1}) h_2(x_{j-1}, x_j),$$

and

$$\frac{h_3(x_{j-1}, x_j)}{(h_1(x_j, x_{j-1}))^2} \geq -\frac{h_2(x_{j-1}, x_j)}{h_1(x_j, x_{j-1})},$$

or

$$\frac{h_3(x_{j-1}, x_j)}{(h_1(x_j, x_{j-1}))^2} + \frac{h_2(x_{j-1}, x_j)}{h_1(x_j, x_{j-1})} \geq 0. \tag{3.5}$$

Next,

$$h_3(x_{j+1}, x_j) = \int_{x_j}^{x_{j+1}} h_2(t, x_j) \Delta t$$

$$\leq \frac{1}{2} \int_{x_j}^{x_{j+1}} (t - x_j)^2 \Delta t$$

$$\leq \frac{h_1(x_{j+1}, x_j)}{2} \int_{x_j}^{x_{j+1}} (t - x_j) \Delta t$$

$$= \frac{1}{2} h_1(x_{j+1}, x_j) h_2(x_{j+1}, x_j),$$

whereupon

$$2\frac{h_3(x_{j+1}, x_j)}{(h_1(x_{j+1}, x_j))^2} \leq \frac{h_2(x_{j+1}, x_j)}{h_1(x_{j+1}, x_j)}. \tag{3.6}$$

Hence, by (3.5), we obtain

$$\frac{h_3(x_{j+1}, x_j)}{(h_1(x_j, x_{j-1}))^2} + \frac{h_2(x_{j+1}, x_j)}{h_1(x_j, x_{j-1})} + \frac{h_2(x_{j+1}, x_j)}{h_1(x_{j+1}, x_j)} - \frac{h_3(x_{j+1}, x_j)}{(h_1(x_j, x_{j-1}))^2} \geq 0. \tag{3.7}$$

Let

$$l(t) = (t - x_j)^2, \quad t \in [x_{j-1}, x_j].$$

Then

$$
\begin{aligned}
h_2(x, x_j) &= \int_{x_j}^{x} (t - x_j)\Delta t \\
&= -\int_{x}^{x_j} (t - x_j)\Delta t \\
&= -\frac{1}{2} \int_{x}^{x_j} ((t - x_j) + (t - x_j))\Delta t \\
&\le -\frac{1}{2} \int_{x}^{x_j} ((t - x_j) + (\sigma(t) - x_j))\Delta t \\
&= -\frac{1}{2} \int_{x}^{x_j} l^{\Delta}(t)\Delta t \\
&= -\frac{1}{2}(l(x_j) - l(x)) \\
&= \frac{1}{2}(x - x_j)^2, \quad x \in [x_{j-1}, x_j].
\end{aligned}
$$

Hence,

$$
\begin{aligned}
h_3(x_{j-1}, x_j) &= \int_{x_j}^{x_{j-1}} h_2(t, x_j)\Delta t \\
&= -\int_{x_{j-1}}^{x_j} h_2(t, x_j)\Delta t \\
&\ge -\frac{1}{2} \int_{x_{j-1}}^{x_j} (t - x_j)^2 \Delta t \\
&= -\frac{1}{2} \int_{x_{j-1}}^{x_j} (x_j - t)(x_j - t)\Delta t \\
&\ge -\frac{1}{2} h_1(x_j, x_{j-1}) \int_{x_{j-1}}^{x_j} (x_j - t)\Delta t
\end{aligned}
$$

$$= -\frac{1}{2} h_1(x_j, x_{j-1}) \int_{x_j}^{x_{j-1}} (t - x_j) \Delta t$$

$$= -\frac{1}{2} h_1(x_j, x_{j-1}) h_2(x_{j-1}, x_j),$$

whereupon

$$2 \frac{h_3(x_{j-1}, x_j)}{(h_1(x_j, x_{j-1}))^2} \geq -\frac{h_2(x_{j-1}, x_j)}{h_1(x_j, x_{j-1})}. \tag{3.8}$$

Next,

$$h_3(x_{j+1}, x_j) = \int_{x_j}^{x_{j+1}} h_2(t, x_j) \Delta t$$

$$\leq \frac{1}{2} \int_{x_j}^{x_{j+1}} (t - x_j)^2 \Delta t$$

$$< \frac{1}{2} h_1(x_{j+1}, x_j) \int_{x_j}^{x_{j+1}} (t - x_j) \Delta t$$

$$= \frac{1}{2} h_1(x_{j+1}, x_j) h_2(x_{j+1}, x_j),$$

from where

$$2 \frac{h_3(x_{j+1}, x_j)}{(h_1(x_{j+1}, x_j))^2} < \frac{h_2(x_{j+1}, x_j)}{h_1(x_{j+1}, x_j)}. \tag{3.9}$$

Now, applying (3.5), (3.6), (3.7), and (3.9), we obtain

$$\left| \frac{h_3(x_{j-1}, x_j)}{(h_1(x_j, x_{j-1}))^2} + \frac{h_2(x_{j+1}, x_j)}{h_1(x_j, x_{j-1})} + \frac{h_2(x_{j+1}, x_j)}{h_1(x_{j+1}, x_j)} - \frac{h_3(x_{j+1}, x_j)}{(h_1(x_{j+1}, x_j))^2} \right|$$

$$> \left| -\frac{h_3(x_{j-1}, x_j)}{(h_1(x_j, x_{j-1}))^2} \right| + \left| \frac{h_3(x_{j+1}, x_j)}{(h_1(x_{j+1}, x_j))^2} \right| \iff$$

$$\frac{h_3(x_{j-1}, x_j)}{(h_1(x_j, x_{j-1}))^2} + \frac{h_2(x_{j+1}, x_j)}{h_1(x_j, x_{j-1})} + \frac{h_2(x_{j+1}, x_j)}{h_1(x_{j+1}, x_j)} - \frac{h_3(x_{j+1}, x_j)}{(h_1(x_{j+1}, x_j))^2}$$

$$> -\frac{h_3(x_{j-1}, x_j)}{(h_1(x_j, x_{j-1}))^2} + \frac{h_3(x_{j+1}, x_j)}{(h_1(x_{j+1}, x_j))^2} \iff$$

$$2 \frac{h_3(x_{j-1}, x_j)}{(h_1(x_j, x_{j-1}))^2} + \frac{h_2(x_{j+1}, x_j)}{h_1(x_j, x_{j-1})} + \frac{h_2(x_{j+1}, x_j)}{h_1(x_{j+1}, x_j)} - 2 \frac{h_3(x_{j+1}, x_j)}{(h_1(x_{j+1}, x_j))^2} > 0.$$

Therefore, the system (3.4) has a unique solution $a_j, j \in \{0, 1, \ldots, m\}$. By (3.2), we get

$$s_2(x) = f(x_j) - s_2^\Delta(x_j-)h_1(x_j, x) - a_{j-1}\frac{h_3(x, x_j)}{h_1(x_j, x_{j-1})}$$

$$+ a_j\left(\frac{h_3(x, x_j)}{h_1(x_j, x_{j-1})} + h_2(x, x_j)\right), \quad x \in [x_{j-1}, x_j], \quad j \in \{1, \ldots, m\}.$$

Finally, applying (3.3), we arrive at

$$s_2(x) = f(x_j) - \frac{f(x_j) - f(x_{j-1})}{h_1(x_j, x_{j-1})}h_1(x_j, x)$$

$$+ a_{j-1}\left(\frac{h_3(x_{j-1}, x_j)}{(h_1(x_j, x_{j-1}))^2}h_1(x_j, x) - \frac{h_3(x, x_j)}{h_1(x_j, x_{j-1})}\right)$$

$$+ a_j\left(-\frac{h_3(x_{j-1}, x_j)}{(h_1(x_j, x_{j-1}))^2}h_1(x_j, x) - \frac{h_2(x_{j-1}, x_j)}{h_1(x_j, x_{j-1})}h_1(x_j, x)\right.$$

$$\left. + \frac{h_3(x, x_j)}{h_1(x_j, x_{j-1})} + h_2(x, x_j)\right), \quad x \in [x_{j-1}, x_j], \quad j \in \{1, \ldots, m\},$$

which is the natural cubic spline of f on the knot set K.

Example 3.18. Let $\mathbb{T} = \mathbb{Z}$ and

$$f(t) = \frac{1+t}{1+t^2}, \quad t \in \mathbb{T}.$$

Take

$$a = x_0 = 0, \quad x_1 = 2, \quad x_2 = b = 4.$$

Then, by (3.4) and using $a_0 = 0$, $a_2 = 0$, we get

$$a_1\left(\frac{h_3(x_0, x_1)}{(h_1(x_1, x_0))^2} + \frac{h_2(x_0, x_1)}{h_1(x_1, x_0)} + \frac{h_2(x_2, x_1)}{h_1(x_2, x_1)} - \frac{h_3(x_2, x_1)}{(h_1(x_2, x_1))^2}\right)$$

$$= \frac{f(x_2) - f(x_1)}{h_1(x_2, x_1)} - \frac{f(x_1) - f(x_0)}{h_1(x_1, x_0)},$$

or

$$a_1\left(\frac{h_3(0, 2)}{(h_1(2, 0))^2} + \frac{h_2(0, 2)}{h_1(2, 0)} + \frac{h_2(4, 2)}{h_1(4, 2)} - \frac{h_3(4, 2)}{(h_1(4, 2))^2}\right) = \frac{f(4) - f(2)}{h_1(4, 2)} - \frac{f(2) - f(0)}{h_1(2, 0)}.$$

Now, we will compute $h_i(\cdot, \cdot)$, $i \in \{1, 2, 3\}$. We have

$$h_1(2, 0) = 2,$$

$$h_2(0, 2) = \int_2^0 h_1(\tau, 2)\Delta\tau = -\int_0^2 (\tau - 2)\Delta\tau$$

$$= -\mu(0)(-2) - \mu(1)(1 - 2) = 3,$$

$$h_3(0, 2) = \int_2^0 h_2(\tau, 2)\Delta\tau = -\int_0^2 h_2(\tau, 2)\Delta\tau$$

$$= -\mu(0)h_2(0, 2) - \mu(1)h_2(1, 2) = -3 - \int_2^1 h_1(\tau, 2)\Delta\tau$$

$$= -3 + \int_1^2 (\tau - 2)\Delta\tau = -3 + \mu(1)(1 - 2) = -3 - 1 = -4,$$

$$h_3(2, 4) = \int_4^2 h_2(\tau, 4)\Delta\tau = -\int_2^4 h_2(\tau, 4)\Delta\tau$$

$$= -\mu(2)h_2(2, 4) - \mu(3)h_2(3, 4)$$

$$= -\int_4^2 h_1(\tau, 4)\Delta\tau - \int_4^3 h_1(\tau, 4)\Delta\tau$$

$$= \int_2^4 (\tau - 4)\Delta\tau + \int_3^4 (\tau - 4)\Delta\tau$$

$$= \mu(2)(2 - 4) + \mu(3)(3 - 4) + \mu(3)(3 - 4)$$

$$= -2 - 1 - 1 = -4,$$

$$h_1(4, 2) = 4 - 2 = 2,$$

$$h_2(4, 2) = \int_2^4 h_1(\tau, 2)\Delta\tau = \int_2^4 (\tau - 2)\Delta\tau = \mu(2)(2 - 2) + \mu(3)(3 - 2) = 1,$$

$$h_3(4, 2) = \int_2^4 h_2(\tau, 2)\Delta\tau = \mu(2)h_2(2, 2) + \mu(3)h_2(3, 2)$$

$$= h_2(3, 2) = \int_2^3 h_1(\tau, 2)\Delta\tau = \mu(2)h_1(2, 2) = 0,$$

$$f(0) = 1, \quad f(2) = \frac{3}{5}, \quad f(4) = \frac{5}{17}.$$

Hence, for a_1, we get the following equation:

$$a_1\left(-\frac{4}{2^2} + \frac{3}{2} + \frac{1}{2} - 0\right) = \frac{\frac{5}{17} - \frac{3}{5}}{2} - \frac{\frac{3}{5} - 1}{2},$$

or

$$a_1 = \frac{-\frac{26}{85}}{2} + \frac{1}{5} = -\frac{13}{85} + \frac{1}{5} = \frac{4}{85}.$$

Now let

$$l_1(t) = \frac{t^2}{2} - \frac{5}{2}t, \quad l_2(t) = \frac{1}{6}t^3 - \frac{3}{2}t^2 + \frac{13}{3}t,$$

$$l_3(t) = \frac{t^2 - 9t}{2}, \quad l_4(t) = \frac{1}{6}t^3 - 5t^2 + \frac{89}{6}t, \quad t \in \mathbb{T}.$$

Then

$$l_1^{\Delta}(t) = \frac{\sigma(t) + t}{2} - \frac{5}{2} = \frac{t + 1 + t}{2} - \frac{5}{2} = t - 2,$$

$$l_2^{\Delta}(t) = \frac{1}{6}\left((\sigma(t))^2 + t\sigma(t) + t^2\right) - \frac{3}{2}(\sigma(t) + t) + \frac{13}{3}$$

$$= \frac{1}{6}\left((t + 1)^2 + t(t + 1) + t^2\right) - \frac{3}{2}(t + 1 + t) + \frac{13}{3}$$

$$= \frac{1}{6}\left(t^2 + 2t + 1 + t^2 + t + t^2\right) - \frac{3}{2}(2t + 1) + \frac{13}{3}$$

$$= \frac{t^2}{2} + \frac{t}{2} + \frac{1}{6} - 3t - \frac{3}{2} + \frac{13}{3}$$

$$= \frac{t^2 - 5t}{2} + \frac{1 - 9 + 26}{6} = \frac{t^2 - 5t}{2} + 3,$$

$$l_3^{\Delta}(t) = \frac{1}{2}(\sigma(t) + t - 9) = \frac{1}{2}(t + 1 + t - 9) = t - 4,$$

$$l_4^{\Delta}(t) = \frac{1}{6}\left((\sigma(t))^2 + t\sigma(t) + t^2\right) - 5(\sigma(t) + t) + \frac{89}{6}$$

$$= \frac{1}{6}\left((t + 1)^2 + t(t + 1) + t^2\right) - 5(t + 1 + t) + \frac{89}{6}$$

$$= \frac{1}{6}\left(3t^2 + 3t + 1\right) - 5(2t + 1) + \frac{89}{6}$$

$$= \frac{t^2}{2} - \frac{9}{2}t + 15 - 5$$

$$= \frac{t^2}{2} - \frac{9}{2}t + 10, \quad t \in \mathbb{T}.$$

From here,

$$h_2(x, 2) = \int_2^x (\tau - 2)\Delta\tau = l_1(\tau)\big|_{\tau=2}^{\tau=x}$$

$$= l_1(x) - l_1(2) = \frac{x^2}{2} - \frac{5}{2}x - 2 + 5$$

$$= \frac{x^2 - 5x}{2} + 3,$$

$$h_3(x, 2) = \int_2^x h_2(\tau, 2)\Delta\tau = \int_2^x l_2^\Delta(\tau)\Delta\tau$$

$$= l_2(x) - l_2(2) = \frac{x^3}{6} - \frac{3x^2}{2} + \frac{13}{3}x - \frac{8}{6} + 6 - \frac{26}{3}$$

$$= \frac{x^3}{6} - \frac{3}{2}x^2 + \frac{13}{3}x - 4,$$

$$h_2(x, 4) = \int_4^x (\tau - 4)\Delta\tau = \int_4^x l_3^\Delta(\tau)\Delta\tau$$

$$= l_3(\tau)|_{\tau=4}^{\tau=x} = \frac{x^2 - 9x}{2} - \frac{16 - 36}{2}$$

$$= \frac{x^2 - 9x}{2} + 10,$$

$$h_3(x, 4) = \int_4^x h_2(\tau, 4)\Delta\tau = \int_4^x l_4^\Delta(\tau)\Delta\tau$$

$$= l_4(x) - l_4(4) = \frac{1}{6}x^3 - 5x^2 + \frac{89}{6} - \frac{64}{6} + 80 - \frac{178}{3}$$

$$= \frac{x^3 - 30x^2 + 89 - 64 + 480 - 356}{6} = \frac{x^3 - 30x^2 + 149}{6}, \quad x \in [2, 4].$$

Therefore,

$$S_2(x) = f(x_1) - \frac{f(x_1) - f(x_0)}{x_1 - x_0} h_1(x_1, x)$$

$$+ a_1\left(-\frac{h_3(x_0, x_1)}{(h_1(x_1, x_0))^2} h_1(x_1, x) - \frac{h_2(x_0, x_1)}{h_1(x_1, x_0)} h_1(x_1, x) + \frac{h_3(x, x_1)}{h_1(x_1, x_0)} + h_2(x, x_1)\right)$$

$$= f(2) - \frac{f(2) - f(0)}{2} h_1(2, x)$$

$$+ \frac{4}{85}\left(-\frac{h_3(0, 2)}{(h_1(2, 0))^2} h_1(2, x) - \frac{h_2(0, 2)}{h_1(2, 0)} h_1(2, x) + \frac{h_3(x, 2)}{h_1(2, 0)} + h_2(x, 2)\right)$$

$$= \frac{3}{5} - \frac{\frac{3}{5} - 1}{2}(2 - x)$$

$$+ \frac{4}{85}\left(\frac{4}{4}(2 - x) - \frac{3}{2}(2 - x) + \frac{1}{2}\left(\frac{x^3}{6} - \frac{3}{2}x^2 + \frac{13}{3}x - 4\right) + \frac{x^2 - 5x}{2} + 3\right)$$

$$= \frac{3}{5} + \frac{1}{5}(2 - x) + \frac{4}{85}\left(-\frac{1}{2}(2 - x) + \frac{x^3}{12} - \frac{3x^2}{4} + \frac{13x}{6} - 2 + \frac{x^2 - 5x}{2} + 3\right)$$

$$= 1 - \frac{x}{5} + \frac{4}{85} \cdot \frac{-12 + 6x + x^3 - 9x^2 + 26x - 24 + 6x^2 - 30x + 36}{12}$$

$$= 1 - \frac{x}{5} + \frac{4}{85} \cdot \frac{x^3 - 3x^2 + 2x}{12}$$

$$= \frac{255 - 51x + x^3 - 3x^2 + 2x}{255}$$

$$= \frac{x^3 - 3x^2 - 49x + 255}{255}, \quad x \in [0,2],$$

and

$$s_2(x) = f(x_2) - \frac{f(x_2) - f(x_1)}{h_1(x_2, x_1)} h_1(x_1, x) + a_1 \left(\frac{h_3(x_1, x_2)}{(h_1(x_2, x))^2} h_1(x_2, x_1) - \frac{h_3(x, x_2)}{h_1(x_2, x_1)} \right)$$

$$= f(4) - \frac{f(4) - f(2)}{h_1(4,2)} h_1(4, x) + \frac{4}{85} \left(\frac{h_3(2,4)}{(h_1(4,2))^2} h_1(4, x) - \frac{h_3(x,4)}{h_1(4,2)} \right)$$

$$= \frac{5}{17} - \frac{\frac{5}{17} - \frac{3}{5}}{2} (4 - x) + \frac{4}{85} \left(-\frac{4}{4}(4 - x) - \frac{x^3 - 30x^2 + 149}{12} \right)$$

$$= \frac{5}{17} - \frac{25 - 81}{170} (4 - x) + \frac{4}{85} \left(\frac{-48 + 12x - x^3 + 30x^2 - 149}{12} \right)$$

$$= \frac{50 - 56(4 - x)}{170} - \frac{x^3 - 30x^2 - 12x + 101}{255}$$

$$= \frac{-174 + 56x}{170} - \frac{x^3 - 30x^2 - 12x + 101}{255}$$

$$= \frac{-522 + 168x - 2x^3 + 60x^2 + 24x - 202}{510}$$

$$= \frac{-2x^3 + 60x^2 + 192x - 724}{510} = \frac{-x^3 + 30x^2 + 96x - 362}{255}, \quad x \in [2,4].$$

In the next example, we compute the natural cubic spline on an interval consisting of two subintervals for a given arbitrary function. We perform the computation of the coefficients with Matlab. Then, we choose two different forms of the arbitrary function and also compare the graphs of the function and its natural cubic spline interpolation.

Example 3.19. Let $\mathbb{T} = 2\mathbb{Z}$ and let $f : \mathbb{T} \to \mathbb{R}$ be a given function. Consider

$$K = \{a = x_0 = -6, \; x_1 = -2, \; x_2 = 2, \; x_3 = 6, \; x_4 = 10 = b\}.$$

In Example 2.28 we have computed

$$h_0(t,s) = 1, \quad h_1(t,s) = t - s,$$

$$h_2(t,s) = \frac{t^2}{2} - (s+1)t + s + \frac{s^2}{2}, \quad t, s \in \mathbb{T}.$$

We also compute

$$h_3(t,s) = \int_s^t h_2(\tau, s) \Delta\tau$$

$$= \int_{s}^{t}\left(\frac{\tau^2}{2} - (1+s)\tau + s + \frac{s^2}{2}\right)\Delta\tau$$

$$= \left(\frac{\tau^3}{6} - \frac{\tau^2}{2} + \frac{2\tau}{3}\right) - (1+s)\left(\frac{\tau^2}{2} - \tau\right) + \left(s + \frac{s^2}{2}\right)\tau\Big|_{\tau=s}^{\tau=t}$$

$$= \frac{t^3}{6} - \frac{2+s}{2}t^2 + \left(\frac{s^2}{2} + 2s + \frac{4}{3}\right)t - \left(\frac{s^3}{6} + s^2 + \frac{4s}{3}\right), \quad t, s \in \mathbb{T}.$$

We have $a_0 = a_4 = 0$. To find a_1, a_2, and a_3, we solve the linear system

$$\begin{bmatrix} u_{11} & u_{12} & 0 \\ u_{21} & u_{22} & u_{23} \\ 0 & u_{32} & u_{33} \end{bmatrix} \begin{bmatrix} a_1 \\ a_2 \\ a_3 \end{bmatrix} = \begin{bmatrix} g_1 \\ g_2 \\ g_3 \end{bmatrix},$$

where

$$u_{jj} = \frac{h_3(x_{j-1}, x_j)}{(h_1(x_j, x_{j-1}))^2} + \frac{h_2(x_{j-1}, x_j)}{h_1(x_j, x_{j-1})} + \frac{h_2(x_{j+1}, x_j)}{h_1(x_{j+1}, x_j)} - \frac{h_3(x_{j+1}, x_j)}{(h_1(x_{j+1}, x_j))^2}, \quad \text{for } j = 1, 2, 3,$$

$$u_{j,j+1} = \frac{h_3(x_{j+1}, x_j)}{(h_1(x_{j+1}, x_j))^2}, \quad \text{for } j = 1, 2,$$

$$u_{j,j-1} = \frac{-h_3(x_{j-1}, x_j)}{(h_1(x_j, x_{j-1}))^2}, \quad \text{for } j = 2, 3,$$

and

$$g_j = \frac{f(x_{j+1}) - f(x_j)}{h_1(x_{j+1}, x_j)} - \frac{f(x_j) - f(x_{j-1})}{h_1(x_j, x_{j-1})}, \quad \text{for } j = 1, 2, 3.$$

Then we construct the cubic spline as

$$s_2(x) = f(x_j) - \frac{f(x_j) - f(x_{j-1})}{h_1(x_j, x_{j-1})} h_1(x_j, x)$$

$$+ a_{j-1}\left(\frac{h_3(x_{j-1}, x_j)}{(h_1(x_j, x_{j-1}))^2} h_1(x_j, x) - \frac{h_3(x, x_j)}{h_1(x_j, x_{j-1})}\right)$$

$$+ a_j\left(-\frac{h_3(x_{j-1}, x_j)}{(h_1(x_j, x_{j-1}))^2} h_1(x_j, x) - \frac{h_2(x_{j-1}, x_j)}{h_1(x_j, x_{j-1})} h_1(x_j, x)\right)$$

$$+ \frac{h_3(x, x_j)}{h_1(x_j, x_{j-1})} + h_2(x, x_j)\right), \quad x \in [x_{j-1}, x_j], \quad j \in \{1, 2, 3, 4\}.$$

For the function $f(x) = \sqrt{x^2 + 2} - x$, the computations are done with Matlab and the graphs of the function and the cubic spline are compared in Figure 3.4. As a second example, the function f is chosen as $f(x) = \frac{2x+5}{x^2+1}$. The computations are done with Matlab and the graphs of the function and the cubic spline are compared in Figure 3.5.

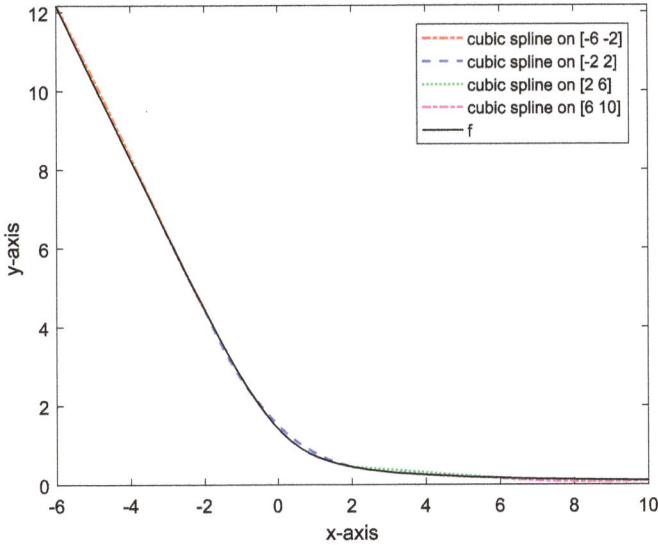

Figure 3.4: The graphs of the natural cubic spline $s_2(x)$ and $f(x) = \sqrt{x^2 + 2} - x$.

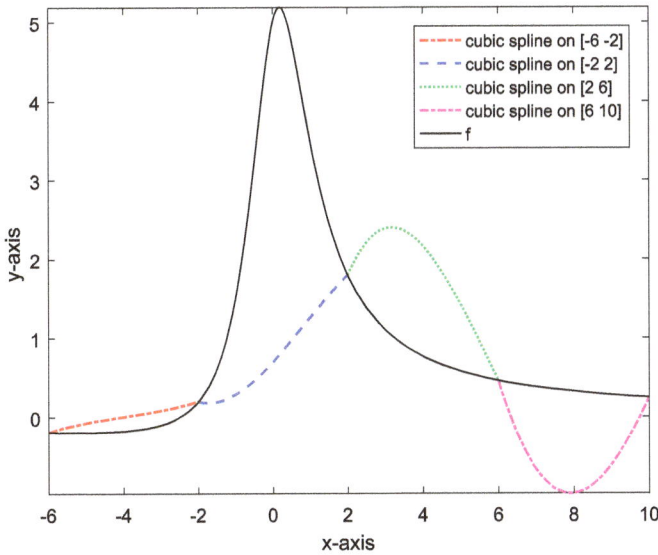

Figure 3.5: The graphs of natural cubic spline $s_2(x)$ and $f(x) = \frac{2x+5}{x^2+1}$.

Exercise 3.20. Let $\mathbb{T} = \{0, \frac{1}{8}, \frac{1}{6}, \frac{1}{3}, \frac{1}{2}, 1\}$, and consider

$$f(t) = 1 + t + \frac{1+t}{1+t^2}, \quad t \in \mathbb{T}, \quad a = x_0 = 0, \quad x_1 = \frac{1}{6}, \quad x_2 = b = 1.$$

Find $s_2(x)$, $x \in [x_{j-1}, x_j]$, $j \in \{1, 2\}$.

3.4 Hermite cubic splines

The construction of a natural cubic spline is complicated and requires the solution of a linear system for the determination of the coefficients. In this section, we will introduce another cubic spline, known as Hermite cubic spline. Its construction is easier and it is based on writing a Hermite interpolation polynomial on each subinterval. Below, we present the definition and the construction of a Hermite cubic spline for an arbitrary time scale.

Let $a, b \in \mathbb{T}$, $a < b$, and consider

$$K = \{a = x_0 \leq \sigma(x_0) < x_1 \leq \sigma(x_1) < \cdots < x_m \leq \sigma(x_m) = b\},$$

as well as $f \in C_{rd}^1([a, b])$.

Definition 3.21. Define the Hermite cubic polynomial as a polynomial in the form

$$s(x) = c_0 + c_1 h_1(x, x_{j-1}) + c_2 h_2(x, x_{j-1}) + c_3 h_3(x, x_{j-1}), \quad x \in [x_{j-1}, x_j], \tag{3.10}$$

$j \in \{1, \ldots, m\}$, where c_0, c_1, c_2, and c_3 are constants so that

$$s(x_j) = f(x_j), \quad s^\Delta(x_j) = f^\Delta(x_j), \quad j \in \{0, 1, \ldots, m\}. \tag{3.11}$$

Now, we will determine the constants c_0, c_1, c_2, and c_3. We have

$$s(x_{j-1}) = c_0 = f(x_{j-1}),$$

and

$$s(x_j) = f(x_{j-1}) + c_1 h_1(x_j, x_{j-1}) + c_2 h_2(x_j, x_{j-1}) + c_3 h_3(x_j, x_{j-1}) = f(x_j),$$

whereupon

$$c_1 h_1(x_j, x_{j-1}) + c_2 h_2(x_j, x_{j-1}) + c_3 h_3(x_j, x_{j-1}) = f(x_j) - f(x_{j-1}).$$

Next,

$$s^\Delta(x) = c_1 + c_2 h_1(x, x_{j-1}) + c_3 h_2(x, x_{j-1}), \quad x \in [x_{j-1}, x_j].$$

From here,

$$s^\Delta(x_{j-1}) = c_1 = f^\Delta(x_{j-1}),$$

and

$$s^\Delta(x_j) = f^\Delta(x_{j-1}) + c_2 h_1(x_j, x_{j-1}) + c_3 h_2(x_j, x_{j-1}) = f^\Delta(x_j).$$

Thus, we get the system

$$c_2 h_2(x_j, x_{j-1}) + c_3 h_3(x_j, x_{j-1}) = f(x_j) - f(x_{j-1}) - f^\Delta(x_{j-1}) h_1(x_j, x_{j-1}),$$
$$c_2 h_1(x_j, x_{j-1}) + c_3 h_2(x_j, x_{j-1}) = f^\Delta(x_j) - f^\Delta(x_{j-1}).$$

From the latter system, we find

$$c_2 = \frac{(f(x_j) - f(x_{j-1}) - f^\Delta(x_{j-1}) h_1(x_j, x_{j-1})) h_2(x_j, x_{j-1})}{(h_2(x_j, x_{j-1}))^2 - h_1(x_j, x_{j-1}) h_3(x_j, x_{j-1})}$$
$$- \frac{(f^\Delta(x_j) - f^\Delta(x_{j-1})) h_3(x_j, x_{j-1})}{(h_2(x_j, x_{j-1}))^2 - h_1(x_j, x_{j-1}) h_3(x_j, x_{j-1})}$$

and

$$c_3 = \frac{(f(x_j) - f(x_{j-1}) - f^\Delta(x_{j-1}) h_1(x_j, x_{j-1})) h_1(x_j, x_{j-1})}{h_3(x_j, x_{j-1}) h_1(x_j, x_{j-1}) - (h_2(x_j, x_{j-1}))^2}$$
$$- \frac{(f^\Delta(x_j) - f^\Delta(x_{j-1})) h_2(x_j, x_{j-1})}{h_3(x_j, x_{j-1}) h_1(x_j, x_{j-1}) - (h_2(x_j, x_{j-1}))^2}.$$

Consequently,

$$s(x) = f(x_{j-1}) + f^\Delta(x_{j-1}) h_1(x, x_{j-1})$$
$$+ \left(\frac{(f(x_j) - f(x_{j-1}) - f^\Delta(x_{j-1}) h_1(x_j, x_{j-1})) h_2(x_j, x_{j-1})}{(h_2(x_j, x_{j-1}))^2 - h_1(x_j, x_{j-1}) h_3(x_j, x_{j-1})} \right.$$
$$\left. - \frac{(f^\Delta(x_j) - f^\Delta(x_{j-1})) h_3(x_j, x_{j-1})}{(h_2(x_j, x_{j-1}))^2 - h_1(x_j, x_{j-1}) h_3(x_j, x_{j-1})} \right) h_2(x, x_{j-1})$$
$$+ \left(\frac{(f(x_j) - f(x_{j-1}) - f^\Delta(x_{j-1}) h_1(x_j, x_{j-1})) h_1(x_j, x_{j-1})}{h_3(x_j, x_{j-1}) h_1(x_j, x_{j-1}) - (h_2(x_j, x_{j-1}))^2} \right.$$
$$\left. - \frac{(f^\Delta(x_j) - f^\Delta(x_{j-1})) h_2(x_j, x_{j-1})}{h_3(x_j, x_{j-1}) h_1(x_j, x_{j-1}) - (h_2(x_j, x_{j-1}))^2} \right) h_3(x, x_{j-1}),$$

whenever $x \in [x_{j-1}, x_j]$, $j \in \{1, \dots, m\}$.

Remark 3.22. We can define the Hermite cubic polynomial in the following way:

$$s(x) = c_0 + c_1 g_1(x, x_{j-1}) + c_2 g_2(x, x_{j-1}) + c_3 g_3(x, x_{j-1}), \quad x \in [x_{j-1}, x_j], \ j \in \{1, \dots, m\},$$

where the constants c_0, c_1, c_2, and c_3 will be determined by the condition (3.11). As above, one can deduct that

$$s(x) = f(x_{j-1}) + f^\Delta(x_{j-1}) g_1(x, x_{j-1})$$
$$+ \left(\frac{(f(x_j) - f(x_{j-1}) - f^\Delta(x_{j-1}) g_1(x_j, x_{j-1})) g_2(x_j, x_{j-1})}{(g_2(x_j, x_{j-1}))^2 - g_1(x_j, x_{j-1}) g_3(x_j, x_{j-1})} \right.$$

$$-\frac{(f^{\Delta}(x_j) - f^{\Delta}(x_{j-1}))g_3(x_j, x_{j-1})}{(g_2(x_j, x_{j-1}))^2 - g_1(x_j, x_{j-1})g_3(x_j, x_{j-1})}\Bigg)g_2(x, x_{j-1})$$

$$+\Bigg(\frac{(f(x_j) - f(x_{j-1}) - f^{\Delta}(x_{j-1})g_1(x_j, x_{j-1}))g_2(x_j, x_{j-1})g_1(x_j, x_{j-1})}{g_3(x_j, x_{j-1})g_1(x_j, x_{j-1}) - (g_2(x_j, x_{j-1}))^2}$$

$$-\frac{(f^{\Delta}(x_j) - f^{\Delta}(x_{j-1}))g_2(x_j, x_{j-1})}{g_3(x_j, x_{j-1})g_1(x_j, x_{j-1}) - (g_2(x_j, x_{j-1}))^2}\Bigg)g_3(x, x_{j-1}),$$

whenever $x \in [x_{j-1}, x_j], j \in \{1, \ldots, m\}$.

We illustrate in a detailed way the construction of Hermite cubic spline in the following example.

Example 3.23. Let $\mathbb{T} = 2^{\mathbb{N}_0}$, and consider

$$a = x_0 = 1, \quad x_1 = 4, \quad x_2 = 16 = b, \quad f(t) = \frac{1}{1+t}, \quad t \in \mathbb{T}.$$

We will find $s(x)$, $x \in [1, 4]$. We have

$$\sigma(t) = 2t, \quad f(x_0) = \frac{1}{2}, \quad f(x_1) = \frac{1}{5},$$

$$f^{\Delta}(t) = -\frac{1}{(1+t)(1+\sigma(t))} = -\frac{1}{(1+t)(1+2t)}, \quad t \in \mathbb{T},$$

$$f^{\Delta}(1) = -\frac{1}{6}, \quad f^{\Delta}(4) = -\frac{1}{45},$$

$$h_1(x, x_0) = x - 1, \quad x \in \mathbb{T}, \quad h_1(x_1, x_0) = 3,$$

$$h_2(x, x_0) = \int_1^x (t-1)\Delta t = \left(\frac{1}{3}t^2 - t\right)\Bigg|_{t=1}^{t=x}$$

$$= \frac{x^2}{3} - x - \left(\frac{1}{3} - 1\right) = \frac{x^2}{3} - x + \frac{2}{3}, \quad x \in \mathbb{T},$$

$$h_2(x_1, x_0) = \frac{16}{3} - 4 + \frac{2}{3} = 6 - 4 = 2,$$

$$h_3(x, x_0) = \int_1^x \left(\frac{t^2}{3} - t + \frac{2}{3}\right)\Delta t = \left(\frac{t^3}{21} - \frac{t^2}{3} + \frac{2}{3}t\right)\Bigg|_{t=1}^{t=x}$$

$$= \frac{x^3}{21} - \frac{x^2}{3} + \frac{2}{3}x - \frac{1}{21} + \frac{1}{3} - \frac{2}{3}$$

$$= \frac{x^3}{21} - \frac{x^2}{3} + \frac{2}{3}x - \frac{8}{21}, \quad x \in \mathbb{T},$$

$$h_3(x_1, x_0) = \frac{64}{21} - \frac{16}{3} + \frac{8}{3} - \frac{8}{21} = \frac{56}{21} - \frac{8}{3} = 0, \quad x \in [1, 4].$$

Hence,

$$c_0 = f(x_0) = \frac{1}{2},$$

$$c_1 = f^\Delta(x_0) = -\frac{1}{6},$$

$$c_2 = \frac{(\frac{1}{5} - \frac{1}{2} + \frac{1}{6} \cdot 3)2}{4 - 3 \cdot 0} - \frac{(-\frac{1}{45} + \frac{1}{6})0}{4 - 3 \cdot 0} = \frac{\frac{2}{5}}{4} = \frac{1}{10},$$

$$c_3 = \frac{(\frac{1}{5} - \frac{1}{2} + \frac{1}{6} \cdot 3)3}{3 \cdot 0 - 2^2} - \frac{(-\frac{1}{45} + \frac{1}{6})2}{3 \cdot 0 - 2^2} = \frac{\frac{3}{5}}{4} - \frac{\frac{13}{45}}{-4} = -\frac{3}{20} + \frac{13}{180} = -\frac{14}{180} = -\frac{7}{90}$$

and

$$s(x) = \frac{1}{2} - \frac{1}{6}(x - 1) + \frac{1}{10}\left(\frac{x^2}{3} - x + \frac{2}{3}\right) - \frac{7}{90}\left(\frac{x^3}{21} - \frac{x^2}{3} + \frac{2}{3}x - \frac{8}{21}\right)$$

$$= \frac{1}{90}\left(45 - 15x + 15 + 3x^2 - 9x + 6 - \frac{x^3}{3} + \frac{7}{3}x^2 - \frac{14}{3}x + \frac{8}{3}\right)$$

$$= \frac{1}{90}\left(-\frac{x^3}{3} = \frac{16}{3}x^2 - \frac{86}{3}x + \frac{206}{3}\right)$$

$$= -\frac{1}{270}(x^3 - 16x^2 + 86x - 206), \quad x \in [1, 4].$$

In the last example, we consider the same time scale and set K as in Example 3.19. We use Matlab to compute the Hermite cubic spline coefficients for an arbitrary function and then take the same two particular functions and compare their graphs with the corresponding Hermite cubic spline.

Example 3.24. We again consider the time scale $\mathbb{T} = 2\mathbb{Z}$ as in Example 3.19. Let $f : \mathbb{T} \to \mathbb{R}$ be a given function and let

$$K = \{a = x_0 = -6, \; x_1 = -2, \; x_2 = 2, \; x_3 = 6, \; x_4 = 10 = b\}.$$

We have already computed the monomials

$$h_0(t, s) = 1,$$

$$h_1(t, s) = t - s,$$

$$h_2(t, s) = \frac{t^2}{2} - (1 + s)t + s + \frac{s^2}{2},$$

$$h_3(t, s) = \frac{t^3}{6} - \frac{2 + s}{2}t + \left(\frac{s^2}{2} + 2s + \frac{4}{3}\right)t - \left(\frac{s^3}{6} + s^2 + \frac{4s}{3}\right), \quad t, s \in \mathbb{T}.$$

To find the Hermite cubic spline for the given function f and the given knot set K, we need to compute the coefficients c_i, $i = 0, 1, 2, 3$ on each interval $[x_{j-1}, x_j]$, $j = 1, 2, 3, 4$. For each $j = 1, 2, 3, 4$, we define and compute

$$A_j = f(x_j) - f(x_{j-1}) - f^\Delta(x_{j-1})h_1(x_j, x_{j-1}),$$

$$B_j = f^\Delta(x_j) - f^\Delta(x_{j-1}),$$

$$D_j = (h_2(x_j, x_{j-1}))^2 - h_1(x_j, x_{j-1})h_3(x_j, x_{j-1}).$$

Then the coefficients of the cubic spline $s_2(x)$ are obtained as

$$c_0 = f(x_{j-1}),$$
$$c_1 = f^\Delta(x_{j-1}),$$
$$c_2 = \frac{A_j h_2(x_j, x_{j-1}) - B_j h_3(x_j, x_{j-1})}{D_j},$$
$$c_3 = \frac{-A_j h_1(x_j, x_{j-1}) + B_j h_1(x_j, x_{j-1})}{D_j},$$

for $j = 1, 2, 3, 4$ and the cubic spline $s_2(x)$ is obtained as

$$s_2(x) = c_0 + c_1 h_1(x, x_{j-1}) + c_2 h_2(x, x_{j-1}) + c_3 h_3(x, x_{j-1}), \quad x \in [x_{j-1}, x_j].$$

As in Example 3.19, we consider first the function $f(x) = \sqrt{x^2 + 2} - x$, $x \in \mathbb{T}$. We compute $f^\Delta(x)$ as

$$f^\Delta(x) = \frac{f(\sigma(x)) - f(x)}{\sigma(x) - x}$$
$$= \frac{\sqrt{x^2 + 4x + 6} - (x + 2) - (\sqrt{x^2 + 2} - x)}{x + 2 - x}$$
$$= \frac{\sqrt{x^2 + 4x + 6} - \sqrt{x^2 + 2} - 2}{2}, \quad x \in \mathbb{T}.$$

The computations of the cubic spline are done with Matlab and the graphs of the function and the cubic spline are compared in Figure 3.6. As a second example, the function f is chosen as $f(x) = \frac{2x+5}{x^2+1}$, $x \in \mathbb{T}$. We compute $f^\Delta(x)$ as

$$f^\Delta(x) = \frac{f(\sigma(x)) - f(x)}{\sigma(x) - x}$$
$$= \frac{\frac{2x+9}{x^2+4x+5} - \frac{2x+5}{x^2+1}}{x + 2 - x}$$
$$= -\frac{2x^2 + 14x + 8}{(x^2 + 1)(x^2 + 4x + 5)}, \quad x \in \mathbb{T}.$$

The computations are done with Matlab and the graphs of the function and the cubic spline are compared in Figure 3.7.

Exercise 3.25. Let $\mathbb{T} = (\frac{1}{3})^{\mathbb{N}_0}$, and consider

$$a = x_0 = \frac{1}{81}, \quad x_1 = \frac{1}{9}, \quad x_2 = b = 1, \quad f(t) = 1 + t - t^2 + \frac{1}{1 + t^2}, \quad t \in \mathbb{T}.$$

Find $s(x)$, $x \in [\frac{1}{81}, \frac{1}{9}]$.

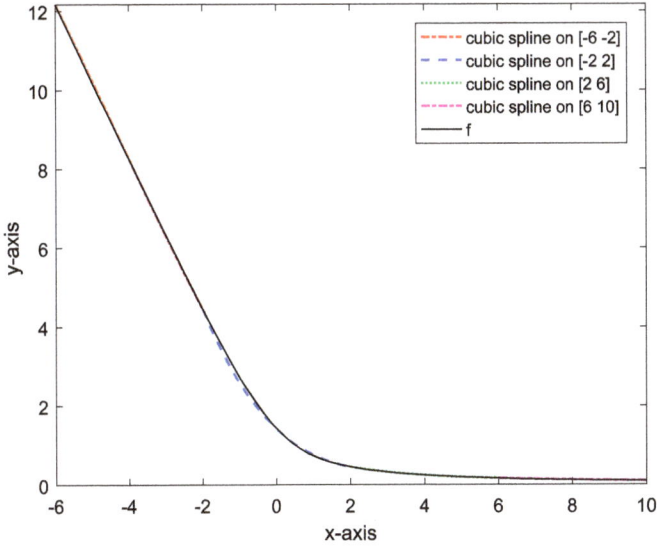

Figure 3.6: The graphs of the Hermite cubic spline $s_2(x)$ and $f(x) = \sqrt{x^2 + 2} - x$.

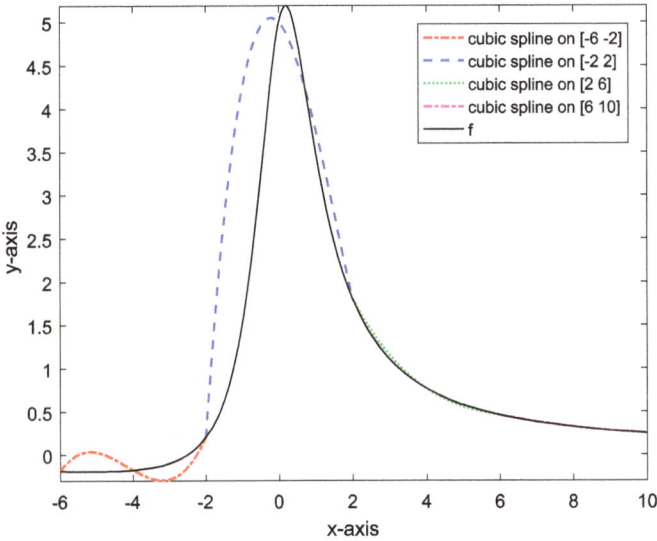

Figure 3.7: The graphs of the Hermite cubic spline $s_2(x)$ and $f(x) = \frac{2x+5}{x^2+1}$.

3.5 Advanced practical problems

Problem 3.26. Let

$$\mathbb{T} = \left\{ 1, \frac{5}{4}, \frac{4}{3}, 2, \frac{5}{2}, 4 \right\},$$

$$a = x_0 = 1, \quad x_1 = \frac{4}{3}, \quad x_2 = \frac{5}{2} = b,$$
$$f(t) = 2 + 3t + t^2, \quad t \in \mathbb{T}.$$

Prove that

$$\sup_{x \in [a,b]} |s_L(x) - f(x)| \leq \frac{133}{6}.$$

Problem 3.27. Let

$$\mathbb{T} = \left\{ 0, \frac{1}{7}, \frac{1}{4}, 1, \frac{7}{6}, \frac{7}{5}, \frac{7}{4}, 2, \frac{5}{2}, 3, \frac{7}{2}, 4 \right\},$$
$$a = x_0 = 1, \quad x_1 = \frac{1}{4}, \quad x_2 = \frac{7}{6}, \quad x_3 = \frac{7}{4}, \quad x_4 = \frac{5}{2}, \quad x_5 = \frac{7}{2}, \quad b = 4,$$

and consider

$$f(t) = t + \frac{1}{t^3 + t^2 + t + 1}, \quad t \in \mathbb{T}.$$

Estimate

$$|s_{\sigma L}(x) - f(x)|, \quad x \in [a, b].$$

Problem 3.28. Let $\mathbb{T} = 2^{\mathbb{N}_0}$, and consider

$$f(t) = 1 + t + e_1(t, 1) + \sin_1(t, 1), \quad t \in \mathbb{T},$$
$$a = x_0 = 1, \quad x_1 = 4, \quad x_2 = b = 16.$$

Find $s_2(x)$, $x \in [x_{j-1}, x_j]$, $j \in \{1, 2\}$.

Problem 3.29. Let $\mathbb{T} = \left(\frac{1}{2}\right)^{\mathbb{N}_0}$,

$$a = x_0 = \frac{1}{256}, \quad x_1 = \frac{1}{64}, \quad x_2 = \frac{1}{8}, \quad x_3 = 1, \quad f(t) = \frac{1+t}{1+t+t^2}, \quad t \in \mathbb{T}.$$

Find $s(x)$, $x \in [\frac{1}{256}, 1]$.

Problem 3.30. Let $a, b \in \mathbb{T}$, $a < b$, and consider

$$K = \{a = x_0 \leq \sigma(x_0) < x_1 \leq \sigma(x_1) < \cdots < x_m \leq \sigma(x_m) = b\},$$

as well as $f \in C_{rd}^1([a, b])$. Find the Hermite cubic spline in the form

$$s(x) = c_0 + c_1 h_1(x, x_{j-1}) + c_2 g_2(x, x_{j-1}) + c_3 g_3(x, x_{j-1}), \quad x \in [x_{j-1}, x_j],$$

so that (3.11) holds.

Problem 3.31. Let $a, b \in \mathbb{T}$, $a < b$, and consider

$$K = \{a = x_0 \le \sigma(x_0) < x_1 \le \sigma(x_1) < \cdots < x_m \le \sigma(x_m) = b\},$$

as well as $f \in C^1_{rd}([a, b])$. Find the Hermite cubic spline in the form

$$s(x) = c_0 + c_1 h_1(x, x_{j-1}) + c_2 h_2(x, x_{j-1}) + c_3 g_3(x, x_{j-1}), \quad x \in [x_{j-1}, x_j],$$

so that (3.11) holds.

Problem 3.32. Let $a, b \in \mathbb{T}$, $a < b$, and consider

$$K = \{a = x_0 \le \sigma(x_0) < x_1 \le \sigma(x_1) < \cdots < x_m \le \sigma(x_m) = b\},$$

as well as $f \in C^1_{rd}([a, b])$. Find the Hermite cubic spline in the form

$$s(x) = c_0 + c_1 h_1(x, x_{j-1}) + c_2 g_2(x, x_{j-1}) + c_3 h_3(x, x_{j-1}), \quad x \in [x_{j-1}, x_j],$$

so that (3.11) holds.

4 The Euler method

Many problems in science and engineering involve a change of one quantity with respect to another, usually with respect to time. Therefore, they are modeled by differential or difference equations, more specifically, initial or boundary value problems.
The differential and difference equations are unified as dynamic equations on time
scales. Especially for the nonlinear equations, an exact analytical solution cannot be
obtained in many cases. The development of efficient numerical methods having high
accuracy has been an extremely important problem in numerical analysis [4, 6, 15].

The basic and most simple method to find an approximate solution of an initial
value problem for a first order differential equation is the so-called Euler method. It
is not very accurate, but the simplicity of its derivation is used in the construction of
more advanced and complicated numerical methods. It is also used to provide the necessary extra information for the application of such more advanced and complicated
methods.

In this chapter, we generalize the Euler method for initial value problems associated with first order dynamic equations on time scales. It can also be adapted to
higher order dynamic equations by transforming the nth order dynamic equation into
a system of n first order dynamic equations.

Let $\mathbb{T}$ be a time scale with forward jump operator σ, backward jump operator ρ,
and delta differentiation operator Δ.

4.1 Analyzing the method

Consider the Cauchy problem

$$\begin{cases} x^{\Delta}(t) = f(t, x(t)), \\ x(t_0) = x_0, \end{cases} \tag{4.1}$$

where $f : \mathbb{T} \times \mathbb{R} \to \mathbb{R}$, $x_0 \in \mathbb{R}$, $t_0 \in \mathbb{T}$ are given, and $x : \mathbb{T} \to \mathbb{R}$ is unknown. Suppose
that $t \in [t_0, \bar{t}]_{\mathbb{T}}$ and take the points

$$t_0 < t_1 < \cdots < t_N = \bar{t}, \quad t_i \in \mathbb{T}, \quad i = 1, \dots, N,$$

where

$$t_i = \begin{cases} \sigma^{l_{i-1}}(t_{i-1}), \ l_{i-1} \in \mathbb{N} & \text{if } t_{i-1} \text{ is right-scattered,} \\ t_{i-1} + q_{i-1}, \ q_{i-1} \in \mathbb{R} & \text{if } t_{i-1} \text{ is right-dense,} \end{cases} \tag{4.2}$$

$i = 1, \dots, N$. Define $\{x_i\}_{i=1}^{N}$ by

https://doi.org/10.1515/9783110787320-004

$$x_{i+1} = x_i + (t_{i+1} - t_i)f(t_i, x_i)$$

$$= x_i + \begin{cases} (\sigma^{l_i}(t_i) - t_i)f(t_i, x_i) & \text{if } t_i \text{ is right-scattered,} \\ q_i f(t_i, x_i) & \text{if } t_i \text{ is right-dense.} \end{cases} \tag{4.3}$$

Formula (4.3) is known as the Euler method for the numerical solution of the initial value problem (4.1) [3].

4.2 Local truncation error

In this section, we derive the local truncation error in the application of the Euler method. To obtain the local truncation error, we consider the cases when t_0 is right-dense and right-scattered separately.

1. Case 1. t_0 is right-scattered.
   In a single step of the Euler method, the computed result is

$$x_1 = x_0 + (\sigma^{l_0}(t_0) - t_0)f(t_0, x(t_0))$$

and it differs from the exact answer $x(t_1) = x(\sigma^{l_0}(t_0))$ by

$$x(t_1) - x_1 = x(\sigma^{l_0}(t_0)) - x(t_0) - (\sigma^{l_0}(t_0) - t_0)f(t_0, x(t_0))$$
$$= x(\sigma^{l_0}(t_0)) - x(t_0) - (\sigma^{l_0}(t_0) - t_0)x^{\Delta}(t_0).$$

If $l_0 = 1$, then $x(t_1) = x_1$. Assuming that x has continuous first and second order delta derivatives, this can be written, using Taylor formula, in the form

$$\int_{t_0}^{\rho(\sigma^{l_0}(t_0))} h_1(\sigma^{l_0}(t_0), \sigma(\tau))x^{\Delta^2}(\tau)\Delta\tau.$$

Another way of writing the error, assuming that the third derivative x^{Δ^3} also exists and is bounded, is

$$h_2(\sigma^{l_0}(t_0), t_0)x^{\Delta^2}(t_0) + O(h_3(\sigma^{l_0}(t_0), t_0)).$$

2. Case 2. t_0 is right-dense.
   In a single step of the Euler method, the computed result

$$x_1 = x_0 + q_0 f(t_0, x(t_0))$$

differs from the exact solution $x(t_1)$ by

$$x(t_1) - x_1 = x(t_0 + q_0) - x(t_0) - q_0 f(t_0, x(t_0))$$
$$= x(t_0 + q_0) - x(t_0) - q_0 x^{\Delta}(t_0).$$

Assuming that x has continuous first and second order derivatives, this can be written in the form

$$\int_{t_0}^{p(t_0+q_0)} h_1(t_0 + q_0, \sigma(\tau)) x^{\Delta^2}(\tau) \Delta\tau.$$

Another way of writing the error, assuming that the third derivative exists and is bounded, is

$$h_2(t_0 + q_0, t_0) x^{\Delta^2}(t_0) + O(h_3(t_0 + q_0, t_0)).$$

Example 4.1. Let $\mathbb{T} = 2^{\mathbb{N}_0}$ and consider the IVP

$$\begin{cases} x^{\Delta}(t) = -\frac{1}{1+2t} x(t), & t > 1, \\ x(1) = \frac{1}{2}. \end{cases}$$

Take

$$t_0 = 1, \quad t_1 = 8, \quad t_2 = 16, \quad t_3 = 32.$$

We have $t_1 = \sigma^3(t_0)$, i. e., $l_0 = 3$. Let

$$g_1(t) = \frac{t^2}{3} - t,$$

$$g_2(t) = \frac{t^3}{21} - \frac{t^2}{3} + \frac{2}{3} t,$$

$$x(t) = \frac{1}{1+t}, \quad t \in \mathbb{T}.$$

Then

$$g_1^{\Delta}(t) = \frac{\sigma(t) + t}{3} - 1 = \frac{2t + t}{3} - 1 = t - 1,$$

$$g_2^{\Delta}(t) = \frac{(\sigma(t))^2 + t\sigma(t) + t^2}{21} - \frac{\sigma(t) + t}{3} + \frac{2}{3} = \frac{4t^2 + 2t^2 + t^2}{21} - \frac{2t + t}{3} + \frac{2}{3} = \frac{t^2}{3} - t + \frac{2}{3},$$

$$x^{\Delta}(t) = -\frac{1}{(1+t)(1+\sigma(t))} = -\frac{1}{(1+t)(1+2t)} = -\frac{1}{2t^2 + 3t + 1},$$

$$x^{\Delta^2}(t) = \frac{2(\sigma(t) + t) + 3}{(2t^2 + 3t + 1)(2(\sigma(t))^2 + 3\sigma(t) + 1)} = \frac{6t + 3}{(2t^2 + 3t + 1)(8t^2 + 6t + 1)}, \quad t \in \mathbb{T},$$

$$x^{\Delta^2}(1) = \frac{9}{6 \cdot 15} = \frac{1}{10}.$$

Hence,

$$x^{\Delta}(t) = -\frac{1}{(1+t)(1+2t)} = -\frac{1}{1+2t}x(t), \quad t > 1,$$

$$x(1) = \frac{1}{2},$$

i. e., x is a solution of the considered IVP. Next,

$$h_2(t, t_0) = \int_1^t (\tau - 1)\Delta\tau = \int_1^t g_1^{\Delta}(\tau)\Delta\tau$$

$$= g_1(t) - g_1(1)$$

$$= \frac{t^2}{3} - t - \left(\frac{1}{3} - 1\right) = \frac{t^2}{3} - t + \frac{2}{3},$$

$$h_2(\sigma^{l_0}(t_0), t_0) = h_2(8, 1) = \frac{64}{3} - 8 + \frac{2}{3} = 22 - 8 = 14,$$

$$h_3(t, t_0) = \int_1^t h_2(\tau, t_0)\Delta\tau = \int_1^t \left(\frac{\tau^2}{3} - \tau + \frac{2}{3}\right)\Delta\tau = \int_1^t g_2^{\Delta}(\tau)\Delta\tau$$

$$= g_2(t) - g_1(1)$$

$$= \frac{t^3}{21} - \frac{t^2}{3} + \frac{2}{3}t - \left(\frac{1}{21} - \frac{1}{3} + \frac{2}{3}\right)$$

$$= \frac{t^3}{21} - \frac{t^2}{3} + \frac{2}{3}t - \left(\frac{1}{21} + \frac{1}{3}\right) = \frac{t^3}{21} - \frac{t^2}{3} + \frac{2}{3}t - \frac{8}{21}, \quad t \in \mathbb{T},$$

$$h_3(\sigma^{l_0}(t_0), t_0) = h_3(8, 1) = \frac{512}{21} - \frac{64}{3} + \frac{16}{3} - \frac{8}{21} = \frac{504}{21} - \frac{48}{3} = 24 - 16 = 8.$$

Therefore the local truncation error is

$$h_2(\sigma^{l_0}(t_0), t_0)x^{\Delta^2}(t_0) + O(h_3(\sigma^{l_0}(t_0), t_0)) = \frac{14}{10} + O(8) = \frac{7}{5} + O(8).$$

Exercise 4.2. Let

$$\mathbb{T} = \left\{0, \frac{1}{8}, \frac{1}{7}, \frac{1}{6}, \frac{1}{5}, \frac{1}{4}, \frac{1}{3}, \frac{1}{2}, 1\right\},$$

$$a = t_0 = 0, \quad t_1 = \frac{1}{6}, \quad t_2 = \frac{1}{4}, \quad t_3 = \frac{1}{2}, \quad t_4 = 1.$$

Consider the IVP

$$\begin{cases} x^{\Delta}(t) = 1 + (x(t))^2, & t > 0, \\ x(0) = 2. \end{cases}$$

Find the local truncation error.

4.3 Global truncation error

We continue with the derivation of the global truncation error of the Euler method. Let $\tilde{x}(t)$ denote the computed solution on the interval $[t_0, \bar{t}]_\mathbb{T}$. That is, at step values t_0, $t_1, \ldots, t_N = \bar{t}$ defined by (4.2), $\tilde{x}$ is computed, using equation (4.3). For "offstep" points, $\tilde{x}(t)$ is defined by linear interpolation, or, equivalently, $\tilde{x}(t)$ is evaluated using a partial step from the most recently computed step values. That is, if $t \in [t_{k-1}, t_k]$, $k = 1, \ldots, N$, then

$$\tilde{x}(t) = x_{k-1} + h_1(t, t_{k-1})f(t_{k-1}, x_{k-1}). \tag{4.4}$$

Define the maximum step size as

$$m = \max_{1 \le i \le N}\{t_i - t_{i-1}\}.$$

Also, let

$$\alpha(t) = x(t) - \tilde{x}(t), \tag{4.5}$$

$$\beta(t) = f(t, x(t)) - f(t, \tilde{x}(t)). \tag{4.6}$$

Suppose that

$$|f(t, x) - f(t, z)| \le L|x - z| \quad \text{for all } t \in \mathbb{T} \text{ and } x, z \in \mathbb{R},$$

where $L > 0$. From (4.5) and (4.6), we have

$$|\beta(t)| \le L|\alpha(t)|, \quad t \in \mathbb{T}.$$

Define $E(t)$, $t \in \mathbb{T}$, so that the exact solution satisfies

$$x(t) = x(t_{k-1}) + h_1(t, t_{k-1})f(t_{k-1}, x(t_{k-1})) + h_2(t, t_{k-1})E(t), \quad t \in [t_{k-1}, t_k], \; t \in \mathbb{T}, \tag{4.7}$$

and assume that $|E(t)| \le p$, $t \in \mathbb{T}$. Subtracting (4.4) from (4.7), we get

$$x(t) - \tilde{x}(t) = x(t_{k-1}) - x_{k-1} + h_1(t, t_{k-1})(f(t_{k-1}, x(t_{k-1})) - f(t_{k-1}, x_{k-1}))$$
$$+ h_2(t, t_{k-1})E(t), \quad t \in \mathbb{T}.$$

Hence, using (4.5) and (4.6), we get

$$\alpha(t) = \alpha(t_{k-1}) + h_1(t, t_{k-1})\beta(t_{k-1}) + h_2(t, t_{k-1})E(t), \quad t \in \mathbb{T},$$

whereupon

$$|\alpha(t)| \le |\alpha(t_{k-1})| + |h_1(t, t_{k-1})|L|\alpha(t_{k-1})| + p|h_2(t, t_{k-1})|$$
$$= (1 + L|h_1(t, t_{k-1})|)|\alpha(t_{k-1})| + |h_2(t, t_{k-1})|p$$
$$\le (1 + L|h_1(t, t_{k-1})|)|\alpha(t_{k-1})| + mp|h_1(t, t_{k-1})|, \quad t \in \mathbb{T}. \qquad (4.8)$$

If $L = 0$, then it follows that

$$|\alpha(t)| \le |\alpha(t_{k-1})| + mph_1(t, t_{k-1}), \quad t \in \mathbb{T}.$$

In particular,

$$|\alpha(t_k)| \le |\alpha(t_{k-1})| + mph_1(t_k, t_{k-1}).$$

Therefore, we deduce

$$|\alpha(t)| \le |\alpha(t_{k-2})| + mp(h_1(t_{k-1}, t_{k-2}) + h_1(t, t_{k-1}))$$
$$= |\alpha(t_{k-2})| + mph_1(t, t_{k-2})$$
$$\le \cdots$$
$$\le |\alpha(t_0)| + mph_1(t, t_0), \quad t \in \mathbb{T}.$$

If $L > 0$, then we have

$$|\alpha(t)| \le (1 + Lh_1(t, t_{k-1}))|\alpha(t_{k-1})| + mph_1(t, t_{k-1})$$
$$= (1 + Lh_1(t, t_{k-1}))|\alpha(t_{k-1})| + \frac{mp}{L}Lh_1(t, t_{k-1}) + \frac{mp}{L} - \frac{mp}{L}, \quad t \in \mathbb{T},$$

i. e.,

$$|\alpha(t)| + \frac{mp}{L} \le (1 + Lh_1(t, t_{k-1}))|\alpha(t_{k-1})| + \frac{mp}{L}(1 + Lh_1(t, t_{k-1}))$$
$$= (1 + Lh_1(t, t_{k-1}))\left(\frac{mp}{L} + |\alpha(t_{k-1})|\right)$$
$$\le e_L(t, t_{k-1})\left(\frac{mp}{L} + |\alpha(t_{k-1})|\right), \quad t \in \mathbb{T}.$$

In particular,

$$|\alpha(t_k)| + \frac{mp}{L} \le e_L(t_k, t_{k-1})\left(\frac{mp}{L} + |\alpha(t_{k-1})|\right).$$

Hence, if $t \in [t_{k-1}, t_k]$, then we get

$$|\alpha(t)| + \frac{mp}{L} \le e_L(t, t_{k-1})\left(\frac{mp}{L} + |\alpha(t_{k-1})|\right)$$
$$\le e_L(t, t_{k-1})e_L(t_{k-1}, t_{k-2})\left(\frac{mp}{L} + |\alpha(t_{k-2})|\right)$$

$$\leq \cdots$$

$$\leq e_L(t, t_0)\left(\frac{mp}{L} + |\alpha(t_0)|\right).$$

Combining the estimates found in the two cases and stating them formally, we have the following result.

Theorem 4.3. *Assuming that f satisfies Lipschitz condition with a constant L, the global error satisfies the bound*

$$|x(t) - \tilde{x}(t)| \leq \begin{cases} |x(t_0) - \tilde{x}(t_0)| + mph_1(t, t_0) & \text{if } L = 0, \\ e_L(t, t_0)(\frac{mp}{L} + |\alpha(t_0)|) - \frac{mp}{L} & \text{if } L > 0. \end{cases} \tag{4.9}$$

Now we consider a sequence of approximations to $x(\bar{t})$. In each of these approximations, a computation using the Euler method is performed, starting from an approximation to $x(t_0)$, and taking a sequence of positive steps. Denote the rth approximation by $\bar{x}_r$.

The only assumption we will make about $\bar{x}_r$, for each specific value of r, is that the initial error $x(t_0) - \bar{x}_r(t_0)$ is bounded by K_r and that the greatest step size is bounded by m_r. It is assumed that $K_r \to 0$ as $r \to \infty$. If $m_r \to 0$, then, by (4.9), we get that

$$|x(\bar{t}) - \bar{x}_r(\bar{t})| \to 0 \quad \text{as } r \to \infty.$$

There are cases when m_r does not tend to zero as $r \to \infty$, for instance, when $\mathbb{T} = 2^{\mathbb{N}_0}$. When $\mathbb{T} = \mathbb{R}$, we have $m_r \to 0$ as $r \to \infty$.

4.4 Numerical examples

In this section, we apply the Euler method to particular examples.

Example 4.4. As a first example, we consider the initial value problem associated with the logistic equation

$$x^{\Delta}(t) = (\alpha \ominus (\alpha x(t)))x(t), \quad x(0) = 2,$$

where $\mathbb{T} = \mathbb{N}_0$ and α is a real number. The logistic equation in both continuous and discrete cases is known to be one of the basic models of the population growth [1]. It is not as simple as the exponential growth model since it takes into account the carrying capacity of the system on which the population of a certain species is studied. However, this model ignores many features and external effects of the population growth. Nevertheless, it is still widely used in population related problems.

Notice that the problem can be written as

$$x^\Delta = \frac{\alpha(1-x)}{1+\alpha\mu(t)x}x, \quad x(0) = 2.$$

We apply the Euler method to the problem with three different step sizes explained below.

Case 1. Let $t \in [0,30]_\mathbb{T}$ and let $t_0 = 0$, $t_i = \sigma^{l_{i-1}}(t_{i-1}) = t_{i-1} + 1$, where $i = 1,\ldots,30$. Then we have $l_i = 1$ and $\sigma(t_{i-1}) - t_{i-1} = 1$ for all $i = 1,\ldots,30$. Hence the computed sequence of values of the solution x is defined as

$$x_i = x_{i-1} + (\sigma(x_{i-1}) - x_{i-1})\frac{\alpha(1-x_{i-1})}{1+\alpha\mu(t_{i-1})x_{i-1}}x_{i-1} = \frac{(1+\alpha)x_{i-1}}{1+\alpha x_{i-1}}, \quad i = 1,\ldots,30.$$

In fact, the exact solution of the problem is obtained as

$$x_i^{(e)}(t) = \frac{(1+\alpha)x_{i-1}^{(e)}(t)}{1+\alpha x_{i-1}^{(e)}(t)},$$

which coincides with the solution obtained by the Euler method with step size $l_i = 1$.

Case 2. Let $t \in [0,30]_\mathbb{T}$ and let $t_0 = 0$, $t_i = \sigma^{l_{i-1}}(t_{i-1}) = t_{i-1} + 2$, where $i = 1,\ldots,15$. In this case, $l_i = 2$ and $\sigma^2(t_{i-1}) - t_{i-1} = 2$ for $i = 0,\ldots,19$. Hence, the computed sequence of values of the solution x is defined as

$$x_i = x_{i-1} + (\sigma^2(t_{i-1}) - t_{i-1})\frac{\alpha(1-x_{i-1})}{1+\alpha\mu(t_{i-1})x_{i-1}}x_{i-1} = \frac{(1+2\alpha)x_{i-1}}{1+2\alpha x_{i-1}}, \quad i = 1,\ldots,15.$$

Case 3. Let $t \in [0,30]_\mathbb{T}$ and let $t_0 = 0$, $t_i = \sigma^{l_{i-1}}(t_{i-1}) = t_{i-1} + 4$, where $i = 1,\ldots,7$. Then we have $l_i = 4$ and $\sigma^{l_{i-1}}(4_{i-1}) - 4_{i-1} = 4$. The computed sequence of values of the solution x is defined as

$$x_i = x_{i-1} + (\sigma^4(t_{i-1}) - t_{i-1})\frac{\alpha(1-x_{i-1})}{1+\alpha\mu(t_{i-1})x_{i-1}}x_{i-1} = \frac{(1+4\alpha)x_{i-1}}{1+4\alpha x_{i-1}}, \quad i = 1,\ldots,7.$$

We denote the computed solution with $l_i = 1$ by $x^{(1)}$, with $l_i = 2$ by $x^{(2)}$, and with $l_i = 4$ by $x^{(4)}$. The exact solution is denoted by $x^{(e)}$. All calculations are done with MATLAB. The values of the approximate and the exact solution are listed in Table 4.1.

In Figures 4.1, 4.2, and 4.3, we compare the graphs of the exact and approximate solutions for the three cases discussed above. In all three figures, the exact solution is represented by the symbol o and the computed solution by the symbol *.

In Figure 4.4 the errors for the three cases discussed above are given. It is obvious that there is no error in Case 1 as stated above, and a small error is present for the Cases 2 and 3.

Table 4.1: The values of $x^{(1)}$, $x^{(2)}$, $x^{(4)}$, and the exact solution $x^{(e)}$ at points of the interval $[0, 30]$.

t	$x^{(e)}$	$x^{(1)}$	$x^{(2)}$	$x^{(4)}$
0.00	2.00000000	2.00000000	2.00000000	2.00000000
1.00	1.20000000	1.20000000		
2.00	1.05882353	1.05882353	1.11111111	
3.00	1.01886792	1.01886792		
4.00	1.00621118	1.00621118	1.02040816	1.05882353
5.00	1.00206186	1.00206186		
6.00	1.00068634	1.00068634	1.00401606	
7.00	1.00022868	1.00022868		
8.00	1.00007621	1.00007621	1.00080064	1.00621118
9.00	1.00002540	1.00002540		
10.00	1.00000847	1.00000847	1.00016003	
11.00	1.00000282	1.00000282		
12.00	1.00000094	1.00000094	1.00003200	1.00068634
13.00	1.00000031	1.00000031		
14.00	1.00000010	1.00000010	1.00000640	
15.00	1.00000003	1.00000003		
16.00	1.00000001	1.00000001	1.00000128	1.00007621
17.00	1.00000000	1.00000000		
18.00	1.00000000	1.00000000	1.00000026	
19.00	1.00000000	1.00000000		
20.00	1.00000000	1.00000000	1.00000005	1.00000847
21.00	1.00000000	1.00000000		
22.00	1.00000000	1.00000000	1.00000001	
23.00	1.00000000	1.00000000		
24.00	1.00000000	1.00000000	1.00000000	1.00000094
25.00	1.00000000	1.00000000		
26.00	1.00000000	1.00000000	1.00000000	
27.00	1.00000000	1.00000000		
28.00	1.00000000	1.00000000	1.00000000	1.00000010
29.00	1.00000000	1.00000000		

Example 4.5. Our second example is an initial value problem associated with a Riccati-type dynamic equation [1]

$$x^{\Delta} = [\Theta(-t)]x^{\sigma} + \frac{x^2}{\mu(t)x - \frac{1}{t}}, \qquad x(t_0) = a. \tag{4.10}$$

The equation can be written as

$$x^{\Delta} = \frac{t}{1 - \mu(t)t}x(\sigma(t)) + \frac{x^2(t)t}{\mu(t)tx - 1}. \tag{4.11}$$

We discuss the problem on two different time scales.

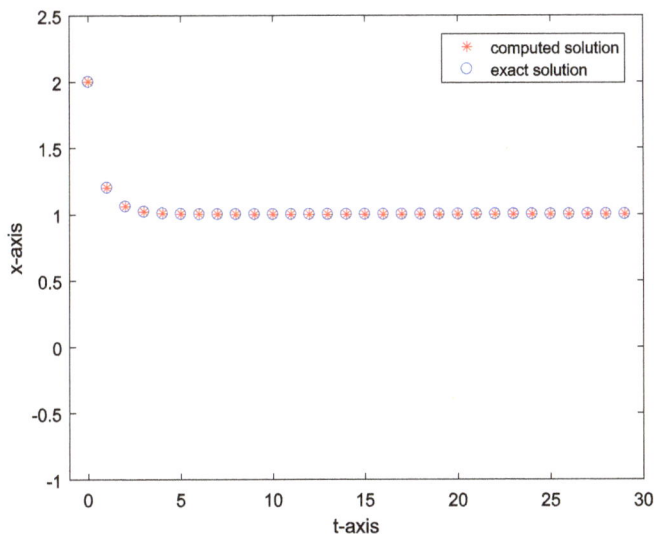

Figure 4.1: Computed and exact values of the solution with step size $l_i = 1$ and $\alpha = 2$.

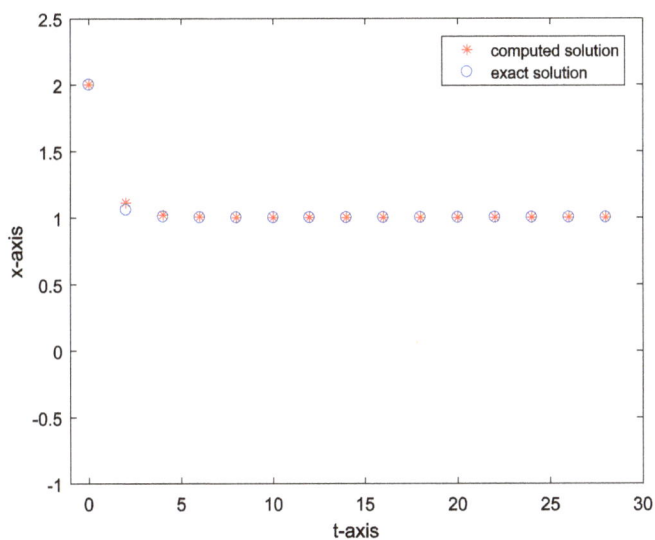

Figure 4.2: Computed and exact values of the solution with step size $l_i = 2$ and $\alpha = 2$.

First, we consider the time scale $\mathbb{T} = [1, 4] \cup \{5, 8, 11, 14, 17, 20, 23, 26, 29, 32\}$. We apply the Euler method for the problem with different choices of the initial value and the step size.

Let $t_0 = 1$, $t_i = t_{i-1} + q$ for $i = 1, \ldots, k = \frac{3}{q}$, $t_{k+1} = 5$ and $t_i = t_{i-1} + 3$ for $i = k + 2, k + 3, \ldots, k + 9$. Since, for $i = 0, \ldots, k - 1$, the points t_i are right-dense, we have

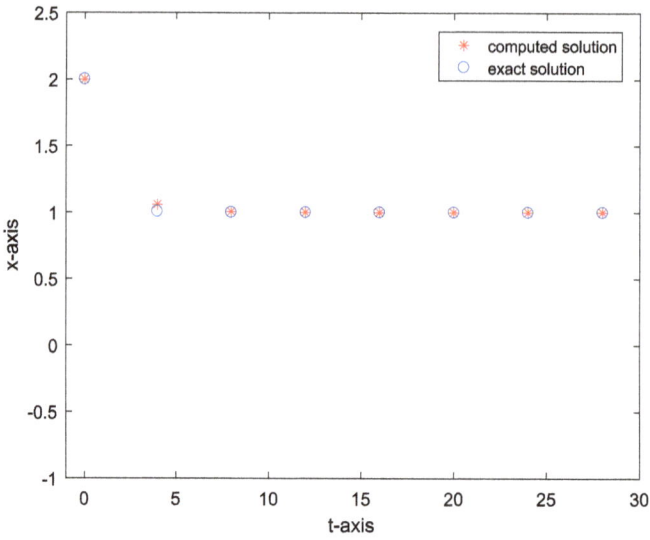

Figure 4.3: Computed and exact values of the solution with step size $l_i = 4$ and $\alpha = 2$.

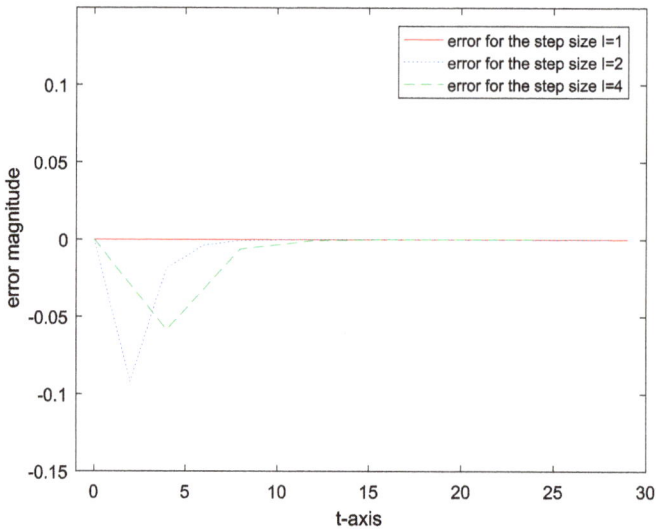

Figure 4.4: The error magnitudes for the logistic equation with step sizes $l_i = 1, 2, 4$ and $\alpha = 2$.

$\sigma(t_i) = t_i$ and $\mu(t_i) = 0$ for these values, and hence, the equation becomes the ordinary differential equation

$$x^\Delta = x' = tx - tx^2. \tag{4.12}$$

The values of x computed by using Euler's method are obtained as

$$x_0 = a, \quad x_i = x_{i-1} + q t_{i-1} x_{i-1} (1 - x_{i-1}), \quad i = 1, 2, \ldots, k. \tag{4.13}$$

Then, for $i = k, \ldots, k + 9$, the points t_i are right-scattered. In this case, we use the dynamic equation (4.11). For $t_k = 4$, we have $\sigma(t_k) = t_k + 1$ and $\mu(t_k) = 1$. Hence,

$$t_{k+1} = 5,$$

$$x_{k+1} = x_k + \mu(t_k) \left[\frac{t_k}{1 - \mu(t_k) t_k} x(\sigma(t_k)) + \frac{x_k^2 t_k}{\mu(t_k) t_k x_k - 1} \right]$$

$$= x_k + \left[\frac{t_k}{1 - t_k} x(t_{k+1}) + \frac{x_k^2 t_k}{t_k x_k - 1} \right],$$

from which we obtain

$$x_{k+1} = \frac{1 - t_k}{1 - 2t_k} \left[\frac{x_k(2 t_k x_k - 1)}{t_k x_k - 1} \right].$$

Finally, for the right-scattered points t_i with $i = k+1, k+2, \ldots, k+8$, we have $\sigma(t_i) = t_i + 3$ and $\mu(t_i) = 3$. Therefore,

$$t_i = t_{i-1} + 3,$$

$$x_i = x_{i-1} + \mu(t_{i-1}) \left[\frac{t_{i-1}}{1 - \mu(t_{i-1}) t_{i-1}} x(\sigma(t_{i-1})) + \frac{x_{i-1}^2 t_{i-1}}{\mu(t_{i-1}) t_{i-1} x_{i-1} - 1} \right]$$

$$= x_{i-1} + 3 \left[\frac{t_{i-1}}{1 - 3 t_{i-1}} x(t_i) + \frac{x_{i-1}^2 t_{i-1}}{3 t_{i-1} x_{i-1} - 1} \right],$$

which gives

$$x_i = \frac{1 - 3 t_{i-1}}{1 - 6 t_{i-1}} \left[\frac{x_{i-1}(6 t_{i-1} x_{i-1} - 1)}{3 t_{i-1} x_{i-1} - 1} \right],$$

where $i = k + 2, k + 3, \ldots, k + 9$.

Notice that the exact solution of the initial value problem (4.10) can be computed. In fact, it is obtained as

$$x^{(e)}(t) = \begin{cases} \dfrac{ae^{\frac{t^2-1}{2}}}{ae^{\frac{t^2-1}{2}} + 1 - a} & \text{if } t \in [1, 4), \\[2ex] x_i(t_i) & \text{if } t_i \in \{4, 5, 8, 11, \ldots, 32\}. \end{cases}$$

We consider two cases for the initial value $x(1) = a$, and for each case we use two different step sizes.

Case 1. Let $x(1) = a = 1.5$. We apply the Euler method with step sizes $q = 0.5$ and $q = 0.25$. The computed solutions for these values of q are compared with the exact solution in Figures 4.5 and 4.6, respectively. In both figures, computed values are shown with $\star$, exact values with o. Finally, the errors for $q = 0.5$ and $q = 0.25$ are shown in Figure 4.7.

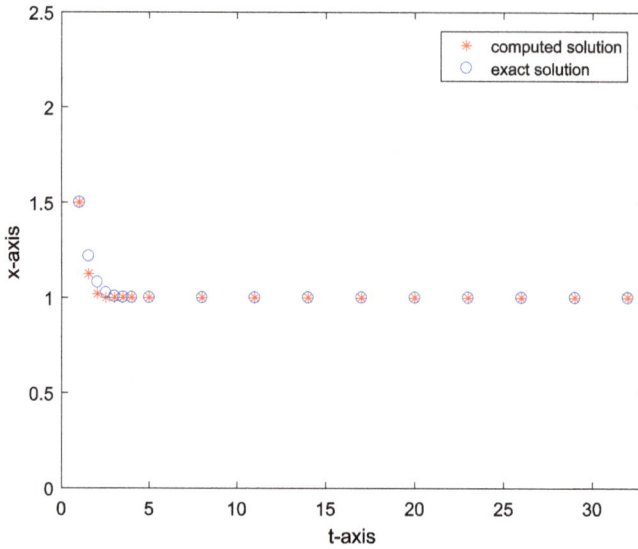

Figure 4.5: Computed and exact values of the solution with step size $q = 0.5$ and $x(1) = 1.5$.

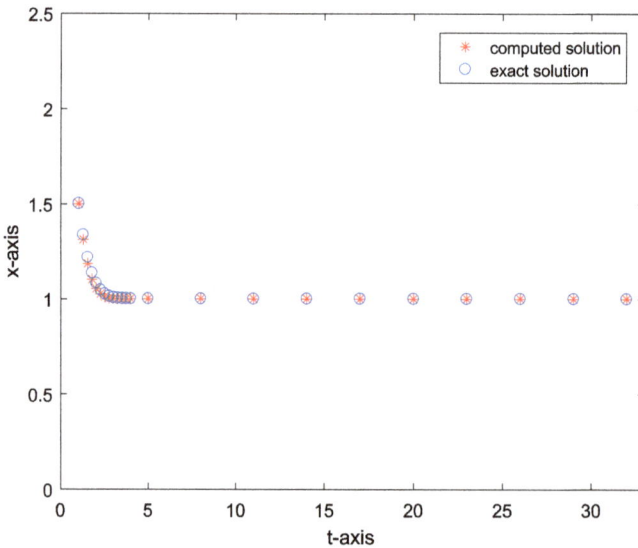

Figure 4.6: Computed and exact values of the solution with step size $q = 0.25$ and $x(1) = 1.5$.

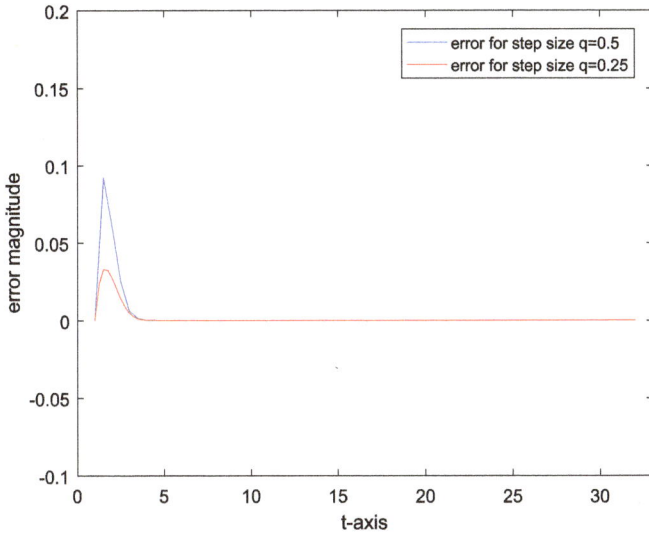

Figure 4.7: The error magnitudes for the cases $q = 0.5$ and $q = 0.25$, and $x(1) = 1.5$.

In Table 4.2, the values of the approximate and exact solution for the initial value $x(0) = 1.5$ are tabulated. The approximate solution obtained with step size $q = 0.25$ is denoted by $x^{(1)}$ and that with the step size $q = 0.5$ is denoted as $x^{(2)}$. The exact solution is denoted as $x^{(e)}$.

Case 2. Let $x(1) = a = 0.5$. The Euler method is applied with the same step sizes, that is, $q = 0.5$ and $q = 0.25$. Figure 4.8 shows the graphs of computed and exact solutions when $q = 0.5$, and Figure 4.9 shows the graphs of computed and exact solutions for $q = 0.25$. In both figures, the symbol $*$ represent the computed and o the exact solution. The errors for the step sizes $q = 0.5$ and $q = 0.25$ are shown in Figure 4.10.

In Table 4.3, the approximate and exact solutions for the initial value $x(1) = 0.5$ are given. The approximate solution obtained with step size $q = 0.25$ is denoted by $x^{(1)}$ and that with the step size $q = 0.5$ is denoted as $x^{(2)}$. The exact solution is denoted as $x^{(e)}$.

We last consider the initial value problem

$$x^{\Delta} = [\Theta(-t)]x^{\sigma} + \frac{x^2}{\mu(t)x - \frac{1}{t}}, \quad x(2) = a, \tag{4.14}$$

on the time scale $\mathbb{T} = \{2, 4, 6\} \cup [7, 10]$.

The discretization is now defined as follows: Take $t_0 = 2$, $t_1 = 4$, $t_2 = 6$, $t_3 = 7$, and $t_i = t_{i-1} + q$. We take two values for q, namely, $q = 0.25$ and 0.20. Accordingly, for $q = 0.25$, we have $i = 5, 6, \ldots, 15$, and for $q = 0.20$, $i = 5, 6, \ldots, 18$.

Table 4.2: The values of the exact solution $x^{(e)}$ and approximate solutions $x^{(1)}$ and $x^{(2)}$ for $x(0) = 1.5$.

t	$x^{(e)}$	$x^{(1)}$	$x^{(2)}$
1.00	1.50000000	1.50000000	1.50000000
1.25	1.33620743	1.31250000	
1.50	1.21716763	1.18432617	1.12500000
1.75	1.13488527	1.10246281	
2.00	1.08035312	1.05304218	1.01953125
2.25	1.04572300	1.02511435	
2.50	1.02474407	1.01063274	0.99961853
2.75	1.01268600	1.00391662	
3.00	1.00614272	1.00121340	1.00009519
3.25	1.00280301	1.00030225	
3.50	1.00120363	1.00005660	0.99995239
3.75	1.00048596	1.00000707	
4.00	1.00018440	1.00000044	1.00003570
5.00	1.00014927	1.00000036	1.00002890
8.00	1.00014376	1.00000034	1.00002783
11.00	1.00014057	1.00000034	1.00002722
14.00	1.00013834	1.00000033	1.00002678
17.00	1.00013663	1.00000033	1.00002645
20.00	1.00013525	1.00000032	1.00002619
23.00	1.00013409	1.00000032	1.00002596
26.00	1.00013310	1.00000032	1.00002577
29.00	1.00013223	1.00000032	1.00002560
32.00	1.00013146	1.00000032	1.00002545

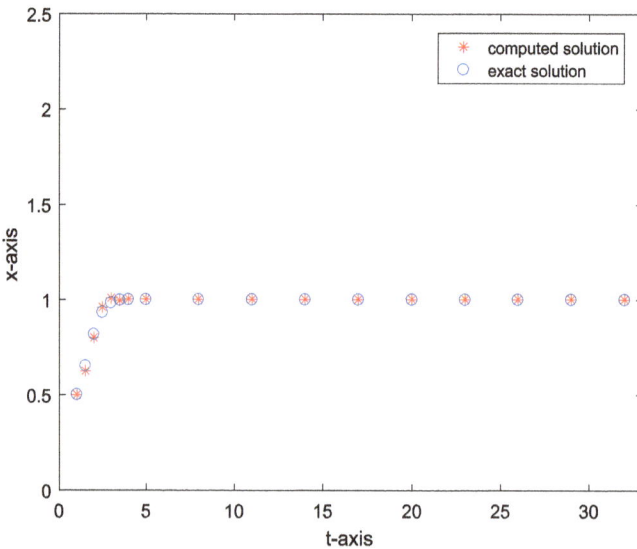

Figure 4.8: Computed and exact values of the solution with step size $q = 0.5$ and $x(1) = 0.5$.

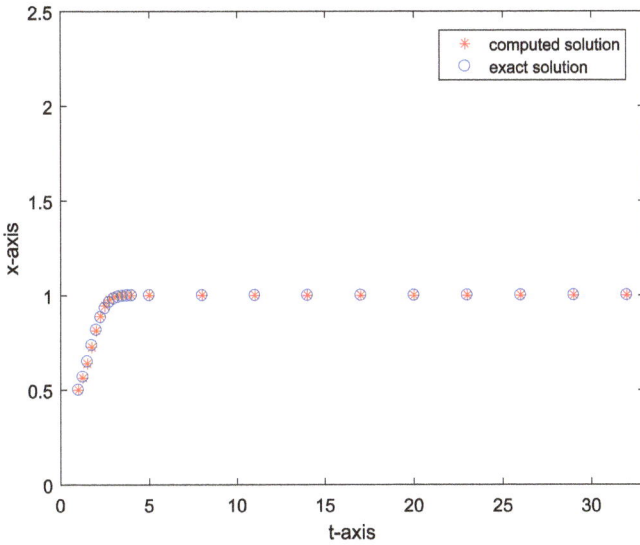

Figure 4.9: Computed and exact values of the solution with step size $q = 0.25$ and $x(1) = 0.5$.

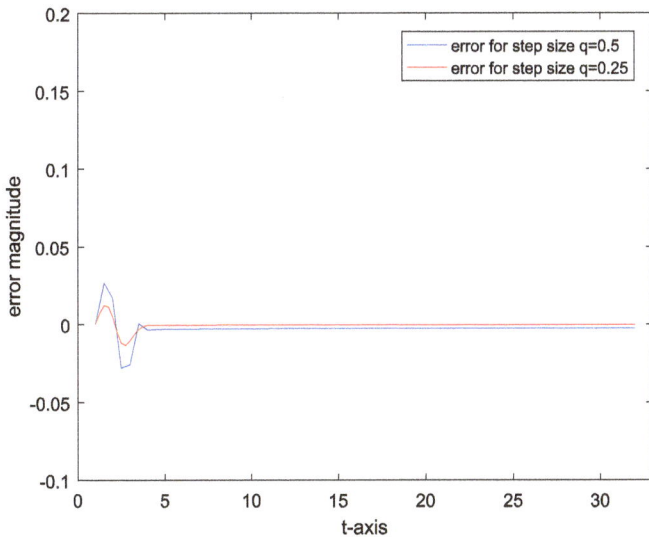

Figure 4.10: The error magnitudes for the cases $q = 0.5$ and $q = 0.25$, and $x(1) = 0.5$.

For the right-scattered points $t_0 = 2$, $t_1 = 4$, $t_2 = 6$, with $\sigma(t_0) = t_0 + 2$, $\sigma(t_1) = t_1 + 2$, we have

$$x_i = \frac{1 - 2t_{i-1}}{1 - 4t_{i-1}} \left[\frac{x_{i-1}(4t_{i-1}x_{i-1} - 1)}{2t_{i-1}x_{i-1} - 1} \right], \quad i = 1, 2.$$

Table 4.3: The values of the exact solution $x^{(e)}$ and approximate solutions $x^{(1)}$ and $x^{(2)}$ for $x(0) = 0.5$.

t	$x^{(e)}$	$x^{(1)}$	$x^{(2)}$
1.00	0.50000000	0.50000000	0.50000000
1.25	0.56985265	0.56250000	
1.50	0.65135486	0.63940430	1.12500000
1.75	0.73715816	0.72586671	
2.00	0.81757448	0.81292231	1.01953125
2.25	0.88403928	0.88896213	
2.50	0.93245331	0.94448564	0.99961853
2.75	0.96377994	0.97725596	
3.00	0.98201379	0.99253685	1.00009519
3.25	0.99168422	0.99809244	
3.50	0.99640640	0.99963938	0.99995239
3.75	0.99640640	0.99995481	
4.00	0.99944722	0.99999717	1.00003570
5.00	0.99955253	0.99999771	1.00002890
8.00	0.99956906	0.99999780	1.00002783
11.00	0.99957863	0.99999785	1.00002722
14.00	0.99958532	0.99999788	1.00002678
17.00	0.99959044	0.99999791	1.00002645
20.00	0.99959457	0.99999793	1.00002619
23.00	0.99959804	0.99999794	1.00002596
26.00	0.99960101	0.99999796	1.00002577
29.00	0.99960362	0.99999797	1.00002560
32.00	0.99960594	0.99999799	1.00002545

For $t_3 = 7$, since $t_2 = 6$ is right-scattered and $\sigma(t_2) = t_2 + 1$, we obtain

$$x_3 = \frac{1-t_2}{1-2t_2}\left[\frac{x_2(t_2 x_2 - 1)}{t_2 x_2 - 1}\right].$$

Finally, for the right-dense points $x_i = x_{i-1} + q$, we calculate

$$x_i = x_{i-1} + q x_{i-1} t_{i-1}(1 - x_{i-1}), \quad i = 1, 2, \ldots, N,$$

where N depends on the choice of q.

We use the Euler method to solve this initial value problem for two different values of the initial condition explained below.

Case 1. In this case, we take the initial condition to be $x(2) = 1.5$ and apply the Euler method with $q = 0.25$ and $q = 0.20$. The approximate solutions obtained for these two values of q are compared with the exact solution in Figures 4.11 and 4.12. The errors for the values $q = 0.25$ and $q = 0.20$ are shown in Figure 4.13.

Table 4.4 shows the approximate and exact solutions obtained by taking the initial value as $x(0) = 1.5$ and the step size as $q = 0.20$. The exact solution is denoted by $x^{(e)}$, the approximate solution obtained for $q = 0.20$ by $x^{(1)}$.

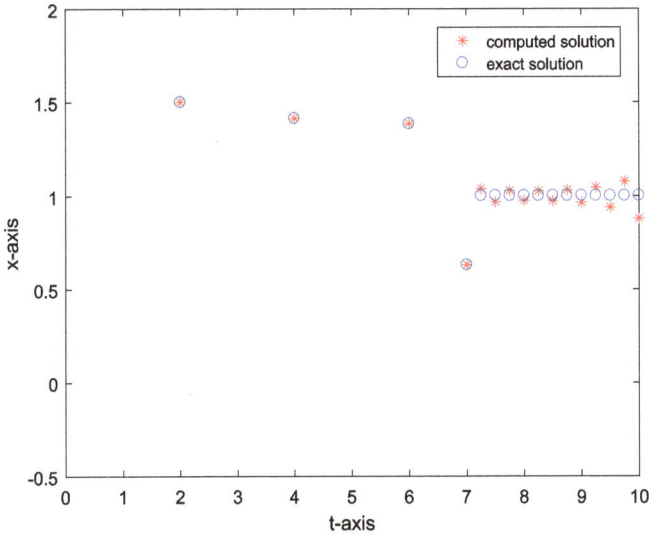

Figure 4.11: Computed and exact values of the solution with step size $q = 0.25$ and $x(2) = 1.5$.

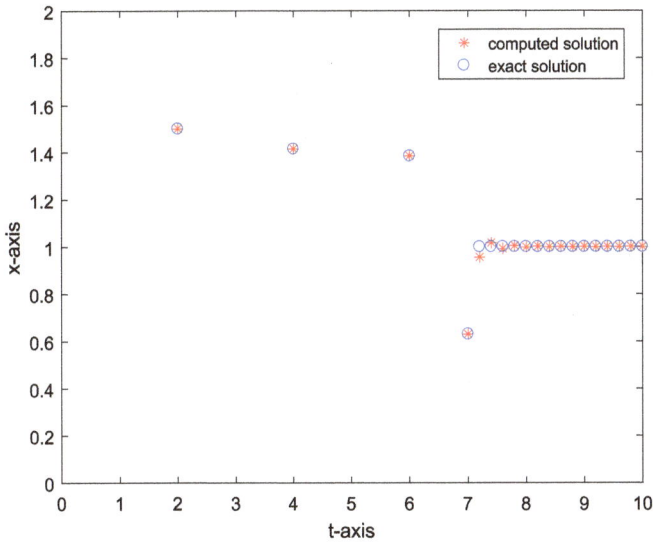

Figure 4.12: Computed and exact values of the solution with step size $q = 0.20$ and $x(2) = 1.5$.

Table 4.5 shows the approximate and exact solutions obtained by taking the initial value as $x(0) = 1.5$ and the step size as $q = 0.25$. The exact solution is denoted by $x^{(e)}$, the approximate solution obtained for $q = 0.25$ by $x^{(2)}$.

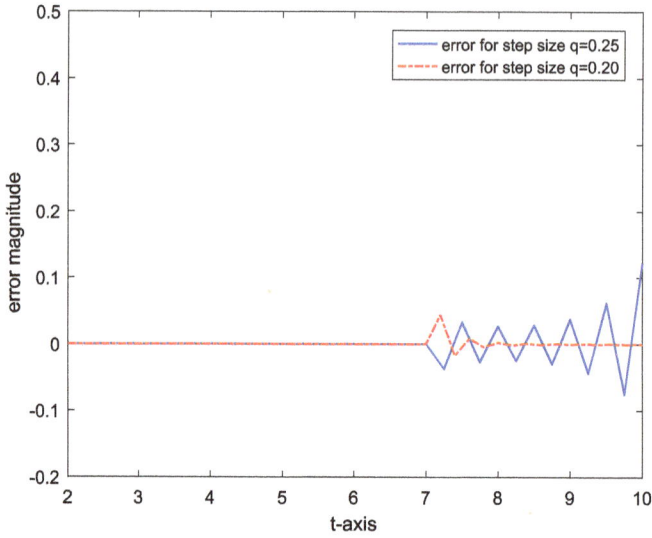

Figure 4.13: The error magnitudes for the cases $q = 0.25$ and $q = 0.20$, and $x(2) = 1.5$.

Table 4.4: The values of the exact solution $x^{(e)}$ and approximate solutions $x^{(1)}$ for $q = 0.20$ and $x(0) = 1.5$.

t	$x^{(e)}$	$x^{(1)}$
2.00	1.50000000	1.50000000
4.00	1.41428571	1.41428571
6.00	1.38398892	1.38398892
7.00	0.62908587	0.62908587
7.20	1.00000000	0.95575745
7.40	1.00000000	1.01664806
7.60	1.00000000	0.99159874
7.80	1.00000000	1.00426137
8.00	1.00000000	0.99758530
8.20	1.00000000	1.00143949
8.40	1.00000000	0.99907533
8.60	1.00000000	1.00062734
8.80	1.00000000	0.99954764
9.00	1.00000000	1.00034343
9.20	1.00000000	0.99972504
9.40	1.00000000	1.00023083
9.60	1.00000000	0.99979677
9.80	1.00000000	1.00018689
10.00	1.00000000	0.99982052

Table 4.5: The values of the exact solution $x^{(e)}$ and approximate solutions $x^{(2)}$ for $q = 0.25$ and $x(0) = 1.5$.

t	$x^{(e)}$	$x^{(2)}$
2.00	1.50000000	1.50000000
4.00	1.41428571	1.41428571
6.00	1.38398892	1.38398892
7.00	0.62908587	0.62908587
7.25	1.00000000	1.03742534
7.50	1.00000000	0.96705322
7.75	1.00000000	1.02679314
8.00	1.00000000	0.97349056
8.25	1.00000000	1.02510394
8.50	1.00000000	0.97202726
8.75	1.00000000	1.02980658
9.00	1.00000000	0.96266125
9.25	1.00000000	1.04353653
9.50	1.00000000	0.93847512
9.75	1.00000000	1.07560659
10.00	1.00000000	0.87738190

Case 2. The initial condition is taken as $x(2) = 0.5$. Computation is done for the same values of q, that is, $q = 0.25$ and $q = 0.20$. The computed solutions for the two values of q are shown in Figures 4.14 and 4.15, and the errors in Figure 4.16.

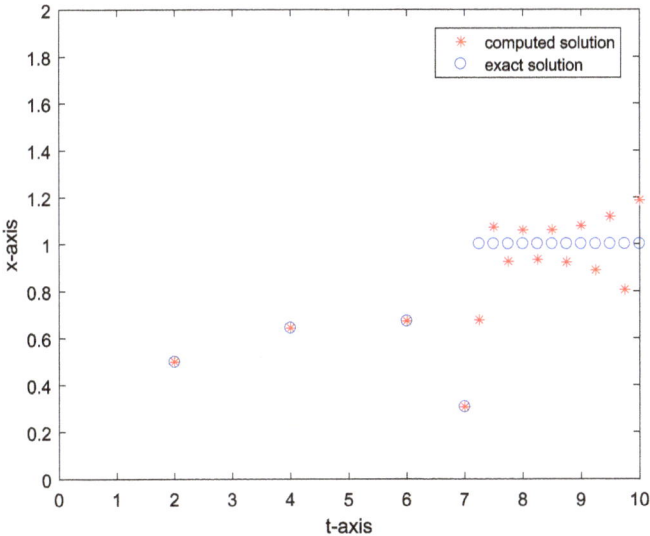

Figure 4.14: Computed and exact values of the solution with step size $q = 0.25$ and $x(2) = 0.5$.

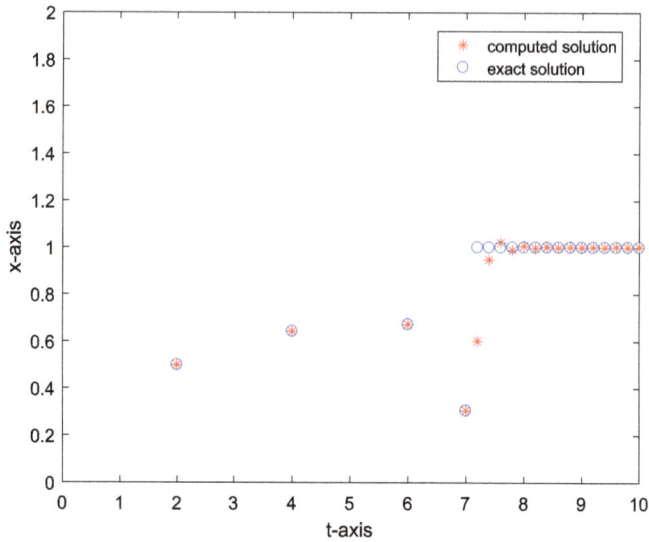

Figure 4.15: Computed and exact values of the solution with step size $q = 0.20$ and $x(2) = 0.5$.

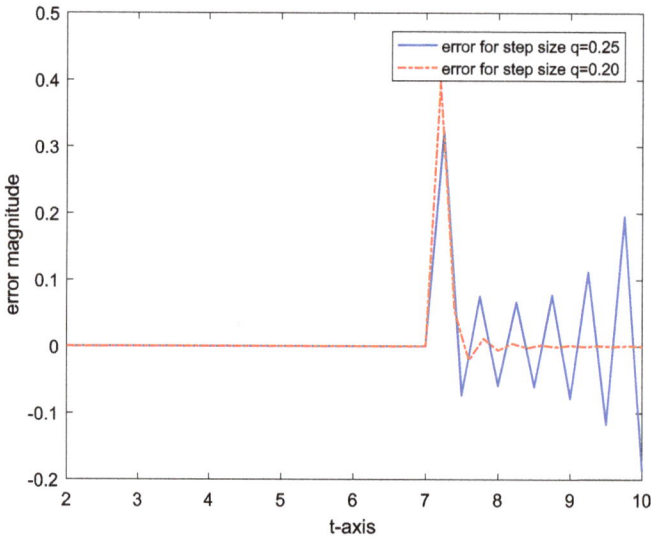

Figure 4.16: The error magnitudes for the cases and $q = 0.25$ and $q = 0.20$, and $x(2) = 0.5$.

As the last example of this chapter, we discuss the Euler method used to solve a second-order dynamic equation.

Example 4.6. In this example, we apply the Euler method to an initial value problem for a second-order dynamic equation given as

$$\begin{cases} x^{\Delta^2}(t) + (t+2)x(t) + 2tx(t) = 0, \\ x(0) = a, \quad x^\Delta(0) = b, \end{cases} \tag{4.15}$$

on the time scale $\mathbb{T} = \alpha\mathbb{N}_0$, where $a, b \in \mathbb{R}$. Let $[t_0, t_f] = [0, 5]$. Note that the exact solution can be computed as follows.

Rewrite the dynamic equation in (4.15) as

$$\left(x^\Delta(t) + 2x(t)\right)^\Delta + t\left(x^\Delta(t) + 2x(t)\right) = 0, \quad t \in \mathbb{T}.$$

Let

$$u(t) = x^\Delta(t) + 2x(t), \quad t \in \mathbb{T},$$

so that we have

$$u^\Delta(t) + tu(t) = 0, \quad t \in \mathbb{T}. \tag{4.16}$$

On the time scale $\alpha\mathbb{N}_0$,

$$u^\Delta(t) = \frac{u(\sigma(t)) - u(t)}{\sigma(t) - t} = \frac{u(t+\alpha) - u(t)}{\alpha}.$$

Then equation (4.16) yields

$$u(t+\alpha) = u(t) - \alpha tu(t) = (1 - \alpha t)u(t), \quad t \in \mathbb{T}.$$

From the initial conditions, we get

$$u(0) = x^\Delta(0) + 2x(0) = b + 2a.$$

Next we consider the dynamic equation

$$x^\Delta(t) + 2x(t) = u(t), \quad t \in \mathbb{T},$$

or

$$x^\Delta(t) = u(t) - 2x(t), \quad t \in \mathbb{T},$$

which gives the exact solution of the initial value problem (4.15) as

$$x(t+\alpha) = x(t) + \alpha\left(u(t) - 2x(t)\right), \quad t \in \mathbb{T},$$

where

$$u(t) = (1 - \alpha t)u(t - \alpha), \quad t \in \mathbb{T},$$

with the initial values

$$u(0) = b + 2a, \quad x(0) = a.$$

To apply the Euler method, we write the second-order dynamic equation in (4.15) as a system of first-order dynamic equations. Let $x_1(t) = x(t)$ and $x_2(t) = x^\Delta(t)$, $t \in \mathbb{T}$. Then the initial value problem (4.15) becomes

$$\begin{cases} x_1^\Delta(t) = x_2(t), \\ x_2^\Delta(t) = -2x_1(t) - (t+2)x_2(t), \quad t \in \mathbb{T}, \\ x_1(0) = a, \quad x_2(0) = b. \end{cases}$$

We choose a constant step size q such that $q \geq \alpha$ and, starting with

$$t_0 = 0, \quad x_{1,0} = a, \quad x_{2,0} = b,$$

we compute the sequence of approximations

$$\begin{cases} t_{i+1} = t_i + q, \\ x_{1,i+1} = x_{1,i} + qx_{2,i}, \\ x_{2,i+1} = x_{2,i} + q(-2x_{1,i} - (t_i + 2)x_{2,i}). \end{cases}$$

We compute the exact $x^{(e)}$ and approximate $x^{(a)}$ solutions for the values $\alpha = 0.2$, $q = 0.4$, $a = 1$, and $b = 1$. The values of the solutions are compared in Table 4.6. The graphs of the solutions are given in Figure 4.17.

Table 4.6: The values of the exact solution $x^{(e)}$ and approximate solutions $x^{(a)}$ for $\alpha = 0.2$, $q = 0.4$, $a = 1$, and $b = 1$.

t	$x^{(e)}$	$x^{(a)}$
0.00	1.00000000	1.00000000
0.40	1.32000000	1.40000000
0.80	1.35072000	1.48000000
1.20	1.15777382	1.30400000
1.60	0.84298652	0.94624000
2.00	0.52296696	0.54567680
2.40	0.27782076	0.23744973
2.80	0.12809919	0.07315282
3.20	0.05262471	0.01565708
3.60	0.02000236	0.00300823
4.00	0.00731252	0.00063614
4.40	0.00263933	0.00011205
4.80	0.00095035	0.00003152

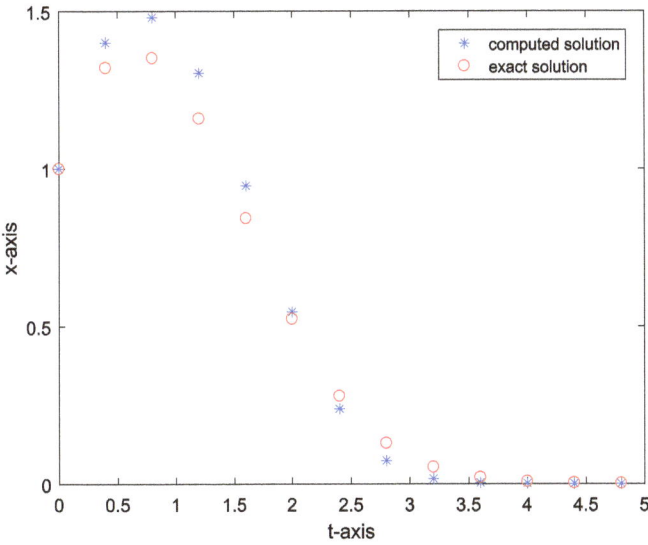

Figure 4.17: The exact solution and approximate solutions for $\alpha = 0.2$, $q = 0.4$, $a = 1$, and $b = 1$.

Next, we compute the exact $x^{(e)}$ and approximate $x^{(a)}$ solutions for the values $\alpha = 0.2$, $q = 0.4$, $a = 0.5$, and $b = -0.3$. The values of the solutions are compared in Table 4.7. The graphs of the solutions are given in Figure 4.18.

It is obvious that the accuracy is very poor, which is natural since the Euler method is of order 1.

Table 4.7: The values of the exact solution $x^{(e)}$ and approximate solutions $x^{(a)}$ for $\alpha = 0.2$, $q = 0.4$, $a = 0.5$, and $b = -0.3$.

t	$x^{(e)}$	$x^{(a)}$
0.00	0.50000000	0.50000000
0.40	0.40400000	0.38000000
0.80	0.34972800	0.35600000
1.20	0.28258883	0.30640000
1.60	0.20117583	0.22121600
2.00	0.12363805	0.12740992
2.40	0.06540532	0.05542200
2.80	0.03009878	0.01707240
3.20	0.01235433	0.00365400
3.60	0.00469430	0.00070206
4.00	0.00171600	0.00014846
4.40	0.00061935	0.00002615
4.80	0.00022301	0.00000735

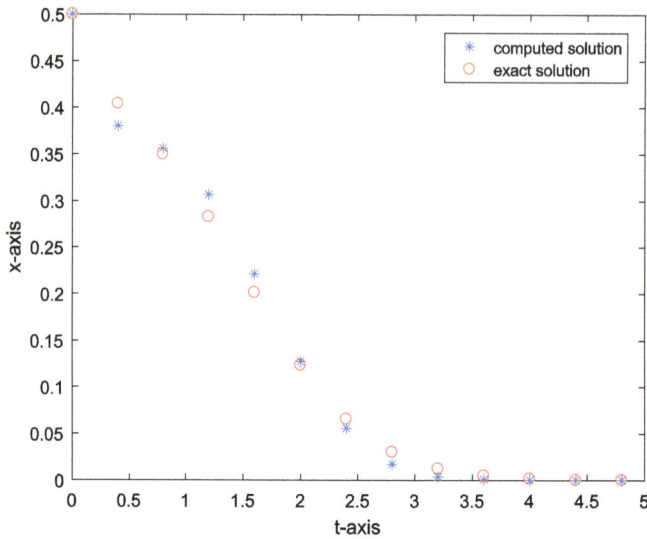

Figure 4.18: The exact solution and approximate solutions for $\alpha = 0.2$, $q = 0.4$, $a = 0.5$, and $b = -0.3$.

4.5 Advanced practical problems

Problem 4.7. Let

$$\mathbb{T} = \left\{-3, -\frac{5}{2}, -2, -1, -\frac{1}{2}, -\frac{1}{4}, -\frac{1}{8}, 0, \frac{1}{8}, \frac{1}{7}, \frac{1}{6}, \frac{1}{5}, 1, 2, 3\right\},$$

$$a = t_0 = -3, \quad t_1 = -2, \quad t_2 = -\frac{1}{4}, \quad t_3 = 0, \quad t_4 = \frac{1}{6}, \quad t_5 = 2, \quad t_6 = 3.$$

Consider the IVP

$$\begin{cases} x^\Delta(t) = \frac{1+x(t)}{1+x(t)+(x(t))^2}, & t > -3, \\ x(-3) = 1. \end{cases}$$

Apply the Euler method and find the local truncation error.

Problem 4.8. Let $\mathbb{T} = 2^{\mathbb{N}_0}$. Apply the Euler method for the following IVP:

$$\begin{cases} x^\Delta(t) = t^2 - (x(t))^2, & t > 1, \\ x(1) = 1, \end{cases}$$

where

$$a = t_0 = 1, \quad t_1 = 2, \quad t_2 = 4, \quad t_3 = 8, \quad t_4 = 16, \quad t_5 = 32.$$

Evaluate the local and global truncation errors.

Problem 4.9. Let

$$\mathbb{T} = \left\{ -\frac{7}{8}, -\frac{3}{4}, 0, \frac{1}{6}, \frac{9}{2}, 5, \frac{11}{2}, 6, 7 \right\}.$$

Apply the Euler method for the following IVP:

$$\begin{cases} x^{\Delta}(t) = t + \frac{x(t)}{1+(x(t))^4}, & t > -\frac{7}{8}, \\ x\left(-\frac{7}{8}\right) = \frac{1}{2}, \end{cases}$$

where

$$a = t_0 - \frac{7}{8}, \quad t_1 = -\frac{3}{4}, \quad t_2 = 0, \quad t_3 = \frac{1}{6}, \quad t_4 = \frac{9}{2}, \quad t_5 = 5.$$

Evaluate the local and global truncation errors.

Problem 4.10. Let

$$\mathbb{T} = \left\{ 0, \frac{1}{6}, \frac{1}{2}, 1, \frac{3}{2}, 2, \frac{7}{3}, \frac{8}{3}, 3, \frac{10}{3}, \frac{11}{3}, \frac{14}{3}, 5 \right\}.$$

Rewrite the following IVP:

$$\begin{cases} x^{\Delta^2}(t) + x(t)x^{\Delta}(t) + 4x(t) = t^2, & t > 0, \\ x(0) = x^{\Delta}(0) = 1, \end{cases}$$

as a first-order system and use the Euler method with

$$a = t_0 = 0, \quad t_1 = \frac{1}{6}, \quad t_2 = \frac{1}{2}, \quad t_3 = 2, \quad t_4 = \frac{8}{3}, \quad t_5 = \frac{10}{3}.$$

Evaluate the local and global truncation errors.

Problem 4.11. Let $\mathbb{T} = 3^{\mathbb{N}_0}$. Rewrite the following IVP:

$$\begin{cases} x^{\Delta^3}(t) + \sin_1(t,1)x^{\Delta^2}(t) + tx(t)x^{\Delta}(t) + (x(t))^2 = t + 1, & t > 1, \\ x^{\Delta^2}(1) = -1, \quad x^{\Delta}(1) = 0, \quad x(1) = 5, \end{cases}$$

as a first-order system and use the Euler method with

$$a = t_0 = 1, \quad t_1 = 9, \quad t_2 = 27, \quad t_3 = 81, \quad t_4 = 243.$$

Evaluate the local and global truncation errors.

Problem 4.12. Let $\mathbb{T} = 4^{\mathbb{N}_0}$. Rewrite the following IVP:

$$\begin{cases} x^{\Delta^4}(t) + \sin_1(t,1)x^{\Delta}(t)x^{\Delta^3}(t) + 2x(t)x^{\Delta^2}(t) + x^{\Delta}(t) + (x(t))^2 = t^2 + t + 1 + e_1(t,1), & t > 1, \\ x^{\Delta^3}(1) = x^{\Delta^2}(1) = 2, \quad x^{\Delta}(1) = 1, \quad x(1) = -1, \end{cases}$$

as a first-order system and use the Euler method with

$$a = t_0 = 1, \quad t_1 = 4, \quad t_2 = 16, \quad t_3 = 64, \quad t_4 = 256.$$

Evaluate the local and global truncation errors.

5 The order 2 Taylor series method – TS(2)

The Euler method is the basic and simplest method to find an approximate solution to an initial value problem. Its derivation uses Taylor series expansion of the dependent variable, which is truncated after the second term. Using more terms from the Taylor series expansion for the unknown function will result in a more accurate approximate solution.

In this chapter we propose the Taylor series method of order 2 for the computation of the approximate solution of initial value problems associated with dynamic equations of first order. We present the derivation, convergence, and error analysis as we did for the Euler method. We also apply the method to some numerical examples.

5.1 Analyzing the method

Suppose that $\mathbb{T}$ is a time scale and that $t_0, t_f \in \mathbb{T}$, $t_f < \infty$. Consider the initial value problem (IVP)

$$\begin{cases} x^\Delta(t) = f(t, x(t)), & t \in [t_0, t_f], \\ x(t_0) = x_0, \end{cases} \tag{5.1}$$

where $x_0 \in \mathbb{R}$. Throughout this chapter, we assume that the following conditions hold:

(H1)
$$\begin{cases} |f(t,x)| \le A, & t \in \mathbb{T}, \ x \in \mathbb{R}, \\ \text{there exist } \Delta_1 f(t,x) \text{ and } \frac{\partial}{\partial x} f(t,x) \text{ such that} \\ |\Delta_1 f(t,x)| \le A, & |\frac{\partial}{\partial x} f(t,x)| \le A, \quad t \in \mathbb{T}, \ x \in \mathbb{R}. \end{cases}$$

(H2)
$$\begin{cases} \text{If } g(t,x) = \Delta_1 f(t,x) + (\int_0^1 \frac{\partial}{\partial x} f(\sigma(t), x + s\mu(t)f(t,x))ds) f(t,x), \\ t \in \mathbb{T}, \ x \in \mathbb{R}, \text{ there exist } \Delta_1 g(t,x) \text{ and } \frac{\partial}{\partial x} g(t,x) \text{ such that} \\ |\Delta_1 g(t,x)| \le A, & |\frac{\partial}{\partial x} g(t,x)| \le A, \quad t \in \mathbb{T}, \ x \in \mathbb{R}, \end{cases}$$

where $A > 0$ is a given constant.

Suppose that $r > 0$, t, $t + r \in [t_0, t_f]$, $\rho^2(t+r) \in [t_0, t_f]$. Then, by the Taylor formula of the second order, we compute

$$x(t + r) = x(t) + h_1(t + r, t)x^\Delta(t) + h_2(t + r, t)x^{\Delta^2}(t)$$

$$+ \int_t^{\rho^2(t+r)} h_2(t + r, \sigma(\tau))x^{\Delta^3}(\tau)\Delta\tau.$$

Let

https://doi.org/10.1515/9783110787320-005

$$R_2(r) = \int_t^{\rho^2(t+r)} h_2(t+r, \sigma(\tau)) x^{\Delta^3}(\tau) \Delta\tau.$$

Then

$$x(t+r) = x(t) + h_1(t+r, t) x^{\Delta}(t) + h_2(t+r, t) x^{\Delta^2}(t) + R_2(r).$$

Assume that $\{t_0 < t_1 < \cdots < t_{m+1} = t_f\}$ is a partition of the interval $[t_0, t_f]$ such that $t_{n+1} = t_n + r_{n+1} \in \mathbb{T}$, $r_{n+1} > 0$, $n \in \{0, \ldots, m\}$. For example, when $\mathbb{T} = 2^{\mathbb{N}_0}$ and $[t_0, t_f] = [1, 16]$, for $t_0 = 1$, $t_1 = 2$, $t_2 = 4$, $t_3 = 8$, $t_4 = 16$, we have $r_1 = 1$, $r_2 = 2$, $r_3 = 4$, $r_4 = 8$. Then

$$x(t_{n+1}) = x(t_n) + h_1(t_{n+1}, t_n) x^{\Delta}(t_n) + h_2(t_{n+1}, t_n) x^{\Delta^2}(t_n) + R_2(r_{n+1})$$
$$= x(t_n) + r_{n+1} x^{\Delta}(t_n) + h_2(t_{n+1}, t_n) x^{\Delta^2}(t_n) + R_2(r_{n+1}) \tag{5.2}$$

for $n \in \{0, \ldots, m\}$. Neglecting the remainder term $R_2(r_{n+1})$, we get the formula

$$x(t_{n+1}) = x(t_n) + r_{n+1} x^{\Delta}(t_n) + h_2(t_{n+1}, t_n) x^{\Delta^2}(t_n). \tag{5.3}$$

Let $x_n = x(t_n)$, $x_n^{\Delta} = x^{\Delta}(t_n)$ and $x_n^{\Delta^2} = x^{\Delta^2}(t_n)$. Then equation (5.3) can be written as

$$x_{n+1} = x_n + r_{n+1} x_n^{\Delta} + h_2(t_{n+1}, t_n) x_n^{\Delta^2}. \tag{5.4}$$

We shall refer to this relation as the order-2 Taylor series method. The value of x_n^{Δ} can be computed from the IVP (5.1) as

$$x_n^{\Delta} = f(t_n, x_n).$$

To determine $x_n^{\Delta^2}$, we have to differentiate both sides of the equation in (5.1). By the Pötzsche chain rule (Appendix C, Theorem C.7), we get

$$(f(t, x(t)))^{\Delta} = \Delta_1 f(t, x(t)) + \left(\int_0^1 \frac{\partial}{\partial x} f(\sigma(t), x(t) + s\mu(t) x^{\Delta}(t)) ds \right) x^{\Delta}(t), \tag{5.5}$$

for $t \in \mathbb{T}^\kappa$, whereupon

$$x^{\Delta^2}(t) = \Delta_1 f(t, x(t)) + \left(\int_0^1 \frac{\partial}{\partial x} f(\sigma(t), x(t) + s\mu(t) x^{\Delta}(t)) ds \right) x^{\Delta}(t), \quad t \in \mathbb{T}^\kappa.$$

Hence,

$$x_n^{\Delta^2} = \Delta_1 f(t_n, x_n) + \left(\int_0^1 \frac{\partial}{\partial x} f(\sigma(t_n), x_n + s\mu(t_n)x_n^\Delta) ds \right) f(t_n, x_n).$$

Therefore, x_{n+1} can be determined by the formula (5.4).

Example 5.1. Consider the IVP

$$\begin{cases} x^\Delta(t) = g(t) + \frac{1}{(x(t))^2+1}, & t \in [t_0, t_f], \\ x(t_0) = x_0, \end{cases} \tag{5.6}$$

where $g : \mathbb{T} \to \mathbb{R}$ is delta differentiable, $|g(t)| \le B$, $|g^\Delta(t)| \le B$ for some positive constant B and $x_0 \in \mathbb{R}$. Here

$$f(t, x(t)) = g(t) + \frac{1}{(x(t))^2 + 1}.$$

Then, if x is a solution of the IVP (5.6), we have

$$x^{\Delta^2}(t) = g^\Delta(t) - \left(\int_0^1 \frac{2(x(t) + s\mu(t)x^\Delta(t))}{(1 + (x(t) + s\mu(t)x^\Delta(t))^2)^2} ds \right) x^\Delta(t)$$

$$= g^\Delta(t) - \int_0^1 \frac{2(x(t) + s\mu(t)(g(t) + \frac{1}{(x(t))^2+1}))}{(1 + (x(t) + s\mu(t)(g(t) + \frac{1}{(x(t))^2+1}))^2)^2} ds$$

$$\times \left(g(t) + \frac{1}{(x(t))^2 + 1} \right), \quad t \in [t_0, t_f].$$

5.2 Convergence of the TS(2) method

In this section, we give the convergence of the Taylor series method derived in the previous section. In the subsequent discussion we use an estimate for the time scales monomials given in the following theorem. We also use the Pötzsche chain rule given in Appendix C, Theorem C.7 for the computation of higher order delta derivatives of the dependent variable x.

Theorem 5.2 ([9]). *For all $i \in \mathbb{N}$, we have the following estimate:*

$$0 \le h_i(t, s) \le \frac{(t - s)^i}{i!}, \quad t \ge s.$$

By the dynamic equation in the IVP (5.1) and Condition (H1), we have

$$|x^\Delta(t)| \le A, \quad t \in [t_0, t_f]. \tag{5.7}$$

From (5.5) and Condition (H2), we obtain

$$|(f(t,x(t)))^{\Delta}| \le |\Delta_1 f(t,x(t))| + \left(\int_0^1 \left| \frac{\partial}{\partial x} f(\sigma(t), x(t) + s\mu(t)x^{\Delta}(t)) \right| ds \right) |x^{\Delta}(t)| \tag{5.8}$$

$$\le A + A^2, \quad t \in [t_0, t_f].$$

Therefore,

$$|x^{\Delta^2}(t)| \le A + A^2, \quad t \in [t_0, t_f]. \tag{5.9}$$

On the other hand, by Theorem 5.2, we have

$$h_2(t+r, \sigma(\tau)) \le \frac{(t+r-\sigma(\tau))^2}{2}$$

$$\le \frac{(t+r-t)^2}{2}$$

$$= \frac{r^2}{2}, \quad \tau \in [t, \rho^2(t+r)], \quad t, t+r \in [t_0, t_f], \, r > 0, \tag{5.10}$$

and

$$h_1(t+r, \sigma(\tau)) \le t+r-\sigma(\tau)$$

$$\le t+r-t$$

$$= r, \quad \tau \in [t, \rho^2(t+r)], \quad t, t+r \in [t_0, t_f], \quad r > 0. \tag{5.11}$$

Applying again the Pötzsche chain rule, we obtain

$$x^{\Delta^3}(t) = \Delta_1 g(t,x(t)) + \left(\int_0^1 \frac{\partial}{\partial x} g(\sigma(t), x(t) + s\mu(t)x^{\Delta}(t)) ds \right) x^{\Delta}(t),$$

$$|x^{\Delta^3}(t)| \le |\Delta_1 g(t,x(t))| + \left(\int_0^1 \left| \frac{\partial}{\partial x} g(\sigma(t), x(t) + s\mu(t)x^{\Delta}(t)) \right| ds \right) |x^{\Delta}(t)|,$$

$$\le A + A^2, \quad t \in [t_0, t_f].$$

Hence, by (5.10) and (5.11), we get

$$|R_2(r)| = \left| \int_t^{\rho^2(t+r)} h_2(t+r, \sigma(\tau)) x^{\Delta^3}(\tau) \Delta\tau \right|$$

$$\le \int_t^{\rho^2(t+r)} h_2(t+r, \sigma(\tau)) |x^{\Delta^3}(\tau)| \Delta\tau$$

$$\le \frac{r^2}{2}(A+A^2)(\rho^2(t+r) - t)$$

$$\leq \frac{r^2}{2}(A + A^2)(t + r - t)$$

$$\leq \frac{r^3}{2}(A + A^2), \quad t, t + r \in [t_0, t_f], \quad r > 0.$$

On the other hand, by (5.9) and (5.11), we have

$$|R_1(r)| = \left| \int_t^{\rho(t+r)} h_1(t + r, \sigma(\tau)) x^{A^2}(\tau) \Delta\tau \right|$$

$$\leq \int_t^{\rho(t+r)} h_1(t + r, \sigma(\tau)) |x^{A^2}(\tau)| \Delta\tau$$

$$\leq r(A + A^2)(\rho(t + r) - t)$$

$$\leq r(A + A^2)(t + r - t)$$

$$\leq r^2(A + A^2), \quad t, t + r \in [t_0, t_f], \quad r > 0,$$

i. e.,

$$R_2(r) = O(r^3), \quad R_1(r) = O(r^2).$$

Now, we denote

$$e_n = x(t_n) - x_n.$$

By the Taylor formula, for $t_{n+1} = t_n + r_{n+1}$, we have

$$x(t_{n+1}) = x(t_n) + r_{n+1}f(t_n, x(t_n)) + h_2(t_{n+1}, t_n)g(t_n, x(t_n)) + R_2(r_{n+1}),$$

and, by (5.4),

$$x_{n+1} = x_n + r_{n+1}f(t_n, x_n) + h_2(t_{n+1}, t_n)g(t_n, x_n),$$

whereupon applying the mean value theorem in the classical case, we obtain

$$x(t_{n+1}) - x_{n+1} = x(t_n) - x_n + r_{n+1}(f(t_n, x(t_n)) - f(t_n, x_n))$$
$$+ h_2(t_{n+1}, t_n)(g(t_n, x(t_n)) - g(t_n, x_n)) + R_2(r_{n+1})$$
$$= x(t_n) - x_n + r_{n+1}\frac{\partial}{\partial x}f(t_n, \zeta)(x(t_n) - x_n)$$
$$+ h_2(t_{n+1}, t_n)\frac{\partial}{\partial x}g(t_n, \eta)(x(t_n) - x_n) + R_2(r_{n+1})$$
$$= e_n + \left(r_{n+1}\frac{\partial}{\partial x}f(t_n, \zeta) + h_2(t_{n+1}, t_n)\frac{\partial}{\partial x}g(t_n, \eta)\right)e_n + R_2(r_{n+1}),$$

where ζ and η are between $x(t_n)$ and x_n. Let

$$A_n = r_{n+1}\frac{\partial}{\partial x}f(t_n, \zeta) + h_2(t_{n+1}, t_n)\frac{\partial}{\partial x}g(t_n, \eta).$$

We have

$$|A_n| \leq r_{n+1}\left|\frac{\partial}{\partial x}f(t_n, \zeta)\right| + h_2(t_{n+1}, t_n)\left|\frac{\partial}{\partial x}g(t_n, \eta)\right|$$

$$\leq Ar_{n+1} + A\frac{r_{n+1}^2}{2}$$

$$= r_{n+1}\left(1 + \frac{r_{n+1}}{2}\right)A.$$

Then,

$$e_0 = 0, \quad e_1 = R_2(r_1), \quad e_{n+1} = (1 + A_n)e_n + R_2(r_{n+1}), \quad n \in \mathbb{N}.$$

In particular,

$$e_2 = (1 + A_1)e_1 + R_2(r_2),$$
$$e_3 = (1 + A_2)e_2 + R_2(r_3)$$
$$\quad = (1 + A_2)((1 + A_1)e_1 + R_2(r_2)) + R_2(r_3)$$
$$\quad = (1 + A_2)(1 + A_1)e_1 + (1 + A_2)R_2(r_2) + R_2(r_3)$$
$$e_4 = (1 + A_3)e_3 + R_2(r_4)$$
$$\quad = (1 + A_3)((1 + A_2)(1 + A_1)e_1 + (1 + A_2)R_2(r_2) + R_2(r_3)) + R_2(r_4)$$
$$\quad = (1 + A_3)(1 + A_2)(1 + A_1)e_1 + (1 + A_3)(1 + A_2)R_2(r_2)$$
$$\quad + (1 + A_3)R_2(r_3) + R_2(r_4),$$

and so on. Let $r_{\max} = \max\{r_1, \ldots, r_{m+1}\}$. Then, using that $0 < r_j \leq t_f - t_0, j \in \{1, \ldots, m+1\}$, and $0 < t_f - t_0 < \infty$, we have that $R_2(r_j) = O(r_{\max}^3), j \in \{1, \ldots, m+1\}$. Since $0 < t_f - t_0 < \infty$ and $t_{j-1} + r_j \in [t_0, t_f], j \in \{1, \ldots, m+1\}$, we have that there exists a constant $0 < B < \infty$ such that $mr_{\max} \leq B$. Then

$$|e_2| \leq \left(1 + r_{\max}\left(1 + \frac{r_{\max}}{2}\right)A\right)R_2(r_{\max}) + R_2(r_{\max}),$$

$$|e_3| \leq \left(1 + r_{\max}\left(1 + \frac{r_{\max}}{2}\right)A\right)^2 R_2(r_{\max})$$

$$+ \left(1 + r_{\max}\left(1 + \frac{r_{\max}}{2}\right)A\right)R_2(r_{\max}) + R_2(r_{\max}),$$

$\cdots$

$$|e_m| \le \sum_{j=0}^{m-1} \left(1 + r_{max}\left(1 + \frac{r_{max}}{2}\right)A\right)^j R_2(r_{max})$$

$$\le \sum_{j=0}^{m-1} e^{r_{max}(1 + \frac{r_{max}}{2})jA} R_2(r_{max})$$

$$\le m e^{mr_{max}(1 + \frac{r_{max}}{2})A} R_2(r_{max})$$

$$\le m e^{B(1 + \frac{r_{max}}{2})A} R_2(r_{max})$$

$$\le m e^{B(1 + \frac{t_f - t_0}{2})A} R_2(r_{max})$$

$$= m r_{max} r_{max}^2 e^{(1 + \frac{t_f - t_0}{2})A} C$$

$$\le r_{max}^2 e^{B(1 + \frac{t_f - t_0}{2})A} BC.$$

Since $t_f < \infty$, we conclude that

$$e_m = O(r_{max}^2),$$

that is, the order of convergence of the order 2 Taylor series method is 2.

5.3 The trapezoid rule

The trapezoid rule is a method that can be deduced from the Taylor series method of order 2. It is an implicit method and its application to nonlinear differential equations requires use of suitable numerical methods.

In this section, we will introduce the trapezoid rule for IVPs associated with the first order dynamic equations on time scales [11].

We start again with the Taylor formula for x^Δ which gives

$$x^\Delta(t + r) = x^\Delta(t) + r x^{\Delta^2}(t) + \int_t^{\rho(t+r)} h_1(t + r, \sigma(\tau)) x^{\Delta^3}(\tau) \Delta\tau$$

$$= x^\Delta(t) + r x^{\Delta^2}(t) + R_1(r),$$

whereupon

$$r x^{\Delta^2}(t) = x^\Delta(t + r) - x^\Delta(t) - R_1(r), \quad t, t + r \in [t_0, t_f], \quad r > 0.$$

We substitute the latter relation into equation (5.2) and find

$$x(t + r) = x(t) + r x^\Delta(t) + h_2(t + r, t) x^{\Delta^2}(t) + R_2(r)$$

$$= x(t) + r x^\Delta(t) + \frac{h_2(t + r, t)}{r}(r x^{\Delta^2}(t)) + R_2(r)$$

$$= x(t) + r x^\Delta(t) + \frac{h_2(t + r, t)}{r}(x^\Delta(t + r) - x^\Delta(t) - R_1(r)) + R_2(r)$$

$$= x(t) + \left(r - \frac{h_2(t+r,t)}{r}\right)x^\Delta(t) + \frac{h_2(t+r,t)}{r}x^\Delta(t+r)$$

$$- \frac{h_2(t+r,t)}{r}R_1(r) + R_2(r)$$

$$= x(t) + \left(r - \frac{h_2(t+r,t)}{r}\right)f(t,x(t)) + \frac{h_2(t+r,t)}{r}f(t+r,x(t+r))$$

$$- \frac{h_2(t+r,t)}{r}R_1(r) + R_2(r), \quad t, t+r \in [t_0,t_f], \quad r > 0.$$

Evaluating this relation at $t = t_n$ and neglecting the remainder terms leads to the trapezoid rule

$$x_{n+1} = x_n + \left(r_{n+1} - \frac{h_2(t_{n+1},t_n)}{r_{n+1}}\right)f(t_n,x_n) + \frac{h_2(t_{n+1},t_n)}{r_{n+1}}f(t_{n+1},x_{n+1}). \qquad (5.12)$$

Definition 5.3. The relation (5.12) will be called trapezoid rule.

Example 5.4. Let $\mathbb{T} = \mathbb{R}$ and $t_{n+1} - t_n = r$ be constant. Then

$$h_2(t_{n+1},t_n) = \frac{1}{2}r^2,$$

and the trapezoid rule takes the form

$$x_{n+1} = x_n + \frac{1}{2}r(f(t_n,x_n) + f(t_{n+1},x_{n+1})),$$

which is the classical trapezoid rule.

Example 5.5. Let $\mathbb{T} = 2^{\mathbb{N}_0}$. Then

$$h_2(t_{n+1},t_n) = \frac{1}{3}(t_{n+1}^2 - t_n^2) = \frac{1}{3}(t_{n+1} - t_n)(t_{n+1} + t_n) = \frac{r_{n+1}}{3}(t_{n+1} + t_n),$$

and the trapezoid rule takes the form

$$x_{n+1} = x_n + \left(h_{n+1} - \frac{1}{3}(t_{n+1} + t_n)\right)f(t_n,x_n) + \frac{1}{3}(t_{n+1} + t_n)f(t_{n+1},x_{n+1}).$$

Exercise 5.6. Let $\mathbb{T} = 2^{\mathbb{N}_0}$. Consider the IVP

$$\begin{cases} x^\Delta(t) = 1 + x(t) + (x(t))^3, \quad t > 1, \\ x(1) = 1 \end{cases}$$

and assume that

$$a = t_0 = 1, \quad t_1 = 2, \quad t_2 = 4, \quad t_3 = 8, \quad t_4 = 16.$$

Write the trapezoid rule.

5.4 Numerical examples

In this section, we will apply the trapezoid rule to specific examples. The first example is a linear dynamic and the second one is a nonlinear dynamic equation. We use MATLAB for the numerical computations and employ the Newton method to find the solution of the implicit relation arising in the second example.

Example 5.7. Let $\mathbb{T} = \mathbb{N}_0$. Consider the IVP associated with the linear dynamic equation

$$\begin{cases} x^\Delta(t) = \frac{1}{t+1}x(t) + \frac{1}{t^2+1}, & t > 0, \\ x(0) = x_0, \end{cases} \tag{5.13}$$

where $t_0 = 0$, $t_f = 20$. The exact solution of this equation has the form

$$x(t) = x_0 e_{\frac{1}{t+1}}(t,0) + \int_0^t e_{\frac{1}{t+1}}(t,\sigma(\tau))\frac{1}{\tau^2+1}\Delta\tau, \quad t \geq 0.$$

On $\mathbb{T} = \mathbb{N}_0$ we have $\sigma(t) = t + 1$, $\mu(t) = 1$, and

$$e_{\frac{1}{t+1}}(t,s) = \prod_{p=s}^{t-1}\left(1 + \frac{1}{p+1}\right) = \frac{s+2}{s+1}\frac{s+3}{s+2}\cdots\frac{t}{t-1}\frac{t+1}{t} = \frac{t+1}{s+1}, \quad t \geq s.$$

The integral $\int_0^t e_{\frac{1}{t+1}}(t,\sigma(\tau))\frac{1}{\tau^2+1}\Delta\tau$, $t \geq 0$ is evaluated as

$$\begin{aligned} \int_0^t e_{\frac{1}{t+1}}(t,\sigma(\tau))\frac{1}{\tau^2+1}\Delta\tau &= \int_0^t e_{\frac{1}{t+1}}(t,\tau+1)\frac{1}{\tau^2+1}\Delta\tau \\ &= \int_0^t \frac{t+1}{\tau+2}\frac{1}{\tau^2+1}\Delta\tau \\ &= (t+1)\sum_{p=0}^{t-1}\frac{1}{(p+2)(p^2+1)}, \quad t \geq 0. \end{aligned}$$

Hence, we obtain the exact solution as

$$x(t_n) = (t_n + 1)\left(x_0 + \sum_{p=0}^{t_n-1}\frac{1}{(p+2)(p^2+1)}\right),$$

for any $t_n \in \mathbb{T}$. Recalling that the monomials on $\mathbb{T}$ have the form

$$h_0(t_{n+1},t_n) = 1,$$
$$h_1(t_{n+1},t_n) = t_{n+1} - t_n = r,$$
$$h_2(t_{n+1},t_n) = \frac{(t_{n+1} - t_n)(t_{n+1} - t_n - 1)}{2} = \frac{r(r-1)}{2},$$

where we assume constant step size $r = t_{n+1} - t_n$, we can write the trapezoid rule for this IVP as

$$x_{n+1} = x_n + \left(r - \frac{r-1}{2}\right)\left(\frac{1}{t_n+1}x_n + \frac{1}{t_n^2+1}\right)$$
$$+ \frac{r-1}{2}\left(\frac{1}{t_{n+1}+1}x_{n+1} + \frac{1}{t_{n+1}^2+1}\right), \quad n \in \left\{0, 1, \ldots, \frac{20}{r}\right\}.$$

Then we have

$$x_{n+1} = \frac{1 + \frac{r+1}{2(t_n+1)}}{1 - \frac{r-1}{2(t_{n+1}+1)}}x_n + \frac{1}{1 - \frac{r-1}{2(t_{n+1}+1)}}\left(\frac{r+1}{2(t_n^2+1)} + \frac{r-1}{2(t_{n+1}^2+1)}\right),$$

where $n \in \{0, 1, \ldots, \frac{20}{r}\}$. Using *MATLAB*, we compute the approximate solution for different values of r and x_0. The approximate and exact values of the solution for $r = 1$ and $r = 2$ with the initial condition $x(0) = 1$ are given in Table 5.1, and those with the initial condition $x(0) = 1$ in Table 5.2. In both tables, $x^{(e)}$ denotes the exact solution, $x^{(1)}$ the approximate solution for $r = 1$, and $x^{(2)}$ the approximate solution for $r = 2$. Since the error is of order r^2, we observe that for $r = 2$ the magnitude of the error is large.

Table 5.1: The values of the exact solution $x^{(e)}$ and approximate solutions $x^{(1)}$ and $x^{(2)}$ for $x(0) = 1$.

t	$x^{(e)}$	$x^{(1)}$	$x^{(2)}$
0.00	1.00000000	1.00000000	1.00000000
1.00	3.00000000	3.00000000	
2.00	5.00000000	5.00000000	4.92000000
3.00	6.86666667	6.86666667	
4.00	8.68333333	8.68333333	8.56601307
5.00	10.47882353	10.47882353	
6.00	12.26375566	12.26375566	12.10199394
7.00	14.04274778	14.04274778	
8.00	15.81809125	15.81809125	15.61077657
9.00	17.59104156	17.59104156	
10.00	19.36234084	19.36234084	19.10920009
11.00	21.13245463	21.13245463	
12.00	22.90168924	22.90168924	22.60263185
13.00	24.67025419	24.67025419	
14.00	26.43829756	26.43829756	26.09328697
15.00	28.20592687	28.20592687	
16.00	29.97322208	29.97322208	29.58224132
17.00	31.74024384	31.74024384	
18.00	33.50703900	33.50703900	33.07007937
19.00	35.27364429	35.27364429	
20.00	37.04008893	37.04008893	36.55714561

Table 5.2: The values of the exact solution $x^{(e)}$ and approximate solutions $x^{(1)}$ and $x^{(2)}$ for $x(0) = 3$.

t	$x^{(e)}$	$x^{(1)}$	$x^{(2)}$
0.00	3.00000000	3.00000000	3.00000000
1.00	7.00000000	7.00000000	
2.00	11.00000000	11.00000000	10.92000000
3.00	14.86666667	14.86666667	
4.00	18.68333333	18.68333333	18.56601307
5.00	22.47882353	22.47882353	
6.00	26.26375566	26.26375566	26.10199394
7.00	30.04274778	30.04274778	
8.00	33.81809125	33.81809125	33.61077657
9.00	37.59104156	37.59104156	
10.00	41.36234084	41.36234084	41.10920009
11.00	45.13245463	45.13245463	
12.00	48.90168924	48.90168924	48.60263185
13.00	52.67025419	52.67025419	
14.00	56.43829756	56.43829756	56.09328697
15.00	60.20592687	60.20592687	
16.00	63.97322208	63.97322208	63.58224132
17.00	67.74024384	67.74024384	
18.00	71.50703900	71.50703900	71.07007937
19.00	75.27364429	75.27364429	
20.00	79.04008893	79.04008893	78.55714561

The exact and approximate solutions are also presented in Figures 5.1, 5.2, 5.3 and 5.4. It can be observed that for large values of r the error increases, and for $r = 1$ the approximate solution matches the exact solution.

The second example is an IVP associated with a nonlinear first-order dynamic equation.

Example 5.8. Let $\mathbb{T} = a\mathbb{N}_0$. Consider the IVP associated with the nonlinear dynamic equation

$$\begin{cases} x^\Delta(t) = \frac{1}{t^2+1} + \frac{t}{(x(t))^2+1}, & t > 0, t \in \mathbb{T}, \\ x(0) = x_0. \end{cases} \tag{5.14}$$

On this time scale, we have $\sigma(t) = t + a$ and $\mu(t) = a$. Let $t_0 = 0$, $t_f = b$, and $t_n = t_{n-1} + r$ for some $r > 0$. The dynamic equation can be written as

$$\frac{x(t+a) - x(t)}{a} = \frac{1}{t^2+1} + \frac{t}{(x(t))^2+1}, \quad t \in \mathbb{T},$$

and hence, the exact solution at any $t_n \in a\mathbb{N}_0$ has the form

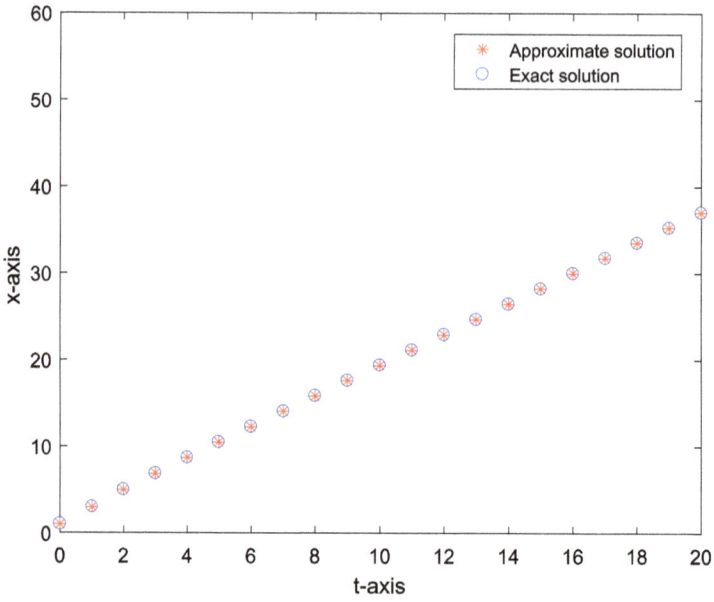

Figure 5.1: Approximate and exact values of the solution of (5.13) with $r = 1$, $x_0 = 1$.

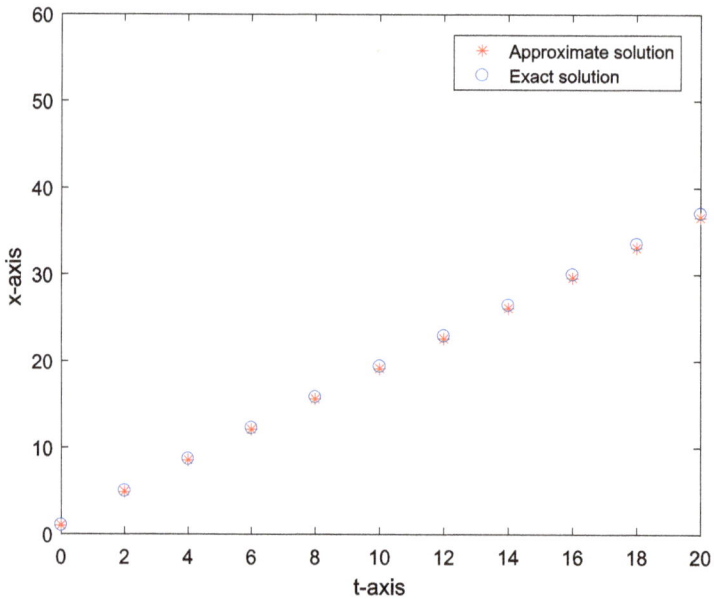

Figure 5.2: Approximate and exact values of the solution of (5.13) with $r = 2$, $x_0 = 1$.

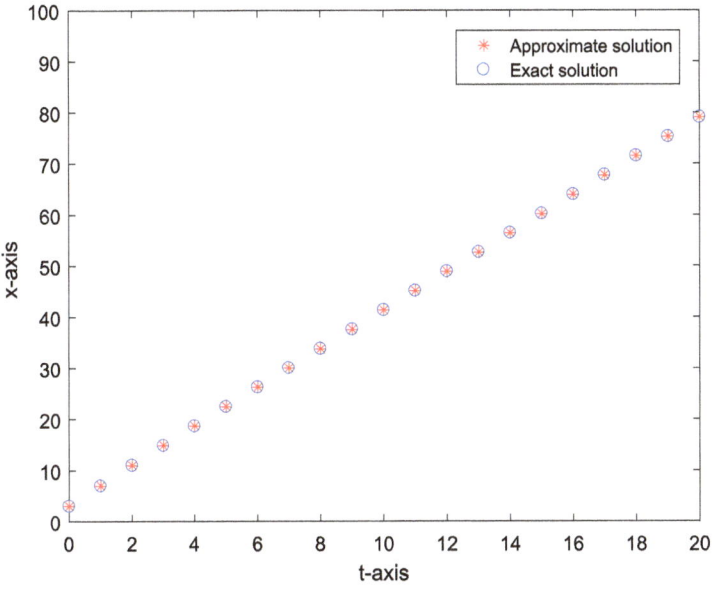

Figure 5.3: Approximate and exact values of the solution of (5.13) with $r = 1$, $x_0 = 3$.

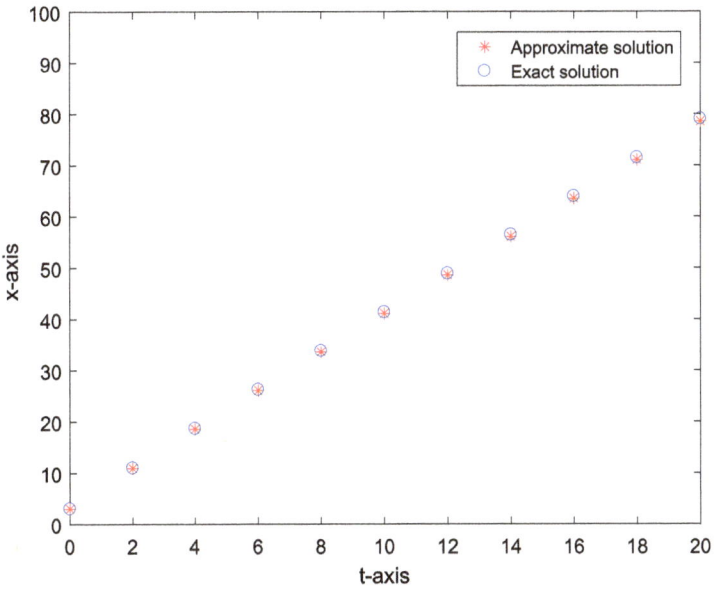

Figure 5.4: Approximate and exact values of the solution of (5.13) with $r = 2$, $x_0 = 3$.

$$x(t_n) = x(t_n - a) + a\left(\frac{1}{(t_n - a)^2 + 1} + \frac{t_n - a}{(x(t_n - a))^2 + 1}\right).$$

Using the monomial h_2, which is given by

$$h_2(t_{n+1}, t_n) = \frac{(t_{n+1} - t_n)(t_{n+1} - t_n - a)}{2} = \frac{r(r - a)}{2},$$

the trapezoid rule for this problem yields

$$x_{n+1} = x_n + \left(r - \frac{r - a}{2}\right)\left(\frac{1}{t_n^2 + 1} + \frac{t_n}{x_n^2 + 1}\right)$$

$$+ \frac{r - a}{2}\left(\frac{1}{t_{n+1}^2 + 1} + \frac{t_{n+1}}{x_{n+1}^2 + 1}\right), \quad n \in \left\{0, 1, \dots, \frac{10}{r}\right\}.$$

This nonlinear recurrence relation can be written as

$$\frac{2x_{n+1}^3 - 2b_n x_{n+1}^2 + 2x_{n+1} - (r - a)t_{n+1} - 2b_n}{2(1 + x_{n+1}^2)} = 0,$$

where

$$b_n = x_n + \frac{r + a}{2}\left(\frac{1}{t_n^2 + 1} + \frac{t_n}{1 + x_n^2 + 1}\right) + \frac{r - a}{2}\frac{1}{t_{n+1}^2 + 1},$$

for $n \in \{0, 1, \dots, \frac{b}{r}\}$. We compute each next term of the sequence $\{x_n\}$, $n \in \{0, 1, \dots, \frac{b}{h}\}$ using Newton's method [14].

The computations are done with *MATLAB* for

$$a = 0.5, \quad r = 0.5, 1, 2, \quad x_0 = 1, \quad b = 20, \quad \text{and}$$
$$a = 0.2, \quad r = 0.4, 0.8, \quad x_0 = 3, \quad b = 8.$$

The approximate and exact solutions for these values of the parameters are compared in Tables 5.3 and 5.4. In Table 5.3, $x^{(e)}$ denotes the exact solution and $x^{(1)}$, $x^{(2)}$, $x^{(3)}$ the approximate solutions for $r = 0.5, 1, 2$, respectively, computed for $a = 0.5$ and $x(0) = 1$. In Table 5.4, $x^{(e)}$ denotes the exact solution and $x^{(1)}$, $x^{(2)}$ the approximate solutions for $r = 0.4, 0.8$, respectively, computed for $a = 0.2$ and $x(0) = 3$.

The graphs of the approximate and exact solutions for the case $a = 0.5, r = 0.5, 1, 2$, and the initial value $x(0) = 1$ are presented in Figures 5.5, 5.6 and 5.7. It is clear that for $r = a$ the computed solution matches the exact solution, and for $r > a$ an error is observed.

The graphs of the approximate and exact solutions for the case $a = 0.2, r = 0.4, 0.8$, and the initial value $x(0) = 3$ are presented in Figures 5.8 and 5.9.

Table 5.3: The values of the exact solution $x^{(e)}$ and the approximate solutions $x^{(1)}$, $x^{(2)}$ and $x^{(3)}$ for $x(0) = 1$ and $a = 0.5$.

t	$x^{(e)}$	$x^{(1)}$	$x^{(2)}$	$x^{(3)}$
0.00	1.00000000	1.00000000	1.00000000	1.00000000
1.00	1.97692308	1.97692308	1.92799782	
2.00	2.59940190	2.59940190	2.57741023	2.59406935
3.00	3.03618334	3.03618334	3.02265747	
4.00	3.42352292	3.42352292	3.41337855	3.44479724
5.00	3.79402385	3.79402385	3.78577255	
6.00	4.15528317	4.15528317	4.14833890	4.17172106
7.00	4.50894103	4.50894103	4.50299225	
8.00	4.85529463	4.85529463	4.85013773	4.86756213
9.00	5.19444102	5.19444102	5.18992800	
10.00	5.52652984	5.52652984	5.52254768	5.53616265
11.00	5.85178922	5.85178922	5.84824944	
12.00	6.17050243	6.17050243	6.16733476	6.17841823
13.00	6.48298140	6.48298140	6.48012939	
14.00	6.78954686	6.78954686	6.78696474	6.79627227
15.00	7.09051545	7.09051545	7.08816578	
16.00	7.38619205	7.38619205	7.38404397	7.39204551
17.00	7.67686568	7.67686568	7.67489354	
18.00	7.96280759	7.96280759	7.96098990	7.96799382
19.00	8.24427079	8.24427079	8.24258946	
20.00	8.52149045	8.52149045	8.51993011	8.52614846

Table 5.4: The values of the exact solution $x^{(e)}$ and the approximate solutions $x^{(1)}$ and $x^{(2)}$ for $x(0) = 3$ and $a = 0.2$.

t	$x^{(e)}$	$x^{(1)}$	$x^{(2)}$
0.00	3.00000000	3.00000000	3.00000000
0.40	3.39586641	3.38940995	
0.80	3.73043213	3.72399602	3.69927048
1.20	3.97566995	3.97120379	
1.60	4.15541211	4.15248125	4.14209378
2.00	4.29534158	4.29335406	
2.40	4.41218246	4.41075566	4.40625031
2.80	4.51582844	4.51474078	
3.20	4.61210777	4.61123132	4.60884799
3.60	4.70451328	4.70377228	
4.00	4.79517350	4.79452174	4.79297384
4.40	4.88539625	4.88480475	
4.80	4.97598082	4.97543101	4.97423099
5.20	5.06740357	5.06688335	
5.60	5.15993195	5.15943337	5.15837967
6.00	5.25369677	5.25321457	
6.40	5.34873916	5.34826990	5.34727656
6.80	5.44504180	5.44458322	
7.20	5.54254995	5.54210061	5.54113241
7.60	5.64118581	5.64074482	
8.00	5.74085838	5.74042521	5.73947000

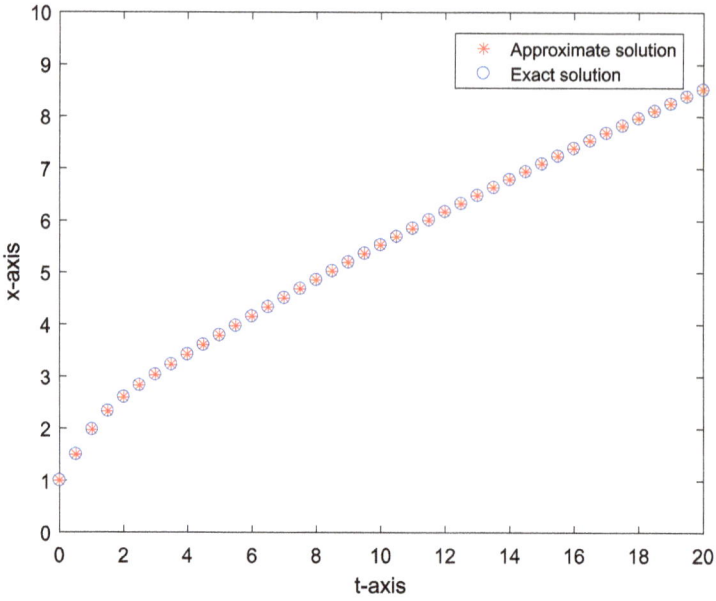

Figure 5.5: Approximate and exact values of the solution of (5.14) with $a = r = 0.5$, $x_0 = 1$.

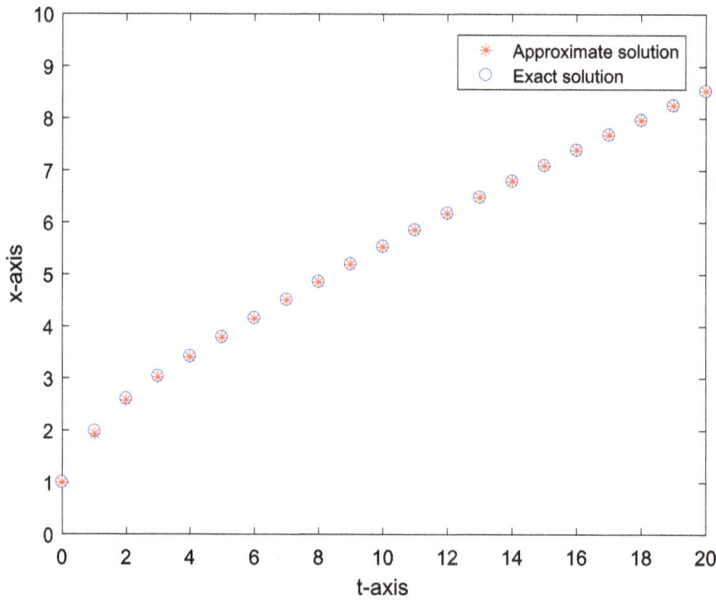

Figure 5.6: Approximate and exact values of the solution of (5.14) with $a = 0.5$, $r = 1$, $x_0 = 1$.

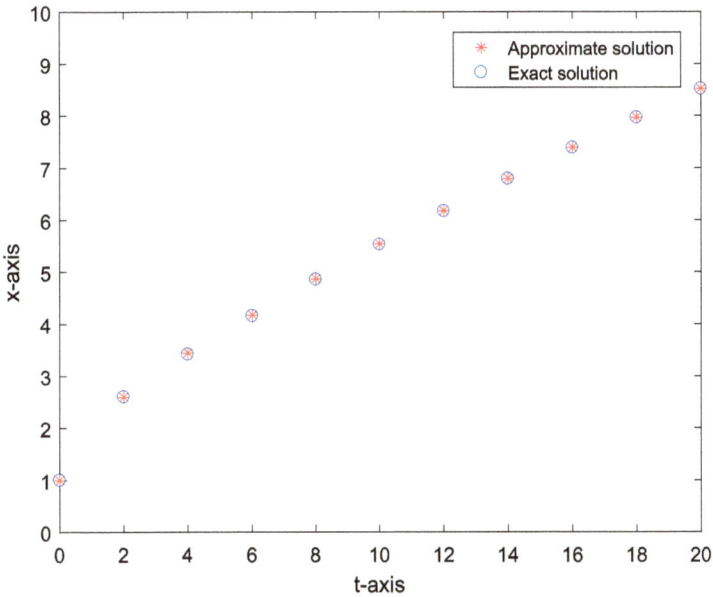

Figure 5.7: Approximate and exact values of the solution of (5.14) with $a = 0.5$, $r = 2$, $x_0 = 1$.

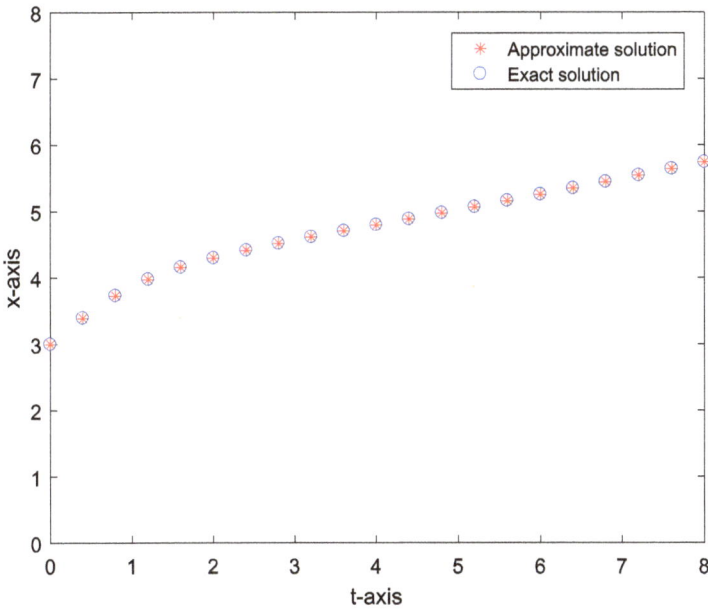

Figure 5.8: Approximate and exact values of the solution of (5.14) with $a = 0.2$, $r = 0.4$, $x_0 = 3$.

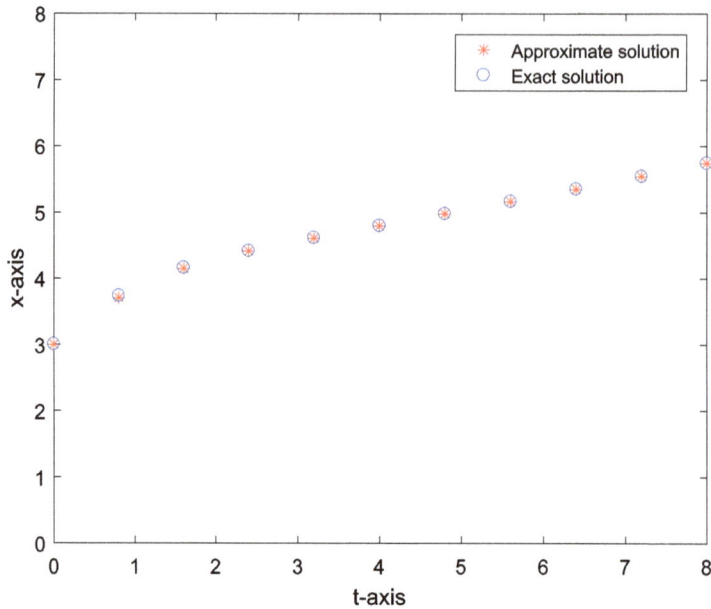

Figure 5.9: Approximate and exact values of the solution of (5.14) with $a = 0.2, r = 0.8, x_0 = 3$.

5.5 Advanced practical problems

Problem 5.9. Let $\mathbb{T} = 2^{\mathbb{N}_0}$. Consider the IVP

$$\begin{cases} x^{\Delta}(t) = 1 + \frac{1+x(t)}{1+(x(t))^2}, & t > 1, \\ x(1) = 1 \end{cases}$$

and assume that

$$a = t_0 = 1, \quad t_1 = 4, \quad t_2 = 16, \quad t_3 = 64, \quad t_4 = b = 216.$$

Write the trapezoid rule.

Problem 5.10. Let

$$\mathbb{T} = \left\{ 0, \frac{1}{8}, \frac{1}{7}, \frac{1}{6}, \frac{1}{5}, \frac{1}{4}, 1, \frac{4}{3}, \frac{7}{3}, \frac{8}{3}, 3, 7, 11 \right\}.$$

Consider the IVP

$$\begin{cases} x^{\Delta}(t) = (1 + x(t))^3, & t > 0, \\ x(0) = 0. \end{cases}$$

Write the trapezoid rule.

Problem 5.11. Let $\mathbb{T} = \mathbb{Z}$. Apply TS(2) for the following IVP

$$\begin{cases} x^\Delta(t) = \sin_1(t, 0) + (x(t))^2, & t > 0, \\ x(0) = 1, \end{cases}$$

where

$$a = t_0 = 0, \quad t_1 = 2, \quad t_2 = 4, \quad t_3 = 5, \quad t_4 = b = 6.$$

Problem 5.12. Let $\mathbb{T} = 2^{\mathbb{N}_0}$. Apply TS(2) for the following IVP:

$$\begin{cases} x^\Delta(t) = \frac{1}{1+(x(t))^2}, & t > 1, \\ x(1) = 1, \end{cases}$$

where

$$a = t_0 = 1, \quad t_1 = 2, \quad t_2 = 8, \quad t_3 = 16, \quad t_4 = b = 32.$$

Problem 5.13. Let

$$\mathbb{T} = \left\{ 0, \frac{1}{3}, \frac{2}{3}, 1, \frac{3}{2}, 2, \frac{17}{8}, \frac{19}{8}, \frac{21}{8}, \frac{11}{4}, 3 \right\}.$$

Apply TS(2) for the following IVP:

$$\begin{cases} x^\Delta(t) = 1 + 3(x(t))^2, & t > 0, \\ x(0) = 1, \end{cases}$$

where

$$a = t_0 = 0, \quad t_1 = \frac{1}{2}, \quad t_2 = 1, \quad t_3 = 2, \quad t_4 = \frac{21}{8}, \quad t_5 = b = 3.$$

6 The order p Taylor series method – TS (p)

The Taylor series method can be extended to an arbitrary order. We already discussed the cases of the first and second order methods in the previous chapters. In this chapter, we give a generalization of the Taylor series method into an arbitrary order $p \geq 2$. The derivation of the method requires some extra conditions on the nonlinear right-hand side function of the dynamic equation. These additional requirements limit the number of problems for which this method can be applied. However, the error in the approximated solution reduces.

Let $\mathbb{T}$ be a time scale and Δ denote the differentiation operator in $\mathbb{T}$ as usual.

6.1 Analyzing the order p Taylor series method

Suppose that $p \in \mathbb{N}$, $p \geq 2$, $t_0, t_f \in \mathbb{T}$, $t_0 < t_f < \infty$, $r > 0$ be such that $t, t + r \in [t_0, t_f]$. Consider the initial value problem (IVP)

$$\begin{cases} x^\Delta(t) = f(t, x(t)), & t \in [t_0, t_f], \\ x(t_0) = x_0, \end{cases} \tag{6.1}$$

where $x_0 \in \mathbb{R}$ is a given constant and the function f satisfies the following conditions:

(H1) $\begin{cases} |f(t,x)| \leq A, \quad t \in \mathbb{T}, \ x \in \mathbb{R}, \\[2mm] \text{there exist } g_k(t, x(t), \ldots, x^{\Delta^k}(t)) = (f(t, x(t)))^{\Delta^k}, \quad k \in \{1, \ldots, p-1\}, \\[2mm] \text{such that } |\frac{\partial f}{\partial y}(t, z)| \leq A, \quad |\Delta_1 g_k(t, y_1, \ldots, y_{k+1})| \leq A, \\[2mm] \text{and } |\frac{\partial}{\partial y_j} g_k(t, y_1, \ldots, y_{k+1})| \leq A, \quad j \in \{1, \ldots, k+1\}, \\[2mm] \text{for any } t \in \mathbb{T} \text{ and for } z, y_j \in \mathbb{R}, \ j \in \{1, \ldots, p-1\}, \\[2mm] \text{where } pe^{t_f - t_0} A < 1 \text{ and } A > 0. \end{cases}$

By the Taylor formula on time scales, we get

$$x(t+r) = x(t) + h_1(t+r, t)x^\Delta(t) + h_2(t+r, t)x^{\Delta^2}(t) + \cdots + h_p(t+r, t)x^{\Delta^p}(t)$$

$$+ \int_t^{\rho^p(t+r)} h_p(t+r, \sigma(u))x^{\Delta^{p+1}}(u)\Delta u.$$

Let

$$R_p(t) = \int_t^{\rho^p(t+r)} h_p(t+r, \sigma(u))x^{\Delta^{p+1}}(u)\Delta u,$$

https://doi.org/10.1515/9783110787320-006

be the remainder term. Let also, $t_0 < t_1 < \cdots < t_{m+1} = t_f$ be a partition of the interval $[t_0, t_f]$ such that $t_{n+1} = t_n + r_{n+1} \in \mathbb{T}$, $r_{n+1} > 0$, $n \in \{0, \ldots, m\}$. Then

$$x(t_{n+1}) = x(t_n) + h_1(t_{n+1}, t_n)x^\Delta(t_n) + h_2(t_{n+1}, t_n)x^{\Delta^2}(t_n)$$
$$+ \cdots + h_p(t_{n+1}, t_n)x^{\Delta^p}(t_n) + R_p(t_{n+1}).$$

Neglecting the remainder term $R_p(t)$, we obtain

$$x(t_{n+1}) = x(t_n) + h_1(t_{n+1}, t_n)x^\Delta(t_n) + h_2(t_{n+1}, t_n)x^{\Delta^2}(t_n) + \cdots + h_p(t_{n+1}, t_n)x^{\Delta^p}(t_n).$$

Set

$$x_n^{\Delta^k} = x^{\Delta^k}(t_n), \quad x_n^{\Delta^k \sigma} = x^{\Delta^k}(\sigma(t_n)), \quad k \in \{0, \ldots, p\}.$$

Thus, we get

$$x_{n+1} = x_n + r_{n+1}x_n^\Delta + h_2(t_{n+1}, t_n)x_n^{\Delta^2} + \cdots + h_p(t_{n+1}, t_n)x_n^{\Delta^p}, \tag{6.2}$$

which will be called the order p Taylor series method. To compute x_{n+1}, we need to determine $x_n^{\Delta^q}$ for $q \in \{1, \ldots, p\}$. From the dynamic equation in the IVP (6.1), we determine x_n^Δ as

$$x_n^\Delta = f(t_n, x_n).$$

Now, we will determine $x_n^{\Delta^2}, \ldots, x_n^{\Delta^p}$. By the generalized Pötzsche chain rule (Appendix C, Theorem C.8), we have

$$(f(t, x(t)))^\Delta = \Delta_1 f(t, x(t)) + \left(\int_0^1 \frac{\partial}{\partial y_1} f(\sigma(t), x(t) + h\mu(t)x^\Delta(t))dh \right)x^\Delta(t)$$
$$= g_1(t, x(t), x^\Delta(t))$$

and, for $q \in \{2, \ldots, p\}$, we compute

$$(f(t, x(t)))^{\Delta^q}$$
$$= (g_{q-1}(t, x(t), x^\Delta(t), \ldots, x^{\Delta^{q-1}}(t)))^\Delta$$
$$= \Delta_1 g_{q-1}(t, x(t), x^\Delta(t), \ldots, x^{\Delta^{q-1}}(t))$$
$$+ \left(\int_0^1 \frac{\partial}{\partial y_1} g_{q-1}(\sigma(t), x(t) + h\mu(t)x^\Delta(t), x^\Delta(t), \ldots, x^{\Delta^{q-1}}(t))dh \right)x^\Delta(t)$$
$$+ \left(\int_0^1 \frac{\partial}{\partial y_2} g_{q-1}(\sigma(t), x(\sigma(t)), x^\Delta(t) + h\mu(t)x^{\Delta^2}(t), \ldots, x^{\Delta^{q-1}}(t))dh \right)x^{\Delta^2}(t)$$

$+ \cdots$

$$+ \left(\int_0^1 \frac{\partial}{\partial y_q} g_{q-1}(\sigma(t), x(\sigma(t)), x^\Delta(\sigma(t)), \ldots, x^{\Delta^{q-1}}(t) + h\mu(t)x^{\Delta^q}(t))dh \right) x^{\Delta^q}(t),$$

for $t \in \mathbb{T}^\kappa$. Therefore, we have

$$x_n^{\Delta^2} = \Delta_1 f(t_n, x_n) + \left(\int_0^1 \frac{\partial}{\partial y_1} f(\sigma(t_n), x_n + h\mu(t_n)x_n^\Delta)dh \right) x_n^\Delta,$$

$$x_n^{\Delta^3} = \Delta_1 g_1(t_n, x_n, x_n^\Delta)$$

$$+ \left(\int_0^1 \frac{\partial}{\partial y_1} g_1(\sigma(t_n), x_n + h\mu(t_n)x_n^\Delta, x_n^\Delta)dh \right) x_n^\Delta$$

$$+ \left(\int_0^1 \frac{\partial}{\partial y_2} g_1(\sigma(t_n), x_n^\sigma, x_n^\Delta + h\mu(t_n)x_n^{\Delta^2})dh \right) x_n^{\Delta^2},$$

$$\vdots$$

$$x_n^{\Delta^{p+1}} = \Delta_1 g_{p-1}(t_n, x_n, x_n^\Delta, \ldots, x_n^{\Delta^{p-1}})$$

$$+ \left(\int_0^1 \frac{\partial}{\partial y_1} g_{p-1}(\sigma(t_n), x_n + h\mu(t_n)x_n^\Delta, x_n^\Delta, \ldots, x_n^{\Delta^{p-1}})dh \right) x_n^\Delta$$

$$+ \left(\int_0^1 \frac{\partial}{\partial y_2} g_{p-1}(\sigma(t_n), x_n^\sigma, x_n^\Delta + h\mu(t_n)x_n^{\Delta^2}, \ldots, x_n^{\Delta^{p-1}})dh \right) x_n^{\Delta^2}$$

$$+ \cdots$$

$$+ \left(\int_0^1 \frac{\partial}{\partial y_p} g_{p-1}(\sigma(t_n), x_n^\sigma, x_n^{\Delta\sigma}, \ldots, x_n^{\Delta^{p-1}} + h\mu(t_n)x_n^{\Delta^p})dh \right) x_n^{\Delta^p},$$

from where we can find $x_n^{\Delta^2}, \ldots, x_n^{\Delta^{p+1}}$.

6.2 Convergence and error analysis of the TS(p) method

Now, we will investigate the convergence of the Taylor series method of order p. We will use the property of the monomials $h_q(t, s)$, $q \in \mathbb{N}_0$, given in Theorem 5.2. By Condition (H1) and the dynamic equation in (6.1), we find

$$|x^\Delta(t)| \leq A, \quad t \in [t_0, t_f]. \tag{6.3}$$

Next, we estimate

$$\left|x^{\Delta^2}(t)\right| \leq \left|\Delta_1 f(t, x(t))\right| + \left(\int_0^1 \left|\frac{\partial}{\partial y_1} f(\sigma(t), x(t) + h\mu(t)x^{\Delta}(t))\right| dh\right) |x^{\Delta}(t)|$$

$$\leq A(1+A), \quad t \in [t_0, t_f],$$ (6.4)

and

$$\left|x^{\Delta^3}(t)\right| \leq \left|\Delta_1 g_1(t, x(t), x^{\Delta}(t))\right|$$

$$+ \left(\int_0^1 \left|\frac{\partial}{\partial y_1} g_1(\sigma(t), x(t) + h\mu(t)x^{\Delta}(t), x^{\Delta}(t))\right| dh\right) |x^{\Delta}(t)|$$

$$+ \left(\int_0^1 \left|\frac{\partial}{\partial y_2} g_1(\sigma(t), x(\sigma(t)), x^{\Delta}(t) + h\mu(t)x^{\Delta^2}(t))\right| dh\right) |x^{\Delta^2}(t)|$$

$$\leq A + A^2 + A(A + A^2) = A(1+A)^2,$$

$$\left|x^{\Delta^4}(t)\right| \leq \left|\Delta_1 g_2(t, x(t), x^{\Delta}(t), x^{\Delta^2}(t))\right|$$

$$+ \left(\int_0^1 \left|\frac{\partial}{\partial y_1} g_2(\sigma(t), x(t) + h\mu(t)x^{\Delta}(t), x^{\Delta}(t), x^{\Delta^2}(t))\right| dh\right) |x^{\Delta}(t)|$$

$$+ \left(\int_0^1 \left|\frac{\partial}{\partial y_2} g_2(\sigma(t), x(\sigma(t)), x^{\Delta}(t) + h\mu(t)x^{\Delta^2}(t), x^{\Delta^2}(t))\right| dh\right) |x^{\Delta^2}(t)|$$

$$+ \left(\int_0^1 \left|\frac{\partial}{\partial y_3} g_2(\sigma(t), x(\sigma(t)), x^{\Delta}(\sigma(t)), x^{\Delta^2}(t) + h\mu(t)x^{\Delta^3}(t))\right| dh\right) |x^{\Delta^3}(t)|$$

$$\leq A + A^2 + A(A + A^2) + A((A + A^2) + A(A + A^2))$$
$$= A + A^2 + 2A(A + A^2) + A^2(A + A^2) = A(1+A)^3,$$

as well as

$$\left|x^{\Delta^5}(t)\right| \leq \left|\Delta_1 g_3(t, x(t), x^{\Delta}(t), x^{\Delta^2}(t), x^{\Delta^3}(t))\right|$$

$$+ \left(\int_0^1 \left|\frac{\partial}{\partial y_1} g_3(\sigma(t), x(t) + h\mu(t)x^{\Delta}(t), x^{\Delta}(t), x^{\Delta^2}(t), x^{\Delta^3}(t))\right| dh\right) |x^{\Delta}(t)|$$

$$+ \left(\int_0^1 \left|\frac{\partial}{\partial y_2} g_3(\sigma(t), x(\sigma(t)), x^{\Delta}(t) + h\mu(t)x^{\Delta^2}(t), x^{\Delta^2}(t), x^{\Delta^3}(t))\right| dh\right) |x^{\Delta^2}(t)|$$

$$+ \left(\int_0^1 \left|\frac{\partial}{\partial y_3} g_3(\sigma(t), x(\sigma(t)), x^{\Delta}(\sigma(t)), x^{\Delta^2}(t) + h\mu(t)x^{\Delta^3}(t), x^{\Delta^3}(t))\right| dh\right) |x^{\Delta^3}(t)|$$

$$+ \left(\int_0^1 \left|\frac{\partial}{\partial y_4} g_3(\sigma(t), x(\sigma(t)), x^{\Delta}(\sigma(t)), x^{\Delta^2}(\sigma(t)), x^{\Delta^3}(t) + h\mu(t)x^{\Delta^4}(t))\right| dh\right) |x^{\Delta^4}(t)|$$

$$\leq A + A^2 + A(A + A^2) + A(1 + A)(A + A^2) + A(A + A^2)(1 + A)^2$$
$$= (A + A^2)(1 + A + A + A^2 + A + 2A^2 + A^3) = A(1 + A)^4, \quad t \in [t_0, t_f],$$

so that we deduce

$$\left|x^{\Delta^p}(t)\right| \leq A(1 + A)^{p-1}, \quad \left|x^{\Delta^{p+1}}(t)\right| \leq A(1 + A)^p, \quad t \in [t_0, t_f].$$

The above estimates can be also proved by induction. We will show that

$$\left|x^{\Delta^p}(t)\right| \leq A(1 + A)^{p-1}, \quad t \in [t_0, t_f]. \tag{6.5}$$

For $n = 1$ and $n = 2$, the estimate is shown in (6.3) and (6.4). Assume that the inequality (6.5) holds for any $p \in \mathbb{N}$. Then, for $n = p + 1$, we have

$$\left|x^{\Delta^{p+1}}\right| \leq \left|\Delta_1 g_{p-1}(t, x, x^{\Delta}, \dots, x^{\Delta^{p-1}})\right|$$
$$+ \int_0^1 \left|\frac{\partial}{\partial y_1} g_{p-1}(\sigma(t), x + h\mu(t)x^{\Delta}, x^{\Delta}, \dots, x^{\Delta^{p-1}})\right| dh |x^{\Delta}|$$
$$+ \int_0^1 \left|\frac{\partial}{\partial y_2} g_{p-1}(\sigma(t), x^\sigma, x^{\Delta} + h\mu(t)x^{\Delta^2}, \dots, x^{\Delta^{p-1}})\right| dh |x^{\Delta^2}|$$
$$+ \cdots$$
$$+ \int_0^1 \left|\frac{\partial}{\partial y_p} g_{p-1}(\sigma(t), x^\sigma, x^{\Delta\sigma}, \dots, x^{\Delta^{p-1}} + h\mu(t)x^{\Delta^p})\right| dh |x^{\Delta^p}|,$$

for $t \in [t_0, t_f]$. Now, Condition (H1) and the induction hypothesis imply that

$$\left|x^{\Delta^{p+1}}\right| \leq A + AA + AA(A + 1) + AA(A + 1)^2 + \cdots + AA(A + 1)^{p-1}$$
$$= A(A + 1) + A^2(A + 1)(1 + (A + 1) + (A + 1)^2 + \cdots + (A + 1)^{p-2})$$
$$= A(A + 1) + A^2(A + 1)\frac{1 - (A + 1)^{p-1}}{1 - (A + 1)}$$
$$= A(A + 1) - A(A + 1) + A(A + 1)(A + 1)^{p-1}$$
$$= A(A + 1)^p,$$

which completes the proof. Moreover, for the remainder terms

$$R_q(r) = \int_t^{\rho^q(t+r)} h_q(t + r, \sigma(u))x^{\Delta^{q+1}}(u)\Delta u, \quad q \in \{1, \dots, p\}, \quad t \in [t_0, t_f],$$

employing the estimate (5.2) and the fact that $\rho^q(t + r) - t \leq t + r - t = r$, we get

$$|R_q(r)| \le \int_t^{\rho^q(t+r)} h_q(t+r, \sigma(u)) |x^{\Delta^{q+1}}(u)| \Delta u$$

$$\le \frac{r^q}{q!} A(1+A)^q (\rho^q(t+r) - t)$$

$$\le \frac{r^{q+1}}{q!} A(1+A)^q, \quad q \in \{1, \dots, p\}. \tag{6.6}$$

Therefore,

$$R_q(r) = O(r^{q+1}), \quad q \in \{1, \dots, p\}.$$

Denote

$$e_n^{\Delta^k} = x^{\Delta^k}(t_n) - x_n^{\Delta^k}, \quad k \in \{0, \dots, p-1\}.$$

Taking into account the fact that

$$x_n^{\Delta} = f(t_n, x_n), \quad x^{\Delta}(t_n) = f(t_n, x(t_n)),$$

$$x_n^{\Delta^q} = g_{q-1}(t_n, x_n, x_n^{\Delta}, \dots, x_n^{\Delta^{q-1}}),$$

$$x^{\Delta^q}(t_n) = g_{q-1}(t_n, x(t_n), x^{\Delta}(t_n), \dots, x^{\Delta^{q-1}}(t_n)), \quad q \in \{2, \dots, p\},$$

we have

$$x(t_{n+1}) = x(t_n) + r_{n+1} f(t_n, x(t_n)) + h_2(t_{n+1}, t_n) g_1(t_n, x(t_n), x^{\Delta}(t_n))$$

$$+ \dots + h_p(t_{n+1}, t_n) g_{p-1}(t_n, x(t_n), \dots, x^{\Delta^{p-1}}(t_n)) + R_p(r_{n+1})$$

and

$$x_{n+1} = x_n + r_{n+1} f(t_n, x_n) + h_2(t_{n+1}, t_n) g_1(t_n, x_n, x_n^{\Delta})$$

$$+ \dots + h_p(t_{n+1}, t_n) g_{p-1}(t_n, x_n, \dots, x_n^{\Delta^{p-1}}).$$

Then

$$x(t_{n+1}) - x_{n+1}$$

$$= (x(t_n) - x_n) + r_{n+1}(f(t_n, x(t_n)) - f(t_n, x_n))$$

$$+ h_2(t_{n+1}, t_n)(g_1(t_n, x(t_n), x^{\Delta}(t_n)) - g_1(t_n, x_n, x_n^{\Delta}))$$

$$+ \dots + h_p(t_{n+1}, t_n)(g_{p-1}(t_n, x(t_n), \dots, x^{\Delta^{p-1}}(t_n)) - g_{p-1}(t_n, x_n, \dots, x_n^{\Delta^{p-1}}))$$

$$+ R_p(r_{n+1}).$$

Note that by Condition (H1), the mean value theorem for f and g_k implies that

$$f(t_n, x(t_n)) - f(t_n, x_n) = \frac{\partial f}{\partial y}(t_n, \xi_1^0)(x(t_n) - x_n) = \frac{\partial f}{\partial y}(t_n, \xi_1^0)e_n,$$

where ξ_1^0 is between $x(t_n)$ and x_n, and $\frac{\partial f}{\partial y}$ stands for the partial derivative with respect to the second variable. Also,

$$g_k(t_n, x(t_n), x^\Delta(t_n), \dots, x^{\Delta^k}(t_n)) - g_k(t_n, x_n, x_n^\Delta, \dots, x_n^{\Delta^k})$$

$$= g_k(t_n, x(t_n), x^\Delta(t_n), \dots, x^{\Delta^k}(t_n)) - g_k(t_n, x_n, x^\Delta(t_n), \dots, x^{\Delta^k}(t_n))$$

$$+ g_k(t_n, x_n, x^\Delta(t_n), \dots, x^{\Delta^k}(t_n)) - g_k(t_n, x_n, x_n^\Delta, \dots, x^{\Delta^k}(t_n))$$

$$+ \cdots + g_k(t_n, x_n, x_n^\Delta, \dots, x_n^{\Delta^{k-1}}, x^{\Delta^k}(t_n)) - g_k(t_n, x_n, x_n^\Delta, \dots, x_n^{\Delta^{k-1}}, x_n^{\Delta^k})$$

$$= \frac{\partial}{\partial y_1} g_k(t_n, \xi_1^k, x^\Delta(t_n), \dots, x^{\Delta^k}(t_n))e_n + \frac{\partial}{\partial y_2} g_k(t_n, x_n, \xi_2^k, \dots, x^{\Delta^k}(t_n))e_n^\Delta$$

$$+ \cdots + \frac{\partial}{\partial y_{k+1}} g_k(t_n, x_n, \dots, x_n^{\Delta^{k-1}}, \xi_{k+1}^k)e_n^{\Delta^k}, \quad k \in \{1, \dots, p-1\},$$

where ξ_j^k is between $x^{\Delta^{j-1}}(t_n)$ and $x_n^{\Delta^{j-1}}$, $j \in \{1, \dots, k+1\}$, and $\frac{\partial}{\partial y_j}$ denotes the partial derivative with respect to the $(j+1)$th variable. Consequently,

$$e_{n+1} = e_n + r_{n+1}\frac{\partial f}{\partial y_1}(t_n, \xi_1^0)e_n$$

$$+ h_2(t_{n+1}, t_n)\left(\frac{\partial g_1}{\partial y_1}(t_n, \xi_1^1, x^\Delta(t_n))e_n + \frac{\partial g_1}{\partial y_2}(t_n, x_n, \xi_2^1)e_n^\Delta\right)$$

$$+ \cdots + h_p(t_{n+1}, t_n)\left(\frac{\partial g_{p-1}}{\partial y_1}(t_n, \xi_1^{p-1}, x^\Delta(t_n), \dots, x^{\Delta^{p-1}}(t_n))e_n\right.$$

$$+ \frac{\partial g_{p-1}}{\partial y_2}(t_n, x_n, \xi_2^{p-1}, \dots, x^{\Delta^{p-1}}(t_n))e_n^\Delta$$

$$\left.+ \cdots + \frac{\partial g_{p-1}}{\partial y_p}(t_n, x_n, x_n^\Delta, \dots, \xi_p^{p-1})e_n^{\Delta^{p-1}}\right) + R_p(r_{n+1}).$$

Let $r_{\max} = \max\{r_1, \dots, r_{m+1}\}$. Since $t_f < \infty$, there is a constant $B > 0$ such that

$$\frac{1}{p!}r_{\max}A(1 + A)^p(e^{r_{\max}}A + 1) \le B.$$

Then

$$|e_{n+1}| \le (1 + h_1(t_{n+1}, t_n) + \cdots + h_p(t_{n+1}, t_n))A(|e_n| + |e_n^\Delta| + \cdots + |e_n^{\Delta^{p-1}}|)$$

$$+ |R_p(r_{n+1})|$$

$$\le \left(1 + r_{\max} + \frac{r_{\max}^2}{2!} + \cdots + \frac{r_{\max}^p}{p!}\right)A(|e_n| + |e_n^\Delta| + \cdots + |e_n^{\Delta^{p-1}}|) + |R_p(r_{n+1})|$$

$$\le e^{r_{\max}}A(|e_n| + |e_n^\Delta| + \cdots + |e_n^{\Delta^{p-1}}|) + |R_p(r_{n+1})|.$$

In a similar way, we make the following estimates:

$$\left|e_{n+1}^{\Delta}\right| \leq e^{r_{max}} A\left(\left|e_n^{\Delta}\right| + \cdots + \left|e_n^{\Delta^{p-1}}\right|\right) + \left|R_{p-1}(r_{n+1})\right|,$$
$$\left|e_{n+1}^{\Delta^2}\right| \leq e^{r_{max}} A\left(\left|e_n^{\Delta^2}\right| + \cdots + \left|e_n^{\Delta^{p-1}}\right|\right) + \left|R_{p-2}(r_{n+1})\right|,$$
$$\cdots$$
$$\left|e_n^{\Delta^{p-2}}\right| \leq e^{r_{max}} A\left|e_n^{\Delta^{p-1}}\right| + \left|R_2(r_{n+1})\right|,$$
$$\left|e_{n+1}^{\Delta^{p-1}}\right| \leq \left|R_1(r_{n+1})\right|.$$

Let

$$B_n = |e_n| + \left|e_n^{\Delta}\right| + \cdots + \left|e_n^{\Delta^{p-1}}\right|.$$

Then

$$B_{n+1} \leq p e^{r_{max}} A B_n + \left|R_1(r_{n+1})\right| + \cdots + \left|R_p(r_{n+1})\right|.$$

Observe that from (6.6) we get

$$\left|R_1(r_{n+1})\right| + \cdots + \left|R_p(r_{n+1})\right| \leq r_{max}^2 A(1 + A) + \frac{r_{max}^3}{2!} A(1 + A)^2 + \cdots + \frac{r_{max}^{p+1}}{p!} A(1 + A)^p$$

$$\leq r_{max}^2 A(1 + A)^p \left(1 + \frac{r_{max}}{2!} + \cdots + \frac{r_{max}^{p-1}}{p!}\right)$$

$$\leq r_{max}^2 A(1 + A)^p \left(1 + r_{max} + \frac{r_{max}^2}{2!} + \cdots + \frac{r_{max}^{p-1}}{(p-1)!}\right)$$

$$\leq r_{max}^2 A(1 + A)^p e^{r_{max}}.$$

Thus,

$$B_{n+1} \leq p e^{r_{max}} A B_n + r_{max}^2 A(1 + A)^p e^{r_{max}}$$
$$\leq p e^{r_{max}} A\left(p e^{r_{max}} A B_{n-1} + r_{max}^2 A(1 + A)^p e^{r_{max}}\right) + r_{max}^2 A(1 + A)^p e^{r_{max}}$$
$$= \left(p e^{r_{max}} A\right)^2 B_{n-1} + \left(p e^{r_{max}} A + 1\right) r_{max}^2 A(1 + A)^p e^{r_{max}}$$
$$\leq \cdots$$
$$\leq \left(p e^{r_{max}} A\right)^{n+1} B_0 + \left(\left(p e^{r_{max}} A\right)^n + \cdots + p e^{r_{max}} A + 1\right) r_{max}^2 A(1 + A)^p e^{r_{max}}$$
$$\leq r_{max}^2 A(1 + A)^p e^{r_{max}} \sum_{j=0}^{\infty} \left(p e^{r_{max}} A\right)^j$$
$$\leq r_{max}^2 A(1 + A)^p e^{t_f - t_0} \sum_{j=0}^{\infty} \left(p e^{t_f - t_0} A\right)^j$$
$$= \frac{1}{1 - p e^{t_f - t_0} A} r_{max}^2 A(1 + A)^p e^{t_f - t_0}.$$

In the last inequality we have used the fact that $B_0 = 0$ and $r_{max} \le t_f - t_0$. Consequently,

$$|e_n| + |e_n^\Delta| + \cdots + |e_n^{\Delta^{p-1}}| = O(r_{max}^2).$$

6.3 The 2-step Adams–Bashforth method – AB(2)

In this section we consider the special case of the Taylor series method of order p, which in the case of $\mathbb{T} = \mathbb{R}$ reduces to the numerical method known as the 2-step Adams–Bashforth method. We shall call this method the 2-step Adams–Bashforth method on time scales.

Consider again the IVP (6.1). Suppose that $r, l > 0$, $t, t + r, t - l \in [t_0, t_f]$, $\rho^2(t + r), \rho(t - l) \in [t_0, t_f]$. Applying the second order Taylor formula, we compute

$$x(t + r) = x(t) + h_1(t + r, t)x^\Delta(t) + h_2(t + r, t)x^{\Delta^2}(t) + R_2(r) \tag{6.7}$$

and applying the first order Taylor formula, we get

$$x^\Delta(t - l) = x^\Delta(t) + h_1(t - l, t)x^{\Delta^2}(t) + R_1(l) = x^\Delta(t) - lx^{\Delta^2}(t) + R_1(l),$$

whereupon

$$x^{\Delta^2}(t) = \frac{1}{l}(x^\Delta(t) - x^\Delta(t - l)) + \frac{1}{l}R_1(l).$$

We put this expression into (6.7) and find

$$x(t + r) = x(t) + rx^\Delta(t) + \frac{h_2(t + r, t)}{l}(x^\Delta(t) - x^\Delta(t - l) + R_1(l)) + R_2(r)$$

$$= x(t) + rf(t, x(t)) + \frac{h_2(t + r, t)}{l}(f(t, x(t)) - f(t - l, x(t - l)))$$

$$+ \frac{h_2(t + r, t)}{l}R_1(l) + R_2(r). \tag{6.8}$$

Assume that $t_0 < t_1 < \cdots < t_{m+1} = t_f$ is a partition of the interval $[t_0, t_f]$ such that $t_{n+1} = t_n + r_{n+1} \in \mathbb{T}$, $r_{n+1} > 0$, $n \in \{0, \ldots, m\}$. Taking $t = t_n$, $r = r_{n+1}$, $l = r_n$ in (6.8), we obtain

$$x(t_{n+1}) = x(t_n) + r_{n+1}f(t_n, x(t_n)) + \frac{h_2(t_{n+1}, t_n)}{r_n}(f(t_n, x(t_n)) - f(t_{n-1}, x(t_{n-1})))$$

$$+ \frac{h_2(t_{n+1}, t_n)}{r_n}R_1(r_n) + R_2(r_{n+1}).$$

Let $x_n = x(t_n)$, $f_n = f(t_n, x(t_n))$. Then, neglecting the remainder terms, we arrive at the 2-step Adams–Bashforth method (AB(2) method).

$$x_{n+1} = x_n + r_{n+1}f_n + \frac{h_2(t_{n+1}, t_n)}{r_n}(f_n - f_{n-1}),$$

or

$$x_{n+1} = x_n + \left(r_{n+1} + \frac{h_2(t_{n+1}, t_n)}{r_n}\right)f_n - \frac{h_2(t_{n+1}, t_n)}{r_n}f_{n-1}. \tag{6.9}$$

Remark 6.1.

1. Note that the 2-step Adams–Bashforth method (6.9) is of order $(1 + O(r_n))O(r_{n+1}^2)$.

2. If $\mathbb{T} = \mathbb{R}$ and $r_n = h$ is constant, then we have $h_2(t_{n+1}, t_n) = \frac{(t_{n+1}-t_n)^2}{2} = \frac{h^2}{2}$ and hence (6.9) takes the form

$$x_{n+1} = x_n + \left(h + \frac{h}{2}\right)f_n - \frac{h}{2}f_{n-1} = x_n + \frac{3h}{2}f_n - \frac{h}{2}f_{n-1},$$

which is the classical 2-step Adams–Bashforth method.

3. The initial condition $x(t_0) = x_0$ provides the first term of the sequence $\{x_n\}$, but one needs the second term x_1 in order to compute the following terms of the sequence. For the computation of x_1, one can use the Euler method on time scales given in Chapter 4 or the trapezoid rule on time scales given in Chapter 5.

6.4 Numerical examples

Below, we apply the method to specific examples of initial value problems associated with nonlinear dynamic equations.

Example 6.2. As a first example, we consider the initial value problem known as the Beverton–Holt model. This model is a population growth model which was initially introduced by R. J. H. Beverton and S. J. Holt to describe the fish population [10]. It has been studied as a continuous and discrete model, that is, both as a differential and a difference equation. Here we consider the unification of Beverton–Holt model as a dynamic equation. In this particular example, we take the time scale as the set of nonnegative integers and get

$$x^\Delta(t) = \frac{\alpha x(t)}{1 + \beta x(t)}, \quad x(0) = x_0, \quad t > 0, \tag{6.10}$$

where α, β are real numbers. Take $\mathbb{T} = \mathbb{N}_0$ and $[t_0, t_f] = [0, 20]$. The monomial h_2 on this time scale has the form

$$h_2(t, s) = \frac{(t - s)(t - s - 1)}{2}, \quad t, s \in \mathbb{T}.$$

If we take constant step size $r_n = h$, then $m = \frac{20}{h}$ and $t_n = nh$ for $n \in \{0, \dots, m\}$. In this case, the AB (2) formula (6.9) takes the form

$$x_{n+1} = x_n + \left(h + \frac{h(h-1)}{2h} \right) \frac{\alpha x_n}{1 + \beta x_n} - \frac{h(h-1)}{2h} \frac{\alpha x_{n-1}}{1 + \beta x_{n-1}}$$
$$= x_n + \frac{3h-1}{2} \frac{\alpha x_n}{1 + \beta x_n} - \frac{h-1}{2} \frac{\alpha x_{n-1}}{1 + \beta x_{n-1}}.$$

Starting with $x_0 = x(0)$, we use the Euler method introduced in Chapter 4 to compute x_1, which gives

$$x_1 = x_0 + h \frac{\alpha x_0}{1 + \beta x_0},$$

and then compute the sequence x_n, $n \in \{2, \ldots, m\}$ by using the AB(2) method.

On the other hand, it is easy to see that the exact solution of the problem can be obtained by writing the dynamic equation in (6.10) as a difference equation, that is,

$$x_0 = x(0),$$
$$x_{n+1} = x_n + \frac{\alpha x_n}{1 + \beta x_n}, \quad n \in \{0, \ldots, 19\}.$$

The approximate and exact solutions are compared in Tables 6.1 and 6.2. In both tables the exact solution is denoted by $x^{(e)}$ and approximate solutions for $h = 1$ and $h = 2$ by $x^{(1)}$ and $x^{(2)}$, respectively. When $h = 1$, the approximate solution is the same as the exact solution. However, for $h = 2$ an error is observed.

Table 6.1: The values of the exact solution $x^{(e)}$ and approximate solutions $x^{(1)}$ and $x^{(2)}$ for $x(0) = 1$, $\alpha = 1.5$, and $\beta = 0.75$.

t	$x^{(e)}$	$x^{(1)}$	$x^{(2)}$
0.00	1.00000000	1.00000000	1.00000000
1.00	1.85714286	1.85714286	
2.00	3.02132196	3.02132196	2.71428571
3.00	4.40895049	4.40895049	
4.00	5.94455919	5.94455919	5.63865546
5.00	7.57815275	7.57815275	
6.00	9.27891347	9.27891347	9.01185991
7.00	11.02763147	11.02763147	
8.00	12.81189858	12.81189858	12.55867971
9.00	14.62337804	14.62337804	
10.00	16.45625923	16.45625923	16.20767199
11.00	18.30635885	18.30635885	
12.00	20.17057940	20.17057940	19.92358857
13.00	22.04657096	22.04657096	
14.00	23.93251289	23.93251289	23.68597760
15.00	25.82696857	25.82696857	
16.00	27.72878606	27.72878606	27.48224143
17.00	29.63702858	29.63702858	
18.00	31.55092475	31.55092475	31.30417723
19.00	33.46983226	33.46983226	
20.00	35.39321087	35.39321087	35.14618456

Table 6.2: The values of the exact solution $x^{(e)}$ and approximate solutions $x^{(1)}$ and $x^{(2)}$ for $x(0) = 2$, $\alpha = 3$, and $\beta = 1$.

t	$x^{(e)}$	$x^{(1)}$	$x^{(2)}$
0.00	2.00000000	2.00000000	2.00000000
1.00	4.00000000	4.00000000	
2.00	6.40000000	6.40000000	6.00000000
3.00	8.99459459	8.99459459	
4.00	11.69443234	11.69443234	11.42857143
5.00	14.45810827	14.45810827	
6.00	17.26403536	17.26403536	17.03940887
7.00	20.09977812	20.09977812	
8.00	22.95759653	22.95759653	22.74434210
9.00	25.83237529	25.83237529	
10.00	28.72057006	28.72057006	28.51162867
11.00	31.61962987	31.61962987	
12.00	34.52766071	34.52766071	34.32066450
13.00	37.44321946	37.44321946	
14.00	40.36518229	40.36518229	40.15915164
15.00	43.29265753	43.29265753	
16.00	46.22492621	46.22492621	46.01940020
17.00	49.16140044	49.16140044	
18.00	52.10159349	52.10159349	51.89633547
19.00	55.04509801	55.04509801	
20.00	57.99156969	57.99156969	57.78645043

The solutions computed with the AB(2) method and the exact solutions for different choices of x_0, α, β, and h are also given in Figures 6.1, 6.2, 6.3 and 6.4.

In the second example we consider the initial value problem in Example 5.8. We aim to compare the two methods used to obtain the approximate solution.

Example 6.3. Consider the initial value problem

$$x^{\Delta}(t) = \frac{1}{1 + t^2} + \frac{t}{1 + (x(t))^2}, \quad x(0) = x_0, \tag{6.11}$$

which is solved by the trapezoid rule in Example 5.8. We take again $\mathbb{T} = a\mathbb{N}_0$ for some $a > 0$ and $[t_0, t_f] = [0, b]$. The monomial h_2 on this time scale has the form $h_2(t, s) = \frac{(t-s)(t-s-a)}{2}$. If we take constant step size $r_n = h$, then $m = \frac{b}{h}$ and $t_n = nh$ for $n \in \{0, \ldots, m\}$. In this case the AB(2) formula (6.9) takes the form

$$x_{n+1} = x_n + \frac{3h - a}{2}\left(\frac{1}{1 + (t_n)^2} + \frac{t_n}{1 + (x_n)^2}\right) - \frac{h - a}{2}\left(\frac{1}{1 + (t_{n-1})^2} + \frac{t_{n-1}}{1 + (x_{n-1})^2}\right).$$

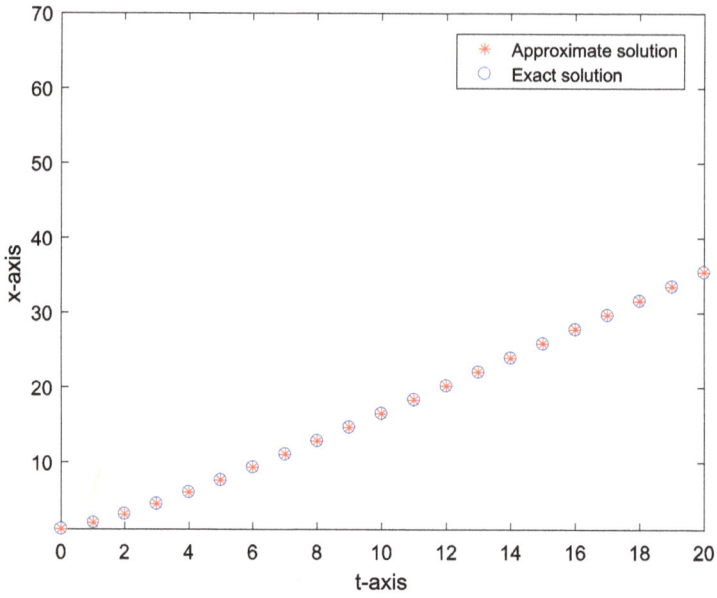

Figure 6.1: Approximate and exact values of the solution with $\alpha = 1.5$, $\beta = 0.75$, $x_0 = 1$, and $h = 1$.

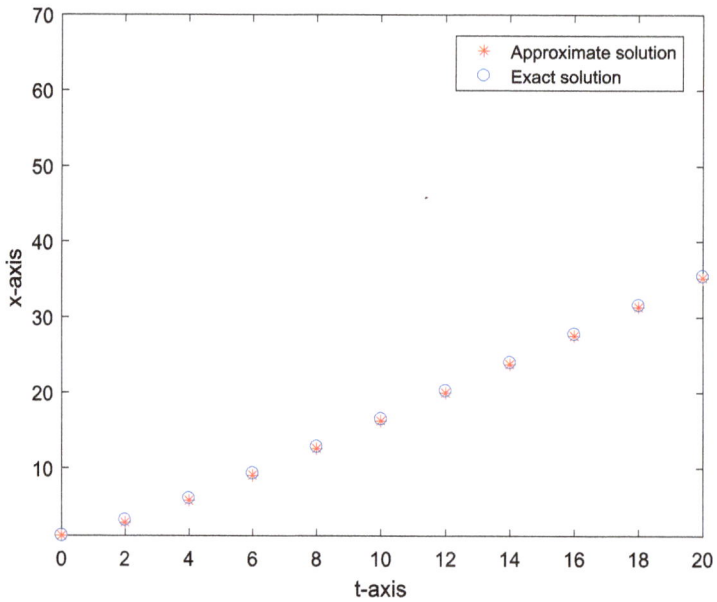

Figure 6.2: Approximate and exact values of the solution with $\alpha = 1.5$, $\beta = 0.75$, $x_0 = 1$, and $h = 2$.

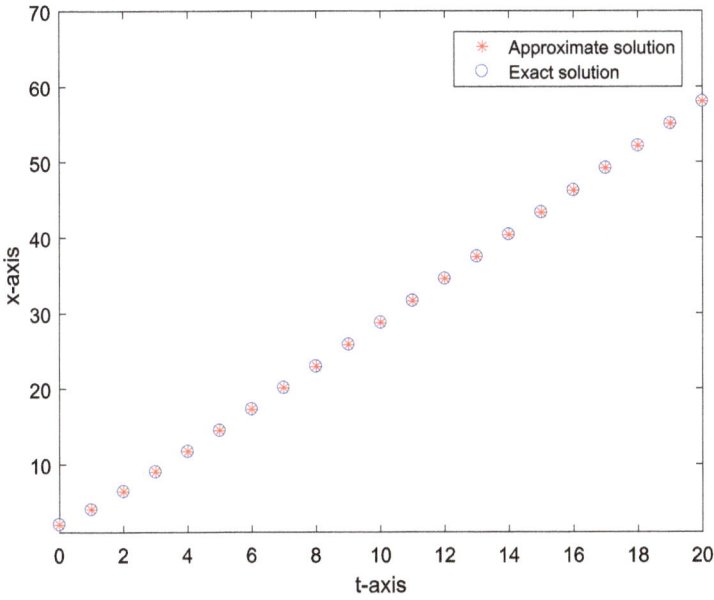

Figure 6.3: Approximate and exact values of the solution with $\alpha = 3$, $\beta = 1$, $x_0 = 2$, and $h = 1$.

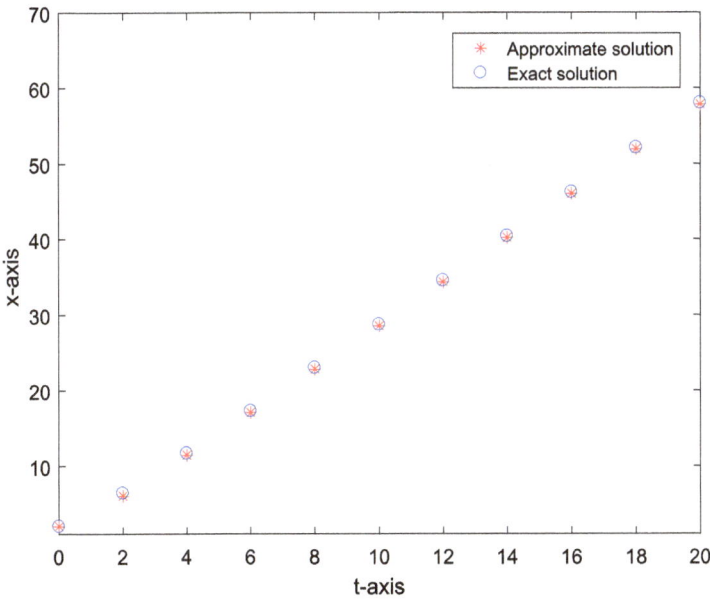

Figure 6.4: Approximate and exact values of the solution with $\alpha = 3$, $\beta = 1$, $x_0 = 2$, and $h = 2$.

Starting with $x_0 = x(0)$, we use the Euler method introduced in Chapter 4 to compute x_1, which gives

$$x_1 = x_0 + h\left(\frac{1}{1+(t_0)^2} + \frac{t_0}{1+(x_0)^2}\right),$$

and then compute the sequence x_n, $n \in \{2, \ldots, m\}$ by using the AB(2) method.

From the discrete structure of the time scale $a\mathbb{N}_0$, the dynamic equation in (6.11) can be written as a difference equation, that is,

$$x_0 = x(0),$$

$$x_{n+1} = x_n + a\left(\frac{1}{1+(an)^2} + \frac{an}{1+(x_n)^2}\right), \quad n \in \{0, \ldots, m\},$$

and hence can be solved analytically on the interval $[0, b]$.

The values of the parameters a, b, x_0, and h are chosen as the same parameters used in Example 5.8. The approximate and exact solutions for $a = 0.5$, $b = 20$, $h = 0.5, 1, 2$, and the initial value $x(0) = 1$ are listed in Table 6.3. In Table 6.4, the approximate and exact solutions for $a = 0.2$, $b = 8$, $h = 0.4, 0.8$, and the initial value $x(0) = 3$ are given.

Table 6.3: The values of the exact solution $x^{(e)}$ and approximate solutions $x^{(1)}$, $x^{(2)}$, and $x^{(3)}$ for $h = 0.5, 1, 2$, $b = 20$, and $x(0) = 1$.

t	$x^{(e)}$	$x^{(1)}$	$x^{(2)}$	$x^{(3)}$
0.00	1.00000000	1.00000000	1.00000000	1.00000000
1.00	1.97692308	1.97692308	1.92799782	
2.00	2.59940190	2.59940190	2.60659764	2.59940190
3.00	3.03618334	3.03618334	3.00140783	
4.00	3.42352292	3.42352292	3.38694224	3.10844893
5.00	3.79402385	3.79402385	3.76145298	
6.00	4.15528317	4.15528317	4.12722174	3.95849454
7.00	4.50894103	4.50894103	4.48475593	
8.00	4.85529463	4.85529463	4.83425866	4.69716122
9.00	5.19444102	5.19444102	5.17594188	
10.00	5.52652984	5.52652984	5.51008813	5.40314291
11.00	5.85178922	5.85178922	5.83703562	
12.00	6.17050243	6.17050243	6.15715160	6.06945724
13.00	6.48298140	6.48298140	6.47081061	
14.00	6.78954686	6.78954686	6.77838010	6.70473486
15.00	7.09051545	7.09051545	7.08021180	
16.00	7.38619205	7.38619205	7.37663730	7.31347165
17.00	7.67686568	7.67686568	7.66796615	
18.00	7.96280759	7.96280759	7.95448563	7.89940697
19.00	8.24427079	8.24427079	8.23646141	
20.00	8.52149045	8.52149045	8.51413872	8.46546300

Table 6.4: The values of the exact solution $x^{(e)}$ and approximate solutions $x^{(1)}$ and $x^{(2)}$, for h = 0.4, 0.8, $x(0)$ = 3, and b = 8.

t	$x^{(e)}$	$x^{(1)}$	$x^{(2)}$
0.00	3.00000000	3.00000000	3.00000000
0.40	3.39586641	3.39586641	
0.80	3.73043213	3.74286015	3.73043213
1.20	3.97566995	3.98499013	
1.60	4.15541211	4.15914713	4.16016040
2.00	4.29534158	4.29522344	
2.40	4.41218246	4.40980634	4.36627079
2.80	4.51582844	4.51217756	
3.20	4.61210777	4.60775159	4.55008030
3.60	4.70451328	4.69978494	
4.00	4.79517350	4.79027335	4.72986961
4.40	4.88539625	4.88044801	
4.80	4.97598082	4.97106244	4.91191419
5.20	5.06740357	5.06256458	
5.60	5.15993195	5.15520366	5.09881334
6.00	5.25369677	5.24909844	
6.40	5.34873916	5.34428194	5.29118354
6.80	5.44504180	5.44073127	
7.20	5.54254995	5.53838787	5.48868810
7.60	5.64118581	5.63717128	
8.00	5.74085838	5.73698868	5.69059353

The solutions computed with the AB(2) method and the exact solutions for the two sets of choices of the parameters a, b, x_0, and h are given in Figures 6.5, 6.6, 6.7, 6.8 and 6.9.

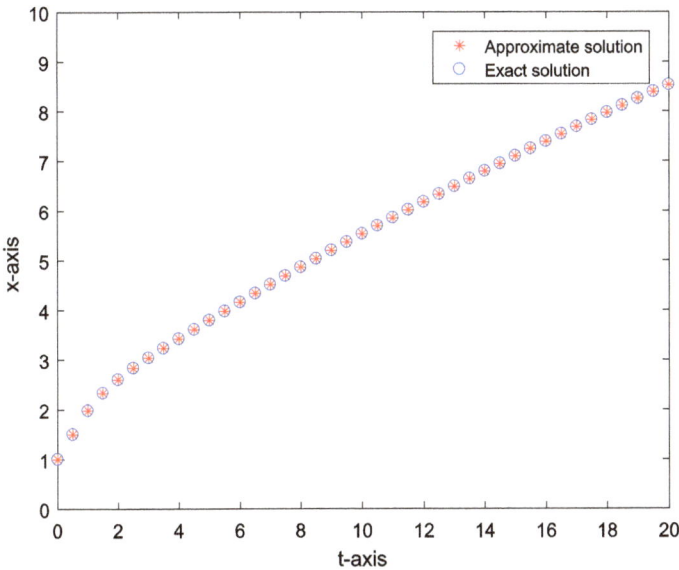

Figure 6.5: Computed and exact values of the solution with a = 0.5, b = 20, x_0 = 1, and h = 0.5.

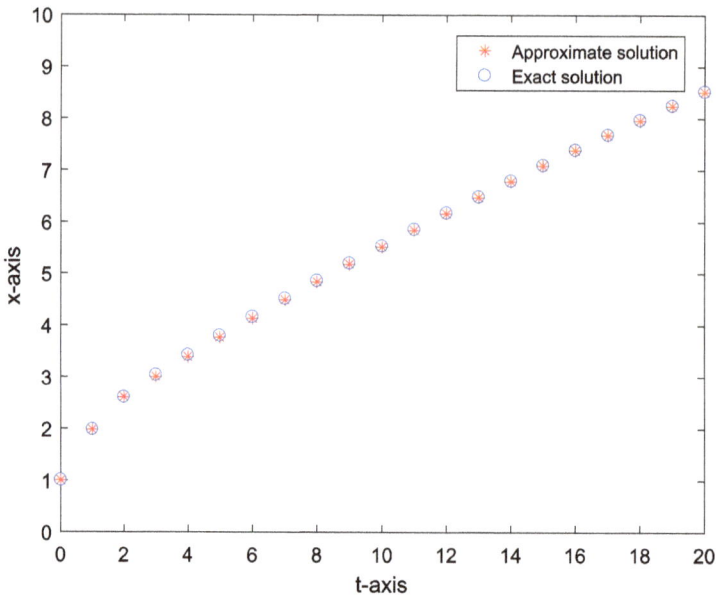

Figure 6.6: Computed and exact values of the solution with $a = 0.5$, $b = 20$, $x_0 = 1$, and $h = 1$.

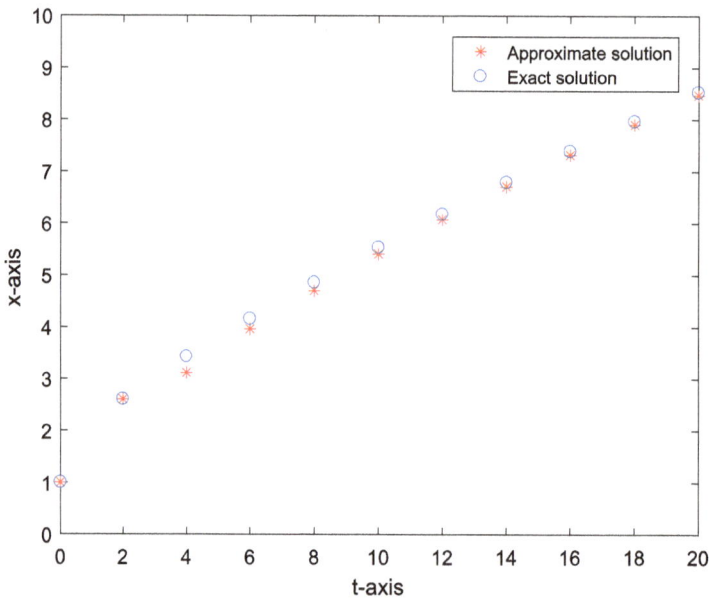

Figure 6.7: Computed and exact values of the solution with $a = 0.5$, $b = 20$, $x_0 = 1$, and $h = 2$.

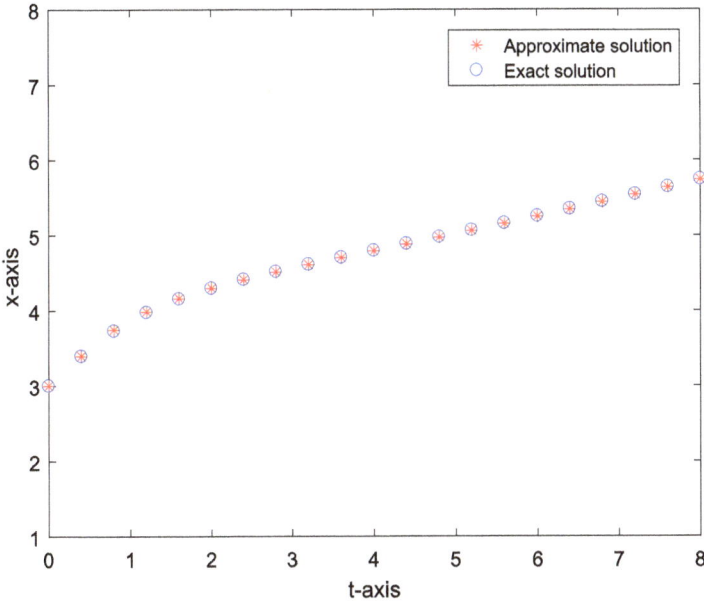

Figure 6.8: Computed and exact values of the solution with $a = 0.2$, $b = 8$, $x_0 = 3$, and $h = 0.4$.

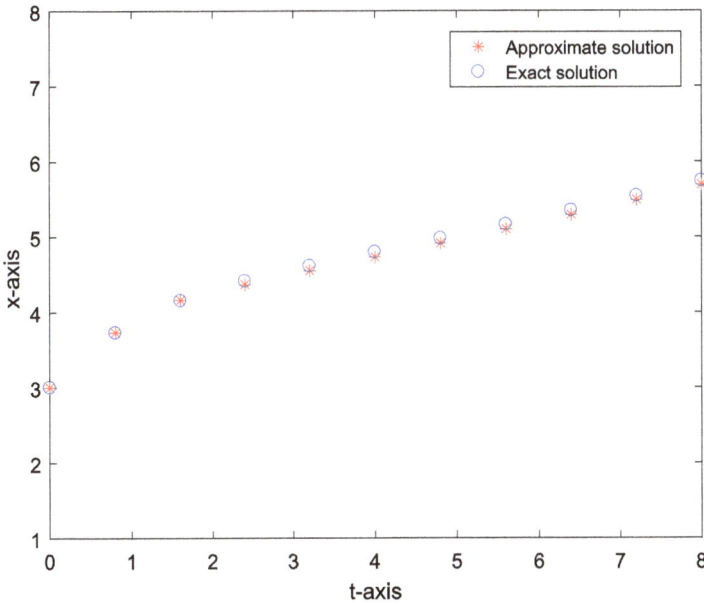

Figure 6.9: Computed and exact values of the solution with $a = 0.2$, $b = 8$, $x_0 = 3$, and $h = 0.8$.

The figures show that there is no significant difference between the exact and computed solutions because *h* is small.

6.5 Advanced practical problems

Problem 6.4. Let $\mathbb{T} = 2^{\mathbb{N}_0}$. Consider the IVP

$$
\begin{cases}
x^{\Delta}(t) = 1 + \frac{1+x(t)}{1+(x(t))^2}, & t > 1, \\
x(1) = 1
\end{cases}
$$

and assume that

$$
a = t_0 = 1, \quad t_1 = 4, \quad t_2 = 16, \quad t_3 = 64, \quad t_4 = b = 216.
$$

Write the 2-step Adams–Bashforth formula.

Problem 6.5. Let

$$
\mathbb{T} = \left\{ 0, \frac{1}{8}, \frac{1}{7}, \frac{1}{6}, \frac{1}{5}, \frac{1}{4}, 1, \frac{4}{3}, \frac{7}{3}, \frac{8}{3}, 3, 7, 11 \right\}.
$$

Consider the IVP

$$
\begin{cases}
x^{\Delta}(t) = (1 + x(t))^3, & t > 0, \\
x(0) = 0.
\end{cases}
$$

Write the 2-step Adams–Bashforth formula.

Problem 6.6. Let $\mathbb{T} = \mathbb{Z}$. Apply AB(2) method for the following IVP:

$$
\begin{cases}
x^{\Delta}(t) = \sin_1(t, 0) + (x(t))^2, & t > 0, \\
x(0) = 1,
\end{cases}
$$

where

$$
a = t_0 = 0, \quad t_1 = 2, \quad t_2 = 4, \quad t_3 = 5, \quad t_4 = b = 6.
$$

Problem 6.7. Let

$$
\mathbb{T} = \left\{ 0, \frac{1}{3}, \frac{2}{3}, 1, \frac{3}{2}, 2, \frac{17}{8}, \frac{19}{8}, \frac{21}{8}, \frac{11}{4}, 3 \right\}.
$$

Apply AB(2) method for the following IVP:

$$\begin{cases} x^{\Delta}(t) = 1 + 3(x(t))^2, & t > 0, \\ x(0) = 1, \end{cases}$$

where

$$a = t_0 = 0, \quad t_1 = \frac{1}{2}, \quad t_2 = 1, \quad t_3 = 2, \quad t_4 = \frac{21}{8}, \quad t_5 = b = 3.$$

7 Linear multistep methods – LMMs

In the previous chapters, the effectiveness of $TS(p)$ methods was shown. For order $p > 1$, these methods have a disadvantage in that they require the right-hand side of the dynamic equation to be differentiable a number of times. This often rules out their use in the real-world applications. The families of linear multistep methods (LMMs) achieve higher order by exploiting x and x^Δ that were computed at the previous k-steps and combining them to generate an approximation of the next step. We begin with two-step methods to describe the strategy.

Suppose that $\mathbb{T}$ is a time scale with forward jump operator σ and delta differentiation operator Δ. Let also, $a, b \in \mathbb{T}$, $a < b$, and $t_j \in [a, b] \subset \mathbb{T}$, $j \in \{0, 1, \ldots, m\}$, so that

$$a = t_0 < t_1 < \cdots < t_m = b.$$

Consider the IVP

$$\begin{cases} x^\Delta(t) = f(t, x(t)), & t \in [a, b], \\ x(a) = x_0, \end{cases} \tag{7.1}$$

where $x_0 \in \mathbb{R}$. Set $r_0 = r_{m+1} = 0$ and

$$t_j = t_{j-1} + r_j, \quad j \in \{1, \ldots, m\}, \quad r = \max_{j \in \{1, \ldots, m\}} r_j.$$

7.1 Two-step methods

For a delta differentiable function z, we need to find constants

$$\alpha_{0j}, \quad \alpha_{1j}, \quad \alpha_{2j}, \quad \beta_{0j}, \quad \beta_{1j}, \quad \beta_{2j}, \quad \gamma_{0j}, \quad \gamma_{1j}, \quad \gamma_{2j},$$

so that

$$\begin{aligned}
&z(t_j + r_{j+1} + r_{j+2}) + \alpha_{1j}z(t_j + r_{j+1}) + \alpha_{0j}z(t_j) \\
&= r_{j+2}(\beta_{2j}z^\Delta(t_j + r_{j+1} + r_{j+2}) + \beta_{1j}z^\Delta(t_j + r_{j+1}) + \beta_{0j}z^\Delta(t_j)) \\
&\quad + r_{j+1}(\gamma_{2j}z^\Delta(t_j + r_{j+1} + r_{j+2}) + \gamma_{1j}z^\Delta(t_j + r_{j+1}) + \gamma_{0j}z^\Delta(t_j)) \\
&\quad + O(r^{p+1}), \quad j \in \{0, \ldots, m-2\},
\end{aligned}$$

where p might be specified in some cases, or one might try to make p as large as possible in others. We choose $z = x$, where x is a solution of the IVP (7.1) and, dropping the $O(r^{p+1})$, we find

$$\begin{aligned}
&x(t_j + r_{j+1} + r_{j+2}) + \alpha_{1j}x(t_j + r_{j+1}) + \alpha_{0j}x(t_j) \\
&= r_{j+2}(\beta_{2j}f(t_j + r_{j+1} + r_{j+2}, x(t_j + r_{j+1} + r_{j+2}))
\end{aligned}$$

https://doi.org/10.1515/9783110787320-007

$$+ \beta_{1j}f(t_j + r_{j+1}, x(t_j + r_{j+1})) + \beta_{0j}f(t_j, x(t_j)))$$
$$+ r_{j+1}(\gamma_{2j}f(t_j + r_{j+1} + r_{j+2}, x(t_j + r_{j+1} + r_{j+2}))$$
$$+ \gamma_{1j}f(t_j + r_{j+1}, x(t_j + r_{j+1})) + \gamma_{0j}f(t_j, x(t_j))), \quad j \in \{0, \dots, m-2\}.$$

Set

$$x_j = x(t_j), \quad f_j = f(t_j, x(t_j)), \quad j \in \{0, 1, \dots, m\}.$$

Then

$$x_{j+2} + \alpha_{1j}x_{j+1} + \alpha_{0j}x_j = r_{j+2}(\beta_{2j}f_{j+2} + \beta_{1j}f_{j+1} + \beta_{0j}f_j)$$
$$+ r_{j+1}(\gamma_{2j}f_{j+2} + \gamma_{1j}f_{j+1} + \gamma_{0j}f_j), \quad j \in \{0, \dots, m-2\}, \qquad (7.2)$$

which will be called a two-step method.

Definition 7.1. A two-step method is said to be explicit (or of explicit type) if $\beta_{2j} = \gamma_{2j} = 0$ for any $j \in \{0, \dots, m-2\}$.

Definition 7.2. A two-step method is said to be implicit if $\beta_{2j} \neq 0$ or $\gamma_{2j} \neq 0$ for some $j \in \{0, \dots, m-2\}$.

We may write (7.2) in the following form:

$$x_{j+2} + \alpha_{1j}x_{j+1} + \alpha_{0j}x_j = r_{j+2}(\beta_{2j}x_{j+2}^\Delta + \beta_{1j}x_{j+1}^\Delta + \beta_{0j}x_j^\Delta)$$
$$+ r_{j+1}(\gamma_{2j}x_{j+2}^\Delta + \gamma_{1j}x_{j+1}^\Delta + \gamma_{0j}x_j^\Delta),$$

where $j \in \{0, \dots, m-2\}$.

7.2 Consistency of two-step methods

Below, we discuss the consistency of two-step methods. To start with, we define the linear difference operator associated with a two-step method and its order of consistency.

Definition 7.3. The linear difference operator associated with the two-step method (7.2) is defined for an arbitrary continuously differentiable function z by

$$\mathcal{L}_{r_{j+1}, r_{j+2}} z(t) = z(t + r_{j+1} + r_{j+2}) + \alpha_{1j}z(t + r_{j+1}) + \alpha_{0j}z(t)$$
$$- r_{j+2}(\beta_{2j}z^\Delta(t + r_{j+1} + r_{j+2}) + \beta_{1j}z^\Delta(t + r_{j+1}) + \beta_{0j}z^\Delta(t))$$
$$- r_{j+1}(\gamma_{2j}z^\Delta(t + r_{j+1} + r_{j+2}) + \gamma_{1j}z^\Delta(t + r_{j+1}) + \gamma_{0j}z^\Delta(t)), \qquad (7.3)$$

for any $j \in \{0, \dots, m-2\}$.

Obviously, for any $a, b \in \mathbb{R}$ and any continuously differentiable functions z, w, we have

$$\mathcal{L}_{r_{j+1},r_{j+2}}(az(t) + bw(t)) = a\mathcal{L}_{r_{j+1},r_{j+2}}z(t) + b\mathcal{L}_{r_{j+1},r_{j+2}}w(t),$$

that is, $\mathcal{L}_{r_{j+1},r_{j+2}}$ is indeed a linear operator.

Definition 7.4. A linear difference operator $\mathcal{L}_{r_{j+1},r_{j+2}}$ is said to be consistent of order p if

$$\mathcal{L}_{r_{j+1},r_{j+2}}z(t) = O(r^{p+1}),$$

with $p > 0$ for any smooth function z.

Definition 7.5. A two-step method is said to be consistent if its difference operator $\mathcal{L}_{r_{j+1},r_{j+2}}$ is consistent of order p for some $p > 0$.

Example 7.6. Consider the Adams–Bashforth method. The associated linear difference operator is

$$\mathcal{L}_{r_{n+1},r_{n=2}}z(t) = z(t + r_{n+1} + r_{n+2}) - z(t + r_{n+1})$$
$$- \left(r_{n+2} + \frac{h_2(t + r_{n+1} + r_{n+2}, t + r_{n+1})}{r_{n+1}} \right) z^\Delta(t + r_{n+1})$$
$$+ \frac{h_2(t + r_{n+1} + r_{n+2}, t + r_{n+1})}{r_{n+1}} z^\Delta(t).$$

By the Taylor expansion, we have

$$z(t + r_{n+1} + r_{n+2}) = z(t) + h_1(t + r_{n+1} + r_{n+2}, t)z^\Delta(t)$$
$$+ h_2(t + r_{n+1} + r_{n+2}, t)z^{\Delta^2}(t) + O(r^3),$$
$$z(t + r_{n+1}) = z(t) + h_1(t + r_{n+1}, t)z^\Delta(t) + h_2(t + r_{n+1}, t)z^{\Delta^2}(t) + O(r^3),$$
$$z^\Delta(t + r_n) = z^\Delta(t) + h_1(t + r_n, t)z^{\Delta^2}(t) + O(r^2).$$

Thus,

$$\mathcal{L}_{r_{n+1},r_{n+2}}z(t) = z(t) + h_1(t + r_{n+1} + r_{n+2}, t)z^\Delta(t) + h_2(t + r_{n+1} + r_{n+2}, t)z^{\Delta^2}(t)$$
$$- z(t) - h_1(t + r_{n+1}, t)z^\Delta(t) - h_2(t + r_{n+1}, t)z^{\Delta^2}(t)$$
$$- \left(r_{n+2} + \frac{h_2(t + r_{n+1} + r_{n+2}, t + r_{n+1})}{r_{n+1}} \right)$$
$$\times (z^\Delta(t) + h_1(t + r_n, t)z^{\Delta^2}(t) + O(r^2))$$
$$+ \frac{h_2(t + r_{n+1} + r_{n+2}, t + r_{n+1})}{r_{n+1}} z^\Delta(t) + O(r^3)$$
$$= r_{n+2}z^\Delta(t) + (h_2(t + r_{n+1} + r_{n+2}, t) - h_2(t + r_{n+1}, t))z^{\Delta^2}(t)$$

$$- \left(r_{n+2} + \frac{h_2(t + r_{n+1} + r_{n+2}, t + r_{n+1})}{r_{n+1}} \right) z^{\Delta}(t)$$

$$- h_1(t + r_n, t) \left(r_{n+2} + \frac{h_2(t + r_{n+1} + r_{n+2}, t + r_{n+1})}{r_{n+1}} \right) z^{\Delta^2}(t)$$

$$+ \frac{h_2(t + r_{n+1} + r_{n+2}, t + r_{n+1})}{r_{n+1}} z^{\Delta}(t) + O(r^2) + O(r^3)$$

$$= \left(h_2(t + r_{n+1} + r_{n+2}, t) - h_2(t + r_{n+1}, t) - r_{n+2} h_1(t + r_{n+1}, t) \right.$$

$$\left. - h_1(t + r_n, t) \frac{h_2(t + r_{n+1} + r_{n+2}, t + r_{n+1})}{r_{n+1}} \right) z^{\Delta^2}(t) + O(r^2) + O(r^3).$$

Hence,

$$\left| \mathcal{L}_{r_{n+1}, r_{n+2}} z(t) \right| \le \left(h_2(t + r_{n+1} + r_{n+2}, t) + h_2(t + r_{n+1}, t) - r_{n+2} h_1(t + r_{n+1}, t) \right.$$

$$\left. + h_1(t + r_n, t) \frac{h_2(t + r_{n+1} + r_{n+2}, t + r_{n+1})}{r_{n+1}} \right) z^{\Delta^2}(t) + O(r^2) + O(r^3)$$

$$\le \left(\frac{(r_{n+1} + r_{n+2})^2}{2} + \frac{r_{n+1}^2}{2} + r_n r_{n+2} + r_n \frac{(r_{n+2})^2}{r_{n+1}} \right) \left| z^{\Delta^2}(t) \right|$$

$$+ O(r^2) + O(r^3)$$

and

$$\mathcal{L}_{r_{n+1}, r_{n+2}} z(t) = O(r^2).$$

Therefore, the Adams–Bashforth method is consistent of order $p = 1$.

Exercise 7.7. Let

$$x_{n+2} = x_{n+1} + r_{n+1} h_3(t_n + r_{n+1} + r_{n+2}, t_n) f_{n+1} - 2 h_2(t_n + r_{n+1}, t_n) f_n.$$

Find the associated linear difference operator and determine its order of consistency.

7.3 Construction of two-step methods

Now, we will give the detailed construction of a two-step method and derive conditions for the method to be consistent with the dynamic equation. For a general two-step method, given by (7.2), the associated linear difference operator is given by (7.3) and the right-hand side can be expanded with the help of the Taylor expansion in the following way:

$$z(t + r_{j+1} + r_{j+2}) = z(t) + h_1(t + r_{j+1} + r_{j+2}, t) z^{\Delta}(t) + h_2(t + r_{j+1} + r_{j+2}, t) z^{\Delta^2}(t)$$

$$+ h_3(t + r_{j+1} + r_{j+2}, t) z^{\Delta^3}(t) + \cdots,$$

$$z(t + r_{j+1}) = z(t) + h_1(t + r_{j+1}, t)z^{\Delta}(t) + h_2(t + r_{j+1}, t)z^{\Delta^2}(t)$$
$$+ h_3(t + r_{j+1}, t)z^{\Delta^3}(t) + \cdots,$$

$$z^{\Delta}(t + r_{j+1} + r_{j+2}) = z^{\Delta}(t) + h_1(t + r_{j+1} + r_{j+2}, t)z^{\Delta^2}(t) + h_2(t + r_{j+1} + r_{j+2}, t)z^{\Delta^3}(t)$$
$$+ h_3(t + r_{j+1} + r_{j+2}, t)z^{\Delta^4}(t) + \cdots,$$

$$z^{\Delta}(t + r_{j+1}) = z^{\Delta}(t) + h_1(t + r_{j+1}, t)z^{\Delta^2}(t) + h_2(t + r_{j+1}, t)z^{\Delta^3}(t)$$
$$+ h_3(t + r_{j+1}, t)z^{\Delta^4}(t) + \cdots.$$

The precise number of terms that have to be retained depends on either the order required or the maximum order possible with the "template" used. Some coefficients in the two-step method may be set zero in order to achieve a method with particular pattern of terms.

First, we focus our attention on the case $p = 1$. We have

$$\mathcal{L}_{r_{j+1}, r_{j+2}} z(t) = z(t) + h_1(t + r_{j+1} + r_{j+2}, t)z^{\Delta}(t) + O(r^2)$$
$$+ \alpha_{1j}(z(t) + h_1(t + r_{j+1}, t)z^{\Delta}(t) + O(r^2)) + \alpha_{0j}z(t)$$
$$- r_{j+2}(\beta_{2j}(z^{\Delta}(t) + O(r)) + \beta_{1j}(z^{\Delta}(t) + O(r)) + \beta_{0j}z^{\Delta}(t))$$
$$- r_{j+1}(\gamma_{2j}(z^{\Delta}(t) + O(r)) + \gamma_{1j}(z^{\Delta}(t) + O(r)) + \gamma_{0j}z^{\Delta}(t))$$
$$= (1 + \alpha_{1j} + \alpha_{0j})z(t)$$
$$+ (h_1(t + r_{j+1} + r_{j+2}, t) + \alpha_{1j}h_1(t + r_{j+1}, t) - r_{j+2}(\beta_{0j} + \beta_{1j} + \beta_{2j})$$
$$- r_{j+1}(\gamma_{2j} + \gamma_{1j} + \gamma_{0j}))z^{\Delta}(t) + O(r^2)$$
$$= (1 + \alpha_{1j} + \alpha_{0j})z(t)$$
$$+ (r_{j+1} + r_{j+2} + \alpha_{1j}r_{j+1} - r_{j+2}(\beta_{0j} + \beta_{1j} + \beta_{2j})$$
$$- r_{j+1}(\gamma_{0j} + \gamma_{1j} + \gamma_{2j}))z^{\Delta}(t) + O(r^2)$$
$$= (1 + \alpha_{1j} + \alpha_{0j})z(t) + (r_{j+1}(1 + \alpha_{1j} - \gamma_{0j} - \gamma_{1j} - \gamma_{2j})$$
$$+ r_{j+2}(1 - \beta_{0j} - \beta_{1j} - \beta_{2j}))z^{\Delta}(t) + O(r^2).$$

Therefore $\mathcal{L}_{r_{j+1}, r_{j+2}} z(t) = O(r^2)$, or the two-step method is consistent of order $p = 1$, if

$$1 + \alpha_{0j} + \alpha_{1j} = 0,$$
$$\beta_{0j} + \beta_{1j} + \beta_{2j} = 1,$$
$$\gamma_{0j} + \gamma_{1j} + \gamma_{2j} = 1 + \alpha_{1j}. \tag{7.4}$$

Definition 7.8. The first, second, and third characteristic polynomials of the two-step method (7.2) are defined to be

$$\rho_{1j}(q) = q^2 + \alpha_{1j}q + \alpha_{0j},$$

$$p_{2j}(q) = \beta_{2j}q^2 + \beta_{1j}q + \beta_{0j},$$
$$p_{3j}(q) = \gamma_{2j}q^2 + \gamma_{1j}q + \gamma_{0j}, \quad q \in \mathbb{R},$$

respectively.

Example 7.9. Consider the following two-step method:

$$x_{j+2} - x_{j+1} + x_j = r_{j+2}(f_{j+2} - f_{j+1}) + r_{j+1}(f_{j+2} + f_{j+1} + f_j).$$

The first, second, and third characteristic polynomials are as follows:

$$p_{1j}(q) = q^2 - q - 1,$$
$$p_{2j}(q) = q^2 - q,$$
$$p_{3j}(q) = q^2 + q + 1.$$

Exercise 7.10. Write the first, second, and third characteristic polynomials of the following two-step method:

$$x_{j+2} + 3x_{j+1} - 4x_j = r_{j+2}(f_{j+2} - f_{j+1} - f_j) + r_{j+1}(f_{j+2} - f_j).$$

Theorem 7.11. *The two-step method (7.2) is consistent with the dynamic equation (7.1) if and only if*

$$p_{ij}(1) = 0, \quad \frac{1}{2}p''_{1j}(1) = p_{2j}(1), \quad p'_{ij}\!\left(\frac{1}{2}\right) = p_{3j}(1). \tag{7.5}$$

Proof. Note that

$$p_{1j}(1) = 1 + \alpha_{0j} + \alpha_{1j}, \quad p'_{1j}(q) = 2q + \alpha_{1j}, \quad p'_{1j}\!\left(\frac{1}{2}\right) = 1 + \alpha_{1j},$$

$$p''_{1j}(q) = 2, \quad \frac{1}{2}p''_{1j}(1) = 1, \quad p_{2j}(1) = \beta_{0j} + \beta_{1j} + \beta_{2j}, \quad p_{3j}(1) = \gamma_{0j} + \gamma_{1j} + \gamma_{2j}.$$

1. Suppose that (7.2) is consistent with the dynamic equation (7.1). Then (7.4) holds and hence (7.5) holds.
2. Suppose that (7.5) holds. Then (7.4) holds.

This completes the proof. □

Example 7.12. Consider the two-step method

$$x_{j+2} - x_j = r_{j+2}(f_{j+2} - f_{j+1} + f_j) + r_{j+1}(-3f_{j+2} + 3f_{j+1} + 2f_j).$$

Here

$$\alpha_{0j} = -1, \quad \alpha_{1j} = 0, \quad \beta_{0j} = 1, \quad \beta_{1j} = -1, \quad \beta_{2j} = 1, \quad \gamma_{0j} = 2, \quad \gamma_{1j} = 3, \quad \gamma_{2j} = -3.$$

Then

$$1 + \alpha_{0j} + \alpha_{1j} = 1 + (-1) + 0 = 0,$$
$$\beta_{0j} + \beta_{1j} + \beta_{2j} = 1 + (-1) + 1 = 1,$$
$$\gamma_{0j} + \gamma_{1j} + \gamma_{2j} = 2 + 3 + (-3) = 2 = 1 + \alpha_{1j}.$$

Therefore (7.4) holds and the considered two-step method is consistent.

Exercise 7.13. Prove that the two-step method

$$x_{j+2} + 3x_{j+1} - 4x_j = r_{j+2}(-4f_{j+2} + 3f_{j+1} + 2f_j) + r_{j+1}(3f_{j+2} + 2f_{j+1} - f_j)$$

is consistent.

Now we will consider the general case $p \geq 1$. We have

$$\mathcal{L}_{r_{j+1},r_{j+2}}z(t)$$
$$= z(t) + h_1(t + r_{j+1} + r_{j+2}, t)z^\Delta(t) + h_2(t + r_{j+1} + r_{j+2}, t)z^{\Delta^2}(t)$$
$$+ \cdots + h_{p+1}(t + r_{j+1} + r_{j+2}, t)z^{\Delta^{p+1}}(t) + O(r^{p+2})$$
$$+ \alpha_{1j}\big(z(t) + h_1(t + r_{j+1}, t)z^\Delta(t) + h_2(t + r_{j+1}, t)z^{\Delta^2}(t)$$
$$+ \cdots + h_{p+1}(t + r_{j+1}, t)z^{\Delta^{p+1}}(t) + O(r^{p+2})\big) + \alpha_{0j}z(t)$$
$$- r_{j+2}\beta_{2j}\big(z^\Delta(t) + h_1(t + r_{j+1} + r_{j+2}, t)z^{\Delta^2}(t) + h_2(t + r_{j+1} + r_{j+2}, t)z^{\Delta^3}(t)$$
$$+ \cdots + h_p(t + r_{j+1} + r_{j+2}, t)z^{\Delta^{p+1}}(t) + O(r^{p+1})\big)$$
$$- r_{j+2}\beta_{1j}\big(z^\Delta(t) + h_1(t + r_{j+1}, t)z^{\Delta^2}(t) + h_2(t + r_{j+1}, t)z^{\Delta^3}(t)$$
$$+ \cdots + h_p(t + r_{j+1}, t)z^{\Delta^{p+1}}(t) + O(r^{p+1})\big) - r_{j+2}\beta_{0j}z^\Delta(t)$$
$$- r_{j+2}\gamma_{2j}\big(z^\Delta(t) + h_1(t + r_{j+1} + r_{j+2}, t)z^{\Delta^2}(t) + h_2(t + r_{j+1} + r_{j+2}, t)z^{\Delta^3}(t)$$
$$+ \cdots + h_p(t + r_{j+1} + r_{j+2}, t)z^{\Delta^{p+1}}(t) + O(r^{p+12})\big)$$
$$- r_{j+2}\gamma_{1j}\big(z^\Delta(t) + h_1(t + r_{j+1}, t)z^{\Delta^2}(t) + h_2(t + r_{j+1}, t)z^{\Delta^3}(t)$$
$$+ \cdots + h_p(t + r_{j+1}, t)z^{\Delta^{p+1}}(t) + O(r^{p+1})\big)$$
$$= (1 + \alpha_{0j} + \alpha_{1j})z(t)$$
$$+ \big(h_1(t + r_{j+1} + r_{j+2}, t) + \alpha_{1j}h_1(t + r_{j+1}, t) - r_{j+2}\beta_{2j} - r_{j+2}\beta_{1j} - r_{j+2}\beta_{0j}$$
$$- r_{j+1}\gamma_{2j} - r_{j+1}\gamma_{1j} - r_{j+1}\gamma_{0j}\big)z^\Delta(t)$$
$$+ \big(h_2(t + r_{j+1} + r_{j+2}, t) + \alpha_{1j}h_2(t + r_{j+1}, t) - r_{j+2}\beta_{2j}h_1(t + r_{j+1} + r_{j+2}, t)$$
$$- r_{j+2}\beta_{1j}h_1(t + r_{j+1}, t) - r_{j+1}\gamma_{2j}h_1(t + r_{j+1} + r_{j+2}, t)$$

$$-r_{j+1}\gamma_{1j}h_1(t + r_{j+1}, t))z^{\Delta^2}(t)$$
$$+ (h_{p+1}(t + r_{j+1} + r_{j+2}, t) + \alpha_{1j}h_{p+1}(t + r_{j+1}, t) - r_{j+2}\beta_{2j}h_p(t + r_{j+1} + r_{j+2}, t)$$
$$- r_{j+2}\beta_{1j}h_p(t + r_{j+1}, t) - r_{j+1}\gamma_{2j}h_p(t + r_{j+1} + r_{j+2}, t)$$
$$- r_{j+1}\gamma_{1j}h_p(t + r_{j+1}, t))z^{\Delta^{p+1}}(t) + O(r^{p+2}).$$

If the coefficients of $h_1, \ldots, h_p$ are equal to zero, then

$$\mathcal{L}_{r_{j+1}, r_{j+2}} z(t) = O(r^{p+1}),$$

i. e., the two-step method is consistent of order p.

7.4 k-Step methods

The most general multistep method has the form

$$x_{j+k} + \alpha_{k-1j}x_{j+k-1} + \alpha_{k-2j}x_{j+k-2} + \cdots + \alpha_{0j}x_j$$
$$= r_{j+k}(\beta_{kkj}f_{j+k} + \beta_{kk-1j}f_{j+k-1} + \cdots + \beta_{k0j}f_j)$$
$$+ r_{j+k-1}(\beta_{k-1kj}f_{j+k} + \beta_{k-1k-1j}f_{j+k-1} + \cdots + \beta_{k-10j}f_j)$$
$$+ \cdots + r_{j+1}(\beta_{1kj}f_{j+k} + \beta_{1k-1j}f_{j+k-1} + \cdots + \beta_{10j}f_j). \tag{7.6}$$

Definition 7.14. The first through the $(k+1)$th characteristic polynomials of the k-step method are defined as

$$\rho_{1j}(q) = q^k + \alpha_{k-1j}q^{k-1} + \alpha_{k-2j}q^{k-2} + \cdots + \alpha_{0j},$$
$$\rho_{2j}(q) = \beta_{kkj}q^k + \beta_{kk-1j}q^{k-1} + \beta_{kk-2j}q^{k-2} + \cdots + \beta_{k0j},$$
$$\cdots$$
$$\rho_{k+1j}(q) = \beta_{1kj}q^k + \beta_{1k-1j}q^{k-1} + \beta_{1k-2j}q^{k-2} + \cdots + \beta_{10j},$$

respectively.

Example 7.15. Consider the 4-step method

$$x_{j+4} - x_{j+3} - x_{j+2} + 2x_{j+1} - 3x_j = r_{j+4}(f_{j+4} - f_{j+3} - f_{j+2}) + r_{j+3}(f_{j+4} + f_j)$$
$$+ r_{j+2}(2f_{j+4} - 3f_{j+3} - f_{j+2} - f_{j+1} - f_j)$$
$$+ r_{j+1}(f_{j+4} + f_{j+3} + f_{j+2} + f_{j+1} + f_j).$$

Then the first through fifth characteristic polynomials are as follows:

$$\rho_{1j}(q) = q^4 - q^3 - q^2 + 2q - 3,$$
$$\rho_{2j}(q) = q^4 - q^3 - q^2,$$

$$\rho_{3j}(q) = -q^4 + 1,$$
$$\rho_{4j}(q) = 2q^4 - 3q^3 - q^2 - q - 1,$$
$$\rho_{5j}(q) = q^4 + q^3 + q^2 + q + 1.$$

Exercise 7.16. Write the first through sixth characteristic polynomials of the following 5-step method:

$$x_{j+5} - x_{j+4} - 2x_{j+3} + 3x_{j+2} - 4x_{j+1} - x_j$$
$$= r_{j+5}(f_{j+5} - f_{j+4} - 2f_{j+3} - 3f_{j+2} - f_{j+1} - f_j) + r_{j+4}(f_{j+1} - f_j)$$
$$+ r_{j+3}(2f_{j+5} + f_{j+4} - f_{j+3} - 2f_{j+2} - 3f_{j+1} - 4f_j)$$
$$+ r_{j+2}(f_{j+5} + 4f_{j+4} - 3f_{j+1} - f_j) + r_{j+1}(f_{j+4} + f_{j+3} - 2f_{j+1} - 4f_j).$$

7.5 Consistency of k-step methods

Definition 7.17. The associated linear difference operator of the k-step method (7.6) is defined as follows:

$$\mathcal{L}_{r_{j+1}, r_{j+2}, \dots, r_{j+k}} z(t)$$
$$= z(t + r_{j+1} + r_{j+2} + \cdots + r_{j+k}) + \alpha_{k-1j} z(t + r_{j+1} + r_{j+2} + \cdots + r_{j+k-1})$$
$$+ \alpha_{k-2j} z(t + r_{j+1} + r_{j+2} + \cdots + r_{j+k-2}) + \cdots + \alpha_{0j} z(t)$$
$$- r_{j+k}(\beta_{kkj} z^{\triangle}(t + r_{j+1} + r_{j+2} + \cdots + r_{j+k})$$
$$+ \beta_{kk-1j} z^{\triangle}(t + r_{j+1} + r_{j+2} + \cdots + r_{j+k-1}) + \cdots + \beta_{k0j} z^{\triangle}(t))$$
$$- r_{j+k-1}(\beta_{k-1kj} z^{\triangle}(t + r_{j+1} + r_{j+2} + \cdots + r_{j+k})$$
$$+ \beta_{k-1k-1j} z^{\triangle}(t + r_{j+1} + r_{j+2} + \cdots + r_{j+k-1}) + \cdots + \beta_{k-10j} z^{\triangle}(t))$$
$$- \cdots - r_{j+1}(\beta_{1kj} z^{\triangle}(t + r_{j+1} + r_{j+2} + \cdots + r_{j+k})$$
$$+ \beta_{1k-1j} z^{\triangle}(t + r_{j+1} + r_{j+2} + \cdots + r f_{j+k-1}) + \cdots + \beta_{10j} z^{\triangle}(t)).$$

Definition 7.18. A linear difference operator $\mathcal{L}_{r_{j+1}, r_{j+2}, \dots, r_{j+k}}$ is said to be consistent of order p if

$$\mathcal{L}_{r_{j+1}, r_{j+2}, \dots, r_{j+k}} z(t) = O(r^{p+1})$$

with $p > 0$ for any smooth function z.

Definition 7.19. A k-step method is said to be consistent if its linear difference operator $\mathcal{L}_{r_{j+1}, r_{j+2}, \dots, r_{j+k}}$ is consistent of order p for some $p > 0$.

Using the Taylor expansion for the right-hand side of $\mathcal{L}_{r_{j+1}, r_{j+2}, \dots, r_{j+k}}$ and collecting the terms in $h_l(\cdot, \cdot)$ of $z, z^\Delta, \dots, z^{\Delta^p}$ and then choosing these terms to be equal to zero, we get that this method has order p. We will illustrate this for the case $p = 1$. We have

$$z(t + r_{j+1} + r_{j+2} + \dots + r_{j+k}) = z(t) + (r_{j+1} + r_{j+2} + \dots + r_{j+k})z^\Delta(t) + O(r^2),$$
$$z(t + r_{j+1} + r_{j+2} + \dots + r_{j+k-1}) = z(t) + (r_{j+1} + r_{j+2} + \dots + r_{j+k-1})z^\Delta(t) + O(r^2),$$
$$z(t + r_{j+1} + r_{j+2} + \dots + r_{j+k-2}) = z(t) + (r_{j+1} + r_{j+2} + \dots + r_{j+k-2})z^\Delta(t) + O(r^2),$$
$$\dots$$
$$z(t + r_{j+1}) = z(t) + r_{j+1}z^\Delta(t) + O(r^2).$$

Then

$$
\begin{aligned}
&\mathcal{L}_{r_{j+1}, r_{j+2}, \dots, r_{j+k}} z(t) \\
&= z(t) + (r_{j+1} + r_{j+2} + \dots + r_{j+k})z^\Delta(t) + O(r^2) \\
&\quad + \alpha_{k-1j}(z(t) + (r_{j+1} + r_{j+2} + \dots + r_{j+k-1})z^\Delta(t) + O(r^2)) \\
&\quad + \alpha_{k-2j}(z(t) + (r_{j+1} + r_{j+2} + \dots + r_{j+k-2})z^\Delta(t) + O(r^2)) \\
&\quad + \dots + \alpha_{0j}z(t) - r_{j+k}(\beta_{kkj}(z^\Delta(t) + O(r)) + \beta_{kk-1j}(z^\Delta(t) + O(r)) + \beta_{k0j}z^\Delta(t)) \\
&\quad - r_{j+k-1}(\beta_{k-1kj}(z^\Delta(t) + O(r)) + \beta_{k-1k-1j}(z^\Delta(t) + O(r)) + \beta_{k-10j}z^\Delta(t)) \\
&\quad - r_{j+k-2}(\beta_{k-2kj}(z^\Delta(t) + O(r)) + \beta_{k-2k-1j}(z^\Delta(t) + O(r)) + \beta_{k-20j}z^\Delta(t)) \\
&\quad + \dots - r_{j+1}(\beta_{1kj}(z^\Delta(t) + O(r)) + \beta_{1k-1j}(z^\Delta(t) + O(r)) + \beta_{10j}z^\Delta(t)) \\
&= (1 + \alpha_{k+-1j} + \alpha_{k-2j} + \dots + \alpha_{kj})z(t) \\
&\quad + r_{j+1}(1 + \alpha_{k-1j} + \alpha_{k-2j} + \dots + \alpha_{1j} - \beta_{1kj} - \beta_{1k-1j} - \dots - \beta_{10j})z^\Delta(t) \\
&\quad + r_{j+2}(1 + \alpha_{k-1j} + \alpha_{k-2j} + \dots + \alpha_{2j} - \beta_{2kj} - \beta_{2k-1j} - \dots - \beta_{20j})z^\Delta(t) \\
&\quad + r_{j+3}(1 + \alpha_{k-1j} + \alpha_{k-2j} + \dots + \alpha_{3j} - \beta_{3kj} - \beta_{3k-1j} - \dots - \beta_{30j})z^\Delta(t) \\
&\quad + \dots + r_{j+k}(1 - \beta_{kkj} - \beta_{kk-1j} - \dots - \beta_{k0j})z^\Delta(t) + O(r^2).
\end{aligned}
$$

Thus,

$$\mathcal{L}_{r_{j+1}, r_{j+2}, \dots, r_{j+k}} z(t) = O(r^2)$$

for any smooth function z if and only if

$$
\begin{aligned}
1 + \alpha_{k-1j} + \alpha_{k-2j} + \dots + \alpha_{0j} &= 0, \\
1 + \alpha_{k-1j} + \alpha_{k-2j} + \dots + \alpha_{1j} &= \beta_{1kj} + \beta_{1k-1j} + \dots + \beta_{10j}, \\
1 + \alpha_{k-1j} + \alpha_{k-2j} + \dots + \alpha_{2j} &= \beta_{2kj} + \beta_{2k-1j} + \dots + \beta_{20j}, \\
&\dots \\
1 + \alpha_{k-1j} &= \beta_{k-1kj} + \beta_{k-1k-1j} + \dots + \beta_{k-10j}, \\
1 &= \beta_{kkj} + \beta_{kk-1j} + \dots + \beta_{k0j}.
\end{aligned}
\tag{7.7}
$$

Example 7.20. Consider the 4-step method

$$x_{j+4} - 7x_{j+3} + 3x_{j+2} + 2x_{j+1} + x_j$$
$$= r_{j+4}(-6f_{j+4} + 3f_{j+3} + f_{j+2} + 2f_{j+1} + f_j)$$
$$+ r_{j+3}(-10f_{j+4} + 3f_{j+3} + 2f_{j+2} - f_{j+1})$$
$$+ r_{j+2}(-8f_{j+4} - f_{j+3} + 3f_{j+2} + 2f_{j+1} + f_j)$$
$$+ r_{j+1}(-4f_{j+4} + f_{j+3} + f_{j+2} + f_{j+1}).$$

Here

$$\alpha_{0j} = 1, \quad \alpha_{1j} = 2, \quad \alpha_{2j} = 3, \quad \alpha_{3j} = -7,$$
$$\beta_{40j} = 1, \quad \beta_{41j} = 2, \quad \beta_{42j} = 1, \quad \beta_{43j} = 3, \quad \beta_{44j} = -6,$$
$$\beta_{30j} = 0, \quad \beta_{31j} = -1, \quad \beta_{32j} = 2, \quad \beta_{33j} = 3, \quad \beta_{34j} = -10,$$
$$\beta_{20j} = 1, \quad \beta_{21j} = 2, \quad \beta_{22j} = 3, \quad \beta_{23j} = -1, \quad \beta_{24j} = -8,$$
$$\beta_{10j} = 0, \quad \beta_{11j} = 1, \quad \beta_{12j} = 1, \quad \beta_{13j} = 1, \quad \beta_{14j} = -4.$$

Hence,

$$1 + \alpha_{0j} + \alpha_{1j} + \alpha_{2j} + \alpha_{3j} + \alpha_{4j} = 1 + 1 + 2 + 3 - 7 = 0,$$
$$1 + \alpha_{1j} + \alpha_{2j} + \alpha_{3j} = 1 + 2 + 3 - 7 = -1,$$
$$\beta_{10j} + \beta_{11j} + \beta_{12j} + \beta_{13j} + \beta_{14j} = 0 + 1 + 1 + 1 - 4 = -1,$$
$$1 + \alpha_{2j} + \alpha_{3j} = 1 + 3 - 7 = -3,$$
$$\beta_{20j} + \beta_{21j} + \beta_{22j} + \beta_{23j} + \beta_{24j} = 1 + 2 + 3 - 1 - 8 = -3,$$
$$1 + \alpha_{3j} = 1 - 7 = -6,$$
$$\beta_{30j} + \beta_{31j} + \beta_{32j} + \beta_{33j} + \beta_{34j} = 0 - 1 + 2 + 3 - 10 = -6,$$
$$\beta_{40j} + \beta_{41j} + \beta_{42j} + \beta_{43j} + \beta_{44j} = 1 + 2 + 1 + 3 - 6 = 1.$$

Therefore the considered 4-step method is consistent of order $p = 1$.

Exercise 7.21. Check if the following 2-step method satisfies the conditions (7.7):

$$x_{j+2} - 4x_{j+1} + 3x_j = r_{j+2}(-3f_{j+2} + 3f_{j+1} + f_j) + r_{j+1}(-6f_{j+2} + 2f_{j+1} + f_j).$$

7.6 Numerical examples

Example 7.22. The first example is an initial value problem associated with the logistic equation studied by Bohner and Peterson in [1]. It is a simple population growth model and on an arbitrary time scale is given as

$$x^{\Delta}(t) = (\alpha \ominus (\alpha x(t)))x(t), \quad x(0) = x_0, \quad t > 0.$$

On $\mathbb{T} = \mathbb{N}_0$, the dynamic equation can be written as

$$x^\Delta = \frac{\alpha(1-x)}{1+\alpha\mu(t)x}x,$$

and we impose an initial condition $x(0) = 2$ and take the interval as $[0, 30]$.

We will apply an explicit two-step method with a constant step size $r_j = r$ for $j = 0, \ldots, n-2$, given as

$$x_{j+2} + 2x_{j+1} - 3x_j = r(-f_{j+1} + 2f_j) + r(2f_{j+1} + f_j),$$

that is, we have $\alpha_1 = 2$, $\alpha_0 = -1$, $\beta_2 = 0$, $\beta_1 = -1$, $\beta_0 = 2$, and $\gamma_2 = 0$, $\gamma_1 = -1$, $\gamma_0 = 1$. It is easy to verify that the consistency condition holds for this two-step method.

When the step size is constant, the method simplifies to

$$x_{j+2} = -2x_{j+1} + 3x_j = r(f_{j+1} + 3f_j).$$

Here $x_0 = x(0) = 2$ and x_1 can be obtained with the Euler method as

$$x_1 = x_0 + r\frac{\alpha(1-x_0)}{1+\alpha r x_0}x_0.$$

The exact solution of the problem is given as

$$x_i^{(e)}(t) = \frac{(1+\alpha)x_{i-1}^{(e)}(t)}{1+\alpha x_{i-1}^{(e)}(t)}.$$

We solve the problem for three cases as follows.

Case 1. Let $t \in [0, 30]_\mathbb{T}$ and $r = 1$, that is, $t_0 = 0$, $t_j = t_{j-1} + 1$, where $j = 1, \ldots, 30$. Hence, the computed sequence of values of the solution x is defined as

$$x_{j+2} = -2x_{j+1} + 3x_j + f_{j+1} + 3f_j,$$

where

$$f_j = \frac{\alpha(1-x_j)}{1+\alpha x_j}x_j,$$

for $j = 1, \ldots, 28$.

Case 2. Let $t \in [0, 30]_\mathbb{T}$ and $r = 2$, that is, $t_0 = 0$, $t_j = t_{j-1} + 2$, where $j = 1, \ldots, 15$. Hence, the computed sequence of values of the solution x is defined as

$$x_{j+2} = -2x_{j+1} + 3x_j + 2(f_{j+1} + 3f_j),$$

where

$$f_j = \frac{\alpha(1-x_j)}{1+2\alpha x_j}x_j,$$

for $j = 1, \ldots, 13$.

Case 3. Let $t \in [0, 30]_{\mathbb{T}}$ and $r = 4$, that is, $t_0 = 0$, $t_j = t_{j-1} + 4$, where $j = 1, \ldots, 8$. Hence, the computed sequence of values of the solution x is defined as

$$x_{j+2} = -2x_{j+1} + 3x_j + 4(f_{j+1} + 3f_j),$$

where

$$f_j = \frac{\alpha(1 - x_j)}{1 + 4\alpha x_j} x_j,$$

for $j = 1, \ldots, 8$.

We denote the computed solution for $r = 1$ by $x^{(1)}$, for $r = 2$ by $x^{(2)}$, and for $r = 4$ by $x^{(4)}$. The exact solution is denoted by $x^{(e)}$. All calculations are done with *MATLAB*. The values of the approximate and the exact solution are listed in Table 7.1.

Table 7.1: The values of $x^{(1)}, x^{(2)}, x^{(4)}$, and the exact solution $x^{(e)}$ at points of the interval $[0, 30]$.

t	$x^{(e)}$	$x^{(1)}$	$x^{(2)}$	$x^{(4)}$
0.00	2.00000000	2.00000000	2.00000000	2.00000000
1.00	1.20000000	1.20000000		
2.00	1.05882353	1.05882353	1.11111111	
3.00	1.01886792	1.01886792		
4.00	1.00621118	1.00621118	1.02040816	1.05882353
5.00	1.00206186	1.00206186		
6.00	1.00068634	1.00068634	1.00401606	
7.00	1.00022868	1.00022868		
8.00	1.00007621	1.00007621	1.00080064	1.00621118
9.00	1.00002540	1.00002540		
10.00	1.00000847	1.00000847	1.00016003	
11.00	1.00000282	1.00000282		
12.00	1.00000094	1.00000094	1.00003200	1.00068634
13.00	1.00000031	1.00000031		
14.00	1.00000010	1.00000010	1.00000640	
15.00	1.00000003	1.00000003		
16.00	1.00000001	1.00000001	1.00000128	1.00007621
17.00	1.00000000	1.00000000		
18.00	1.00000000	1.00000000	1.00000026	
19.00	1.00000000	1.00000001		
20.00	1.00000000	0.99999998	1.00000005	1.00000847
21.00	1.00000000	1.00000005		
22.00	1.00000000	0.99999985	1.00000001	
23.00	1.00000000	1.00000044		
24.00	1.00000000	0.99999868	1.00000000	1.00000094
25.00	1.00000000	1.00000396		
26.00	1.00000000	0.99998812	1.00000000	
27.00	1.00000000	1.00003563		
28.00	1.00000000	0.99989310	1.00000000	1.00000010
29.00	1.00000000	1.00032068		
30.00	1.00000000	0.99903792	1.00000000	

In Figures 7.1, 7.2, and 7.3 we compare the graphs of the exact and approximate solutions for the three cases discussed above. In all three figures, the exact solution is represented by the symbol o and the computed solution by the symbol *.

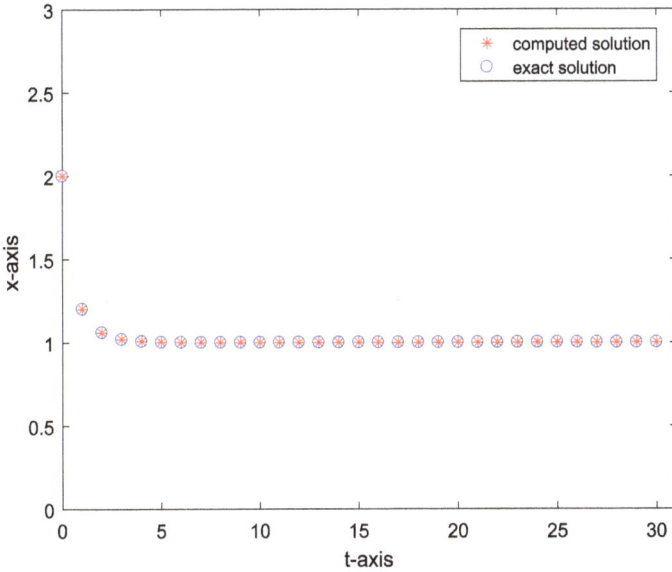

Figure 7.1: Computed and exact values of the solution with step size $r = 1$ and $\alpha = 2$.

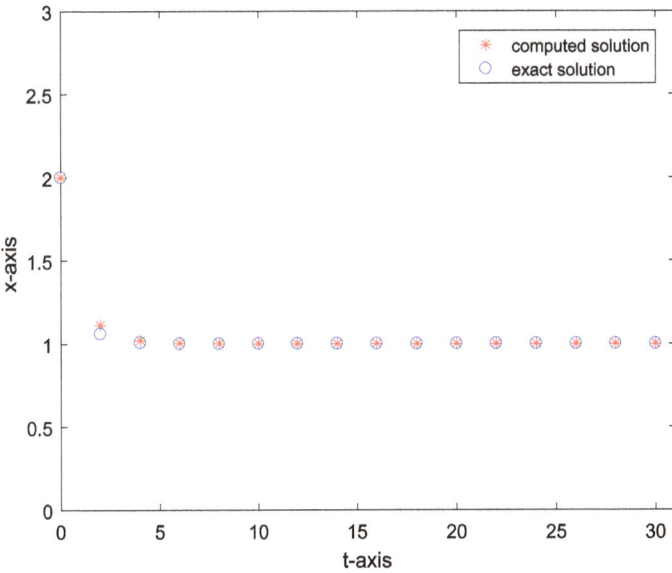

Figure 7.2: Computed and exact values of the solution with step size $r = 2$ and $\alpha = 2$.

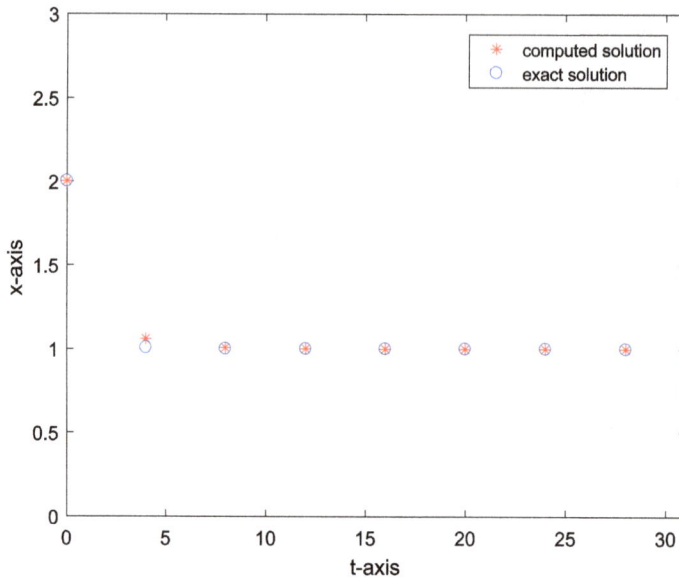

Figure 7.3: Computed and exact values of the solution with step size $r = 4$ and $\alpha = 2$.

In Figure 7.4, the errors for the three cases discussed above are given. It is obvious that there is almost no error in Case 1 as stated above, and a small error is present for the Cases 2 and 3.

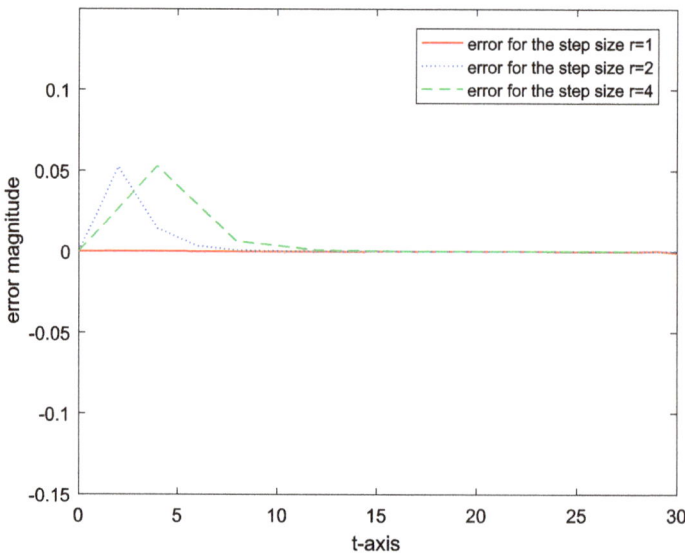

Figure 7.4: The error magnitudes for the logistic equation with step sizes $r = 1, 2, 4$ and $\alpha = 2$.

7.7 Advanced practical problems

Problem 7.23. Let

$$x_{n+2} = 3x_{n+1} + 2r_{n+1}h_2(t_n + r_{n+1} + r_{n+2}, t_n)f_{n+1} - 2h_3(t_n + r_{n+1}, t_n)f_n.$$

Write the associated linear difference operator and determine its order of consistency.

Problem 7.24. Write the first, second, and third characteristic polynomials of the following two-step method:

$$x_{j+2} + 4x_j = r_{j+2}(f_{j+2} + 3f_j) + r_{j+1}(f_{j+2} + 4f_{j+1} + 4f_j).$$

Problem 7.25. Prove that the two-step method

$$x_{j+2} - x_j = r_{j+2}f_{j+2} + r_{j+1}(6f_{j+2} - 2f_{j+1} - 3f_j)$$

is consistent.

Problem 7.26. Prove that the two-step method

$$x_{j+2} + x_{j+1} - 2x_j = r_{j+2}(f_{j+2} + 3f_{j+1} - 4f_j) + r_{j+1}(5f_{j+2} - 2f_{j+1} - f_j)$$

is consistent.

Problem 7.27. Write the first, second, third and fourth characteristic polynomials of the following 3-step method.

$$x_{j+3} - 2x_{j+2} + x_{j+1} - 3x_j = r_{j+3}(f_{j+3} + 2f_{j+2} - 4f_{j+1} - 3f_j)$$
$$+ r_{j+2}(-f_{j+3} - f_{j+2} - f_j)$$
$$+ r_{j+1}(2f_{j+3} - f_{j+2} + f_{j+1} - f_j).$$

Problem 7.28. Check if the following 5-step method satisfies conditions (7.7):

$$x_{j+5} - 3x_{j+4} + x_{j+3} - x_{j+2} + x_{j+1} - x_j$$
$$= r_{j+5}(f_{j+5} - f_{j+4} + f_{j+3} - f_{j+2} + f_{j+1} - f_j)$$
$$+ r_{j+4}(2f_{j+5} + f_{j+4} - f_{j+3} - f_{j+2} - f_{j+1} - f_j)$$
$$+ r_{j+3}(-f_{j+5} - 2f_{j+4} + 2f_{j+3} + f_{j+2} + f_{j+1} - f_j)$$
$$+ r_{j+2}(f_{j+5} - 7f_{j+4} + 6f_{j+3} - f_{j+2} + 3f_{j+1} - 4f_j)$$
$$+ r_{j+1}(7f_{j+5} - 10f_{j+1} + 4f_j).$$

8 Runge–Kutta methods – RKMs

The Runge–Kutta methods are among the most widely used methods in the numerical solutions of initial value problems for ordinary differential equations [4]. They are one-step methods which consist of several stages. Depending on the choice of the parameters involved in their construction, the Runge–Kutta methods can be implicit or explicit as the linear multistep methods.

In this chapter, we will discuss the Runge–Kutta methods on time scales. We will present the construction of one-, two-, and three-stage methods in details and briefly introduce a generalization of the idea for s-stage methods.

As in the previous chapters, we suppose that $\mathbb{T}$ is a time scale with forward jump operator σ and delta differentiation operator Δ. Let $a, b \in \mathbb{T}$, $a < b$, $m \in \mathbb{N}$, and

$$a = t_0 < t_1 < \cdots < t_m = b,$$

as well as $(t_{j-1}, t_j) \cap \mathbb{T} \neq \emptyset$, $j \in \{1, \ldots, m\}$. Denote

$$t_{j+1} = t_j + r_{j+1}, \quad j \in \{0, 1, \ldots, m-1\}, \quad r = \max_{j \in \{1, \ldots, m\}} r_j.$$

Consider the following IVP:

$$\begin{cases} x^\Delta(t) = f(t, x(t)), & t \in [a, b], \\ x(t_0) = x_0, \end{cases} \tag{8.1}$$

where $x_0 \in \mathbb{R}$, and assume that

$$(\text{H1}) \quad \begin{cases} |f(t, x)| \leq A, & t \in \mathbb{T}, \ x \in \mathbb{R}, \\ \text{there exist } \Delta_1 f(t, x) \text{ and } \frac{\partial}{\partial x} f(t, x) \text{ such that} \\ |\Delta_1 f(t, x)| \leq A, \quad |\frac{\partial}{\partial x} f(t, x)| \leq A, \quad t \in \mathbb{T}, \ x \in \mathbb{R}, \end{cases}$$

for some positive constant A.

8.1 One-stage methods

First, we will discuss the one-stage Runge–Kutta method. Define

$$x_{n+1} = x_n + b_1 r_{n+1} f_n,$$

where $x_n = x(t_n)$, $f_n = f(t_n, x_n)$, $b_1 \in \mathbb{R}$ will be determined below. By Pötzsche chain rule, for $t \in [a, b]$, we have

https://doi.org/10.1515/9783110787320-008

$$(f(t,x(t)))^\Delta = \Delta_1 f(t,x(t)) + \left(\int_0^1 \frac{\partial}{\partial x} f(\sigma(t), x(t) + s\mu(t)f(t,x(t)))ds \right) f(t,x(t)),$$

where x is a solution of the IVP (8.1). Set, for $t \in [a,b]$,

$$g(t,x(t)) = \Delta_1 f(t,x(t)) + \left(\int_0^1 \frac{\partial}{\partial x} f(\sigma(t), x(t) + s\mu(t)f(t,x(t)))ds \right) f(t,x(t)).$$

Then

$$x^{\Delta^2}(t) = g(t,x(t)), \quad t \in [a,b].$$

Now, applying the Taylor formula, we arrive at

$$\begin{aligned}
x(t_{n+1}) &= x(t_n) + r_{n+1}x^\Delta(t_n) + h_2(t_{n+1},t_n)x^{\Delta^2}(t_n) + O(r^3) \\
&= x(t_n) + r_{n+1}f(t_n,x(t_n)) + h_2(t_{n+1},t_n)g(t_n,x(t_n)) + O(r^3) \\
&= x_n + r_{n+1}f_n + h_2(t_{n+1},t_n)g(t_n,x_n) + O(r^3).
\end{aligned}$$

Hence,

$$\begin{aligned}
x(t_{n+1}) - x_{n+1} &= x_n + r_{n+1}f_n + h_2(t_{n+1},t_n)g(t_n,x_n) + O(r^3) - x_n - b_1 r_{n+1}f_n \\
&= r_{n+1}(1 - b_1)f_n + h_2(t_{n+1},t_n)g(t_n,x_n) + O(r^3).
\end{aligned}$$

Therefore, the method will be consistent of order $p = 1$ if we choose $b_1 = 1$. Thus, for $b_1 = 1$, we get the Euler method. Note that the Euler method is the only first-order one-stage explicit RKM.

8.2 Two-stage methods

In this section, we continue with the discussion of two-stage methods. In addition to (H1), suppose that

$$\text{(H2)} \quad \frac{\partial f}{\partial x}(t,x) \neq 0, \quad (t,x) \in [a,b] \times \mathbb{R}.$$

Define $t_{n+1}^1 \in \mathbb{T}$ as $t_{n+1}^1 = t_n + r_{n+1}^1$ where $t_n < t_n + r_{n+1}^1 < t_n + r_{n+1}$ and

$$x_n = x(t_n), \quad x_{n+1} = x_n + r_{n+1}^1(b_1 k_1 + b_2 k_2),$$

$$k_1 = f(t_n,x_n), \quad k_2 = a_1 f(t_{n+1}^1,x_n) + a_2 f\left(t_n,x_n + \frac{h_2(t_{n+1}^1,t_n)}{r_{n+1}^1} \cdot \frac{g(t_n,x_n)}{f_x(t_n,x_n)}\right),$$

where $a_1, a_2, b_1, b_2 \in \mathbb{R}$ will be determined below. This scheme will be called the two-stage Runge–Kutta method. We have

$$x(t_{n+1}^1) = x(t_n + r_{n+1}^1)$$

$$= x(t_n) + r_{n+1}^1 x^\Delta(t_n) + h_2(t_{n+1}^1, t_n)g(t_n, x_n) + O(r^3)$$

$$= x_n + r_{n+1}^1 f(t_n, x_n) + h_2(t_{n+1}^1, t_n)g(t_n, x_n) + O(r^3),$$

$$f(t_{n+1}^1, x_n) = f(t_n, x_n) + r_{n+1}^1 \Delta_1 f(t_n, x_n) + O(r^2),$$

$$f\left(t_n, x_n + \frac{h_2(t_{n+1}^1, t_n)}{r_{n+1}^1} \cdot \frac{g(t_n, x_n)}{f_x(t_n, x_n)}\right)$$

$$= f(t_n, x_n) + \frac{h_2(t_{n+1}^1, t_n)}{r_{n+1}^1} \cdot \frac{g(t_n, x_n)}{f_x(t_n, x_n)} f_x(t_n, x_n) + O(r^2)$$

$$= f(t_n, x_n) + g(t_n, x_n)\frac{h_2(t_{n+1}^1, t_n)}{r_{n+1}^1} + O(r^2),$$

$$k_2 = a_1 f(t_{n+1}^1, x_n) + a_2 f\left(t_n, x_n + \frac{h_2(t_{n+1}^1, t_n)}{r_{n+1}^1} \cdot \frac{g(t_n, x_n)}{f_x(t_n, x_n)}\right)$$

$$= a_1(f(t_n, x_n) + r_{n+1}^1 \Delta_1 f(t_n, x_n) + O(r^2))$$

$$+ a_2\left(f(t_n, x_n) + g(t_n, x_n)\frac{h_2(t_{n+1}^1, t_n)}{r_{n+1}^1} + O(r^2)\right)$$

$$= (a_1 + a_2)f(t_n, x_n) + a_1 r_{n+1}^1 \Delta_1 f(t_n, x_n) + a_2 g(t_n, x_n)\frac{h_2(t_{n+1}^1, t_n)}{r_{n+1}^1} + O(r^2),$$

$$b_1 k_1 + b_2 k_2 = (b_1 + b_2(a_1 + a_2))f(t_n, x_n)$$

$$+ a_1 b_2 r_{n+1}^1 \Delta_1 f(t_n, x_n) + a_2 b_2 g(t_n, x_n)\frac{h_2(t_{n+1}^1, t_n)}{r_{n+1}^1} + O(r^2),$$

and

$$x_{n+1} = x_n + (b_1 + b_2(a_1 + a_2))r_{n+1}^1 f(t_n, x_n)$$

$$+ a_1 b_2(r_{n+1}^1)^2 \Delta_1 f(t_n, x_n) + a_2 b_2 g(t_n, x_n)h_2(t_{n+1}^1, t_n) + O(r^3).$$

Hence,

$$x(t_{n+1}^1) - x_{n+1} = x_n + r_{n+1}^1 f(t_n, x_n) + h_2(t_{n+1}^1, t_n)g(t_n, x_n)$$

$$- x_n - (b_1 + b_2(a_1 + a_2))r_{n+1}^1 f(t_n, x_n)$$

$$- a_1 b_2(r_{n+1}^1)^2 \Delta_1 f(t_n, x_n) - a_2 b_2 g(t_n, x_n)h_2(t_{n+1}^1, t_n) + O(r^3)$$

$$= (1 - (b_1 + b_2(a_1 + a_2)))r^1_{n+1}f(t_n, x_n)$$
$$- a_1 b_2(r^1_{n+1})^2 \Delta_1 f(t_n, x_n) + (1 - a_2 b_2)g(t_n, x_n)h_2(t^1_{n+1}, t_n) + O(r^3).$$

Therefore, we can say the following:

1. If

$$\begin{cases} b_1 + b_2(a_1 + a_2) = 1, \\ a_1 b_2 \neq 0 \end{cases} \quad \text{or} \quad \begin{cases} b_1 + b_2(a_1 + a_2) = 1, \\ a_2 b_2 \neq 1, \end{cases}$$

then the method is consistent of order $p = 1$.

2. If

$$\begin{cases} b_1 + b_2(a_1 + a_2) = 1, \\ a_1 b_2 = 0, \\ a_2 b_2 = 1, \end{cases}$$

which is equivalent to

$$\begin{cases} a_1 = b_1 = 0 \\ a_2 b_2 = 1, \end{cases}$$

then the method is consistent of order $p = 2$.

8.3 Three-stage methods

In this section, we derive the three-stage methods and their consistency conditions. Suppose that (H1) and (H2) hold and $t^1_{n+1} = t_n + r^1_{n+1} \in \mathbb{T}, t_n < t_n + r^1_{n+1} < t_n + r_{n+1}$. Let

$$k_1 = f(t_n, x_n),$$
$$k_2 = a_{11}f(t^1_{n+1}, x_n) + a_{12}f\left(t_n, x_n + \frac{\Delta_1 f(t_n, x_n)}{f_x(t_n, x_n)}r^1_{n+1}\right),$$
$$k_3 = a_{21}f(t^1_{n+1}, x_n) + a_{22}f\left(t_n, x_n + \frac{h_2(t^1_{n+1}, t_n)g(t_n, x_n)}{f_x(t_n, x_n)r^1_{n+1}}\right),$$
$$x_{n+1} = x_n + r^1_{n+1}(b_1 k_1 + b_2 k_2 + b_3 k_3),$$

where $a_{11}, a_{12}, a_{21}, a_{22}, b_1, b_2, b_3 \in \mathbb{R}$ will be determined below. This scheme will be called the three-stage Runge–Kutta method. We have

$$x(t^1_{n+1}) = x_n + r^1_{n+1}f(t_n, x_n) + h_2(t^1_{n+1}, t_n)g(t_n, x_n) + O(r^3),$$
$$f(t_n + r^1_{n+1}, x_n) = f(t_n, x_n) + r^1_{n+1}\Delta_1 f(t_n, x_n) + O(r^2),$$
$$f\left(t_n, x_n + \frac{\Delta_1 f(t_n, x_n)}{f_x(t_n, x_n)}r^1_{n+1}\right) = f(t_n, x_n) + \frac{\Delta_1 f(t_n, x_n)}{f_x(t_n, x_n)}r^1_{n+1}f_x(t_n, x_n) + O(r^2),$$

$$= f(t_n, x_n) + r_{n+1}^1 \Delta_1 f(t_n, x_n) + O(r^2),$$

$$f\left(t_n, x_n + \frac{h_2(t_{n+1}^1, t_n)g(t_n, x_n)}{f_x(t_n, x_n)}\right) = f(t_n, x_n) + \frac{h_2(t_{n+1}^1, t_n)g(t_n, x_n)}{f_x(t_n, x_n)r_{n+1}^1} f_x(t_n, x_n) + O(r^2)$$

$$= f(t_n, x_n) + \frac{h_2(t_{n+1}^1, t_n)g(t_n, x_n)}{r_{n+1}^1} + O(r^2).$$

Thus,

$$k_2 = a_{11}f(t_{n+1}^1, x_n) + a_{12}f\left(t_n, x_n + \frac{\Delta_1 f(t_n, x_n)}{f_x(t_n, x_n)} r_{n+1}^1\right)$$

$$= a_{11}f(t_n, x_n) + a_{11}r_{n+1}^1\Delta_1 f(t_n, x_n) + a_{12}f(t_n, x_n) + a_{12}r_{n+1}^1\Delta_1 f(t_n, x_n) + O(r^2)$$

$$= (a_{11} + a_{12})f(t_n, x_n) + (a_{11} + a_{12})r_{n+1}^1\Delta_1 f(t_n, x_n) + O(r^2),$$

$$b_2 k_2 = b_2(a_{11} + a_{12})f(t_n, x_n) + b_2(a_{11} + a_{12})r_{n+1}^1\Delta_1 f(t_n, x_n) + O(r^2),$$

$$k_3 = a_{21}f(t_{n+1}^1, x_n) + a_{22}f\left(t_n, x_n + \frac{\Delta_1 f(t_n, x_n)}{f_x(t_n, x_n)} r_{n+1}^1\right)$$

$$= a_{21}f(t_n, x_n) + a_{21}r_{n+1}^1\Delta_1 f(t_n, x_n)$$

$$+ a_{22}f(t_n, x_n) + a_{12}\frac{h_2(t_{n+1}^1, t_n)g(t_n, x_n)}{r_{n+1}^1} + O(r^2)$$

$$= (a_{21} + a_{22})f(t_n, x_n) + a_{21}r_{n+1}^1\Delta_1 f(t_n, x_n) + a_{22}\frac{h_2(t_{n+1}^1, t_n)g(t_n, x_n)}{r_{n+1}^1} + O(r^2),$$

$$b_3 k_3 = b_3(a_{21} + a_{22})f(t_n, x_n) + a_{21}r_{n+1}^1\Delta_1 f(t_n, x_n)$$

$$+ b_3 a_{22}\frac{h_2(t_{n+1}^1, t_n)g(t_n, x_n)}{r_{n+1}^1} + O(r^2),$$

and

$$b_1 k_1 + b_2 k_2 + b_3 k_3 = (b_1 + b_2(a_{11} + a_{12}) + b_3(a_{21} + a_{22}))f(t_n, x_n)$$

$$+ (b_2(a_{11} + a_{12}) + b_3 a_{21})r_{n+1}^1\Delta_1 f(t_n, x_n)$$

$$+ b_3 a_{22}\frac{h_2(t_{n+1}^1, t_n)g(t_n, x_n)}{r_{n+1}^1} + O(r^2),$$

$$r_{n+1}^1(b_1 k_1 + b_2 k_2 + b_3 k_3) = (b_1 + b_2(a_{11} + a_{12}) + b_3(a_{21} + a_{22}))r_{n+1}^1 f(t_n, x_n)$$

$$+ (b_2(a_{11} + a_{12}) + b_3 a_{21})(r_{n+1}^1)^2\Delta_1 f(t_n, x_n)$$

$$+ b_3 a_{22}h_2(t_{n+1}^1, t_n)g(t_n, x_n) + O(r^3),$$

$$x_{n+1} = x_n + (b_1 + b_2(a_{11} + a_{12}) + b_3(a_{21} + a_{22}))r_{n+1}^1 f(t_n, x_n)$$

$$+ (b_2(a_{11} + a_{12}) + b_3 a_{21})(r_{n+1}^1)^2\Delta_1 f(t_n, x_n)$$

$$+ b_3 a_{22}h_2(t_{n+1}^1, t_n)g(t_n, x_n) + O(r^3).$$

Hence,

$$
\begin{aligned}
x(t_{n+1}^1) - x_{n+1} ={}& x_n + r_{n+1}^1 f(t_n, x_n) + h_2(t_{n+1}^1, t_n)g(t_n, x_n) \\
& - x_n - (b_1 + b_2(a_{11} + a_{12}) + b_3(a_{21} + a_{22}))r_{n+1}^1 f(t_n, x_n) \\
& - (b_2(a_{11} + a_{12}) + b_3 a_{21})(r_{n+1}^1)^2 \Delta_1 f(t_n, x_n) \\
& - b_3 a_{22} h_2(t_{n+1}^1, t_n)g(t_n, x_n) + O(r^3) \\
={}& (1 - (b_1 + b_2(a_{11} + a_{12}) + b_3(a_{21} + a_{22})))r_{n+1}^1 f(t_n, x_n) \\
& - (b_2(a_{11} + a_{12}) + b_3 a_{21})(r_{n+1}^1)^2 \Delta_1 f(t_n, x_n) \\
& + (1 - b_3 a_{22}))h_2(t_{n+1}^1, t_n)g(t_n, x_n) + O(r^3).
\end{aligned}
$$

Therefore, we can say the following:

1. If

$$
\begin{cases}
b_1 + b_2(a_{11} + a_{12}) + b_3(a_{21} + a_{22}) = 1, \\
b_2(a_{11} + a_{12}) + b_3 a_{21} \neq 0
\end{cases}
$$

or

$$
\begin{cases}
b_1 + b_2(a_{11} + a_{12}) + b_3(a_{21} + a_{22}) = 1, \\
b_3 a_{22} \neq 1,
\end{cases}
$$

then the three-stage method is consistent of order $p = 1$.

2. If

$$
\begin{cases}
b_1 + b_2(a_{11} + a_{12}) + b_3(a_{21} + a_{22}) = 1, \\
b_2(a_{11} + a_{12}) + b_3 a_{21} = 0, \\
b_3 a_{22} = 1,
\end{cases}
$$

then the three-stage method is consistent of order $p = 2$.

8.4 *s*-Stage methods

Here, we briefly give the general form of an s-stage Runge–Kutta method. Suppose that (H1) and (H2) hold and $t_{n+1}^1 = t_n + r_{n+1}^1 \in \mathbb{T}$, $t_n < t_n + r_{n+1}^1 < t_n + r_{n+1}$. The general s-stage RK method can be written in the form

$$
x_{n+1} = x_n + r_{n+1}^1 \sum_{j=1}^{s} b_j k_j,
$$

where

$$
k_1 = f(t_n, x_n), \quad k_j = k_j(f, \Delta_1 f, f_x), \quad j \in \{2, \dots, s\},
$$

are determined so that the method is consistent of order $p > 0$. To determine the coefficients $b_j, j \in \{1, \dots, s\}$, we use the Taylor expansions of $f, x, k_j, j \in \{1, \dots, s\}$, of order $p > 0$ so that the considered s-stage method is consistent of order 1 or 2, or so on, up to p.

8.5 Numerical examples

In this section, we solve examples with the Runge–Kutta methods. In the first two, the time scale is chosen as the set of real numbers and the initial value problem is treated with both two- and three-stage Runge–Kutta methods. The third example is the initial value problem solved by using the trapezoid rule in Example 5.7.

Example 8.1. Consider the initial value problem

$$\begin{cases} x^\Delta(t) = -\frac{x^2 t}{1+t^2}, & t \in [1,5] \subset \mathbb{R}, \\ x(1) = 1. \end{cases}$$

(8.2)

where $\mathbb{T} = \mathbb{R}$. We will use a two-stage Runge–Kutta method given by

$$x_n = x(t_n), \qquad x_{n+1} = x_n + r_{n+1}^1(b_1 k_1 + b_2 k_2),$$

$$k_1 = f(t_n, x_n), \qquad k_2 = a_1 f(t_{n+1}^1, x_n) + a_2 f\left(t_n, x_n + \frac{h_2(t_{n+1}^1, t_n)}{r_{n+1}^1} \cdot \frac{g(t_n, x_n)}{f_x(t_n, x_n)}\right),$$

where $a_1, a_2, b_1, b_2 \in \mathbb{R}$ are chosen according to the consistency conditions given as follows.

In order that the method is consistent of order $p = 1$, we should have

$$\begin{cases} b_1 + b_2(a_1 + a_2) = 1, \\ a_1 b_2 \neq 0 \end{cases} \quad \text{or} \quad \begin{cases} b_1 + b_2(a_1 + a_2) = 1, \\ a_2 b_2 \neq 1. \end{cases}$$

The method is consistent of order $p = 2$ if

$$\begin{cases} a_1 = b_1 = 0 \\ a_2 b_2 = 1. \end{cases}$$

We take

$$t_0 = 1, \quad x_0 = x(1) = 1,$$

and

$$t_{n+1}^1 = t_n + r_{n+1}^1 = t_n + r,$$

where the step size r is constant. Note that on the time scale $\mathbb{T} = \mathbb{R}$ we have $\sigma(t) = t$, $\mu(t) = 0$ for all $t \in \mathbb{R}$, and $x^\Delta(t) = x'(t)$, $t \in \mathbb{R}$. For the given initial value problem, we have

$$f(t,x) = -\frac{x^2 t}{1+t^2}, \quad f_x(t,x) = -\frac{2xt}{1+t^2},$$

$$g(t,x) = \frac{\partial}{\partial t}f(t,x) = \frac{\partial}{\partial t}\left(-\frac{x^2 t}{1+t^2}\right) = -\frac{(2xx't + x^2)(1+t^2) - 2tx^2 t}{(1+t^2)^2}$$

$$= -\frac{(-2xt\frac{x^2 t}{1+t^2}(1+t^2) + x^2(1+t^2) - 2tx^2 t}{(1+t^2)^2} = \frac{2x^3 t - x^2 + x^2 t^2}{(1+t^2)^2}, \quad t \in \mathbb{R}.$$

We also have

$$h_2(t_{n+1}^1, t_n) = \frac{(t_{n+1}^1 - t_n)^2}{2},$$

so that

$$x_n + \frac{h_2(t_{n+1}^1, t_n)}{r_{n+1}^1} \cdot \frac{g(t_n, x_n)}{f_x(t_n, x_n)} = x_n + \frac{r^2}{2r}\left(\frac{\frac{2x_n^3 t_n - x_n^2 + x_n^2 t_n^2}{(1+t_n^2)^2}}{-\frac{2x_n t_n}{1+t_n^2}}\right) = x_n - \frac{r}{2}\left(\frac{2x_n^3 t_n - x_n^2 + x_n^2 t_n^2}{2x_n t_n(1+t_n^2)}\right).$$

Then the two stage Runge–Kutta method becomes

$$k_1 = -\frac{x_n^2 t_n}{1+t_n^2}, \quad k_2 = -a_1\frac{x_n^2 t_{n+1}^1}{1+(t_{n+1}^1)^2} - a_2\frac{(x_n - \frac{r}{2}(\frac{2x_n^3 t_n - x_n^2 + x_n^2 t_n^2}{2x_n t_n(1+t_n^2)}))^2 t_n}{1+t_n^2},$$

$$x_{n+1} = x_n + r(b_1 k_1 + b_2 k_2),$$

where $x_0 = 1$, $t_0 = 1$ and $t_{n+1}^1 = t_n + r$ for $n = 0, \ldots, \frac{4}{r}$.

We note that the differential equation in the given initial value problem is a separable first-order equation and the exact solution is obtained as follows. The separation of the variables gives

$$-\frac{dx}{x^2} = \frac{tdt}{1+t^2}, \quad t \in \mathbb{R},$$

and hence, upon integration,

$$\frac{1}{x} = \frac{1}{2}\ln(1+t^2) + C, \quad t \in \mathbb{R},$$

where C is an arbitrary constant to be determined from the initial condition $x(1) = 1$. Since the general solution is

$$x(t) = \frac{1}{C + \ln(\sqrt{1+t^2})}, \quad t \in \mathbb{R},$$

the initial condition gives $C = 1 - \ln\sqrt{2}$, so that the exact solution of the given initial value problem is

$$x(t) = \frac{1}{1 - \ln\sqrt{2} + \ln(\sqrt{1+t^2})}, \quad t \in \mathbb{R}.$$

We perform the computation with *MATLAB* for different values of the step size r and the parameters a_1, a_2, b_1, b_2.

First, we take a two-step Runge–Kutta method, which is consistent of order $p = 1$, by choosing $a_1 = 0.5$, $a_2 = 0.5$, $b_1 = 0$, $b_2 = 1$, so that $b_1 + b_2(a_1 + a_2) = 1$ and $a_1 b_2 = 0.5 \neq 0$. We choose two different step sizes $r = 0.4$ and $r = 0.2$. The approximate solution and the exact solution for $r = 0.4$ are compared in Table 8.1, and their graphs are shown in Figure 8.1, and for $r = 0.2$, a similar comparison is given in Table 8.2 and Figure 8.2.

Table 8.1: The values of the exact solution $x^{(e)}$ and approximate solution $x^{(a)}$ for $a_1 = 0.5$, $a_2 = 0.5$, $b_1 = 0$, $b_2 = 1$, and $r = 0.4$.

t	$x^{(e)}$	$x^{(a)}$
1.00	1.00000000	1.00000000
1.40	0.83610569	0.82440541
1.80	0.72689842	0.71475735
2.20	0.65112992	0.64034943
2.60	0.59597683	0.58661715
3.00	0.55410290	0.54593378
3.40	0.52119030	0.51397716
3.80	0.49458255	0.48813533
4.20	0.47257163	0.46674406
4.60	0.45401491	0.44869523
5.00	0.43812096	0.43322319

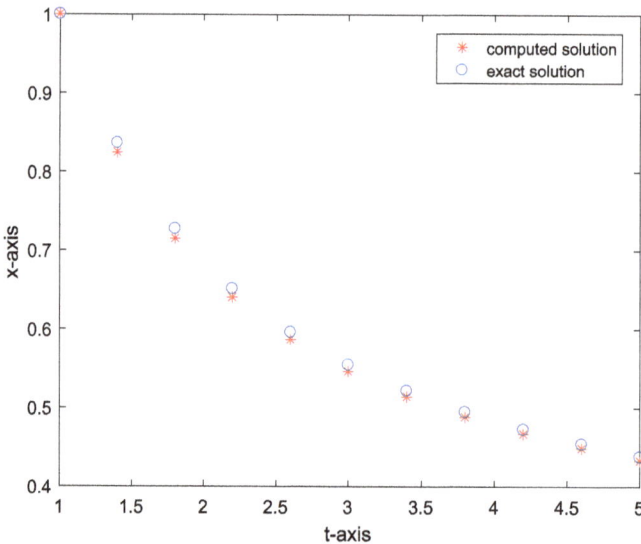

Figure 8.1: Approximate and exact solutions for $a_1 = 0.5$, $a_2 = 0.5$, $b_1 = 0$, $b_2 = 1$, and $r = 0.4$.

Table 8.2: The values of the exact solution $x^{(e)}$ and approximate solution $x^{(a)}$ for $a_1 = 0.5$, $a_2 = 0.5$, $b_1 = 0$, $b_2 = 1$, and $r = 0.2$.

t	$x^{(e)}$	$x^{(a)}$
1.00	1.00000000	1.00000000
1.20	0.90956601	0.90569467
1.40	0.83610569	0.83064648
1.60	0.77621269	0.77023829
1.80	0.72689842	0.72090282
2.00	0.68580268	0.68000476
2.20	0.65112992	0.64561723
2.40	0.62152805	0.61632553
2.60	0.59597683	0.59108015
2.80	0.57370048	0.56909231
3.00	0.55410290	0.54976102
3.20	0.53672088	0.53262209
3.40	0.52119030	0.51731229
3.60	0.50722171	0.50354384
3.80	0.49458255	0.49108614
4.00	0.48308407	0.47975238
4.20	0.47257163	0.46938976
4.40	0.46291742	0.45987213
4.60	0.45401491	0.45109446
4.80	0.44577464	0.44296859
5.00	0.43812096	0.43542003

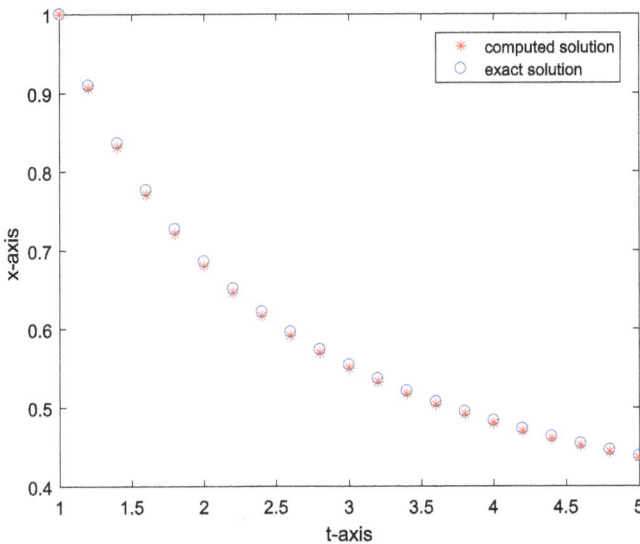

Figure 8.2: Approximate and exact solutions for $a_1 = 0.5$, $a_2 = 0.5$, $b_1 = 0$, $b_2 = 1$, and $r = 0.2$.

Second, we take a two-step Runge–Kutta method, consistent of order $p = 2$, by choosing $a_1 = 0$, $a_2 = 2$, $b_1 = 0$, $b_2 = 0.5$, so that the consistency condition is satisfied. We choose again two different step sizes $r = 0.4$ and $r = 0.2$. The approximate solution and the exact solution for $r = 0.4$ are compared in Table 8.3, and their graphs are shown in Figure 8.3, and for $r = 0.2$, a similar comparison is presented in Table 8.4 and Figure 8.4. The results of the numerical computation show that, as expected, with the second choice of the parameters providing consistency order $p = 2$, we obtain a higher accuracy.

Table 8.3: The values of the exact solution $x^{(e)}$ and approximate solution $x^{(a)}$ for $a_1 = 0$, $a_2 = 2$, $b_1 = 0$, $b_2 = 0.5$, and $r = 0.4$.

t	$x^{(a)}$	$x^{(e)}$
1.00	1.00000000	1.00000000
1.40	0.83800000	0.83610569
1.80	0.73096768	0.72689842
2.20	0.65606530	0.65112992
2.60	0.60108796	0.59597683
3.00	0.55910166	0.55410290
3.40	0.52597162	0.52119030
3.80	0.49911903	0.49458255
4.20	0.47686763	0.47257163
4.60	0.45808680	0.45401491
5.00	0.44198867	0.43812096

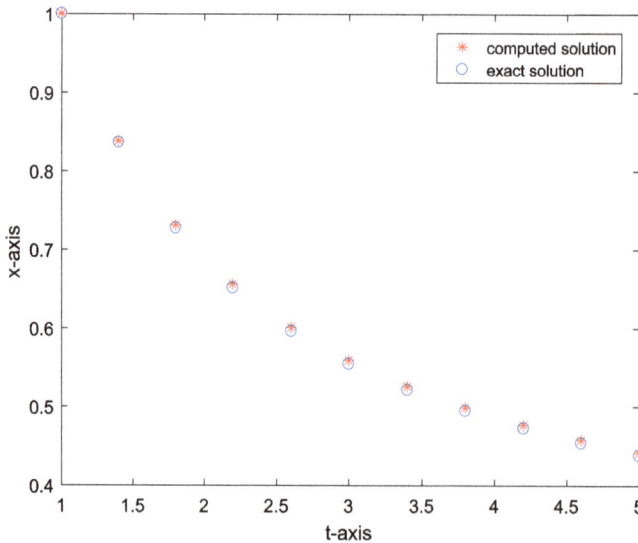

Figure 8.3: Approximate and exact solutions for $a_1 = 0$, $a_2 = 2$, $b_1 = 0$, $b_2 = 0.5$, and $r = 0.4$.

Table 8.4: The values of the exact solution $x^{(e)}$ and approximate solution $x^{(a)}$ for $a_1 = 0$, $a_2 = 2$, $b_1 = 0$, $b_2 = 0.5$, and $r = 0.2$.

t	$x^{(a)}$	$x^{(e)}$
1.00	1.00000000	1.00000000
1.20	0.90975000	0.90956601
1.40	0.83662799	0.83610569
1.60	0.77701742	0.77621269
1.80	0.72789574	0.72689842
2.00	0.68691711	0.68580268
2.20	0.65230745	0.65112992
2.40	0.62273252	0.62152805
2.60	0.59718493	0.59597683
2.80	0.57489755	0.57370048
3.00	0.55528005	0.55410290
3.20	0.53787296	0.53672088
3.40	0.52231457	0.52119030
3.60	0.50831701	0.50722171
3.80	0.49564869	0.49458255
4.00	0.48412150	0.48308407
4.20	0.47358118	0.47257163
4.40	0.46390012	0.46291742
4.60	0.45497191	0.45401491
4.80	0.44670715	0.44577464
5.00	0.43903017	0.43812096

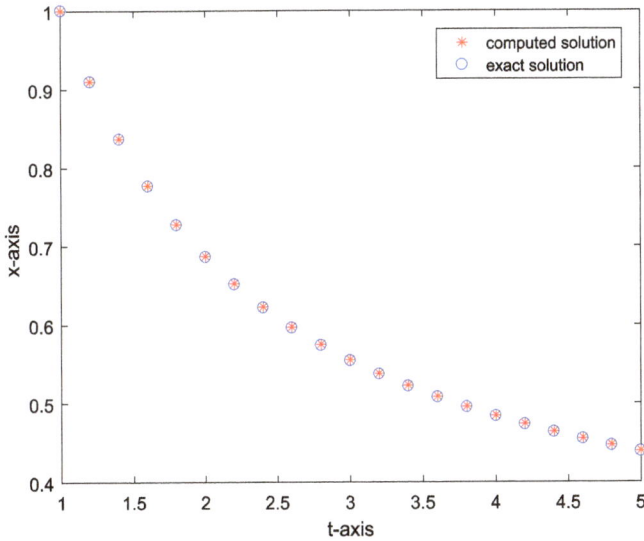

Figure 8.4: Approximate and exact solutions for $a_1 = 0$, $a_2 = 2$, $b_1 = 0$, $b_2 = 0.5$, and $r = 0.2$.

Example 8.2. We consider again the initial value problem from Example 8.1, that is,

$$\begin{cases} x^\Delta(t) = -\frac{x^2 t}{1+t^2}, & t \in [1,5] \subset \mathbb{R}, \\ x(1) = 1, \end{cases} \tag{8.3}$$

where $\mathbb{T} = \mathbb{R}$. We will use a three-stage Runge–Kutta method given by

$$k_1 = f(t_n, x_n),$$

$$k_2 = a_{11} f(t^1_{n+1}, x_n) + a_{12} f\left(t_n, x_n + \frac{\Delta_1 f(t_n, x_n)}{f_x(t_n, x_n)} r^1_{n+1}\right),$$

$$k_3 = a_{21} f(t^1_{n+1}, x_n) + a_{22} f\left(t_n, x_n + \frac{h_2(t^1_{n+1}, t_n) g(t_n, x_n)}{f_x(t_n, x_n) r^1_{n+1}}\right),$$

$$x_{n+1} = x_n + r^1_{n+1}(b_1 k_1 + b_2 k_2 + b_3 k_3),$$

where $a_{11}, a_{12}, a_{21}, a_{22}, b_1, b_2, b_3 \in \mathbb{R}$ will be chosen according to the following condition for consistency of order 2:

$$\begin{cases} b_1 + b_2(a_{11} + a_{12}) + b_3(a_{21} + a_{22}) = 1, \\ b_2(a_{11} + a_{12}) + b_3 a_{21} = 0, \\ b_3 a_{22} = 1, \end{cases}$$

As in the previous example, we take

$$t_0 = 1, \quad x_0 = x(1) = 1, \quad \text{and} \quad t^1_{n+1} = t_n + r^1_{n+1} = t_n + r,$$

where the step size r is constant. From Example 8.1 we have

$$f(t, x) = -\frac{x^2 t}{1+t^2}, \quad f_x(t, x) = -\frac{2xt}{1+t^2},$$

$$g(t, x) = \frac{2x^3 t - x^2 + x^2 t^2}{(1+t^2)^2}, \quad x, t \in \mathbb{R},$$

and compute

$$\Delta_1 f(t, x) = f_t(t, x) = -\frac{x^2(1-t^2)}{(1+t^2)^2}, \quad x, t \in \mathbb{R}.$$

Employing also

$$h_2(t^1_{n+1}, t_n) = \frac{(t^1_{n+1} - t_n)^2}{2},$$

we get

$$x_n + \frac{\Delta_1 f(t_n, x_n)}{f_x(t_n, x_n)} r^1_{n+1} = x_n + r\frac{x_n(1 - t_n^2)}{2t_n(1 + t_n^2)},$$

$$x_n + \frac{h_2(t_{n+1}^1, t_n)g(t_n, x_n)}{f_x(t_n, x_n)r_{n+1}^1} = -\frac{r}{2}\left(\frac{2x_n^3 t_n - x_n^2 + x_n^2 t_n^2}{2x_n t_n(1 + t_n^2)}\right).$$

We choose the parameters involved in the three-stage Runge–Kutta method as follows:

$$b_1 = 0, \quad b_2 = \frac{1}{2}, \quad b_3 = 1,$$

$$a_{11} = 1, \quad a_{12} = 1, \quad a_{21} = -1, \quad a_{22} = 1,$$

so that we have

$$\begin{cases} b_1 + b_2(a_{11} + a_{12}) + b_3(a_{21} + a_{22}) = 1 + \frac{1}{2}(1 + 1) + (-1 + 1) = 1, \\ b_2(a_{11} + a_{12}) + b_3 a_{21} = \frac{1}{2}(1 + 1) + (-1) = 0, \\ b_3 a_{22} = 1, \end{cases}$$

and the condition for consistency of order two holds. Then the three-stage Runge–Kutta method becomes

$$k_1 = -\frac{x_n^2 t_n}{1 + t_n^2},$$

$$k_2 = -\frac{x_n^2 t_{n+1}^1}{1 + (t_{n+1}^1)^2} - \frac{t_n}{1 + t_n^2}\left(x_n + r\frac{x_n(1 - t_n^2)}{2t_n(1 + t_n^2)}\right)^2,$$

$$k_3 = \frac{x_n^2 t_{n+1}^1}{1 + (t_{n+1}^1)^2} - \frac{(x_n - \frac{r}{2}(\frac{2x_n^3 t_n - x_n^2 + x_n^2 t_n^2}{2x_n t_n(1 + t_n^2)}))^2 t_n}{1 + t_n^2},$$

$$x_{n+1} = x_n + r\left(\frac{1}{2}k_2 + k_3\right),$$

where $x_0 = 1$, $t_0 = 1$ and $t_{n+1}^1 = t_n + r$ for $n = 0, \ldots, \frac{4}{r}$.

Recall also that the exact solution of the given initial value problem is obtained as

$$x(t) = \frac{1}{1 - \ln\sqrt{2} + \ln(\sqrt{1 + t^2})}, \quad t \in \mathbb{R}.$$

We perform the computation with *MATLAB* for 3 different values of the step size r, that is, $r = 0.2, 0.4$, and 0.8. The approximate solution and the exact solution for $r = 0.2$ are compared in Table 8.5 and their graphs are shown in Figure 8.5, and for $r = 0.4$, a similar comparison is presented in Table 8.6 and Figure 8.6. Finally, the approximate solution and the exact solution for $r = 0.8$ are compared in Table 8.7 and their graphs are shown in Figure 8.7.

The results of the numerical computation show that, as expected, the stage-three Runge–Kutta method provides a better approximation than a two-stage method.

Table 8.5: The values of the exact solution $x^{(e)}$ and approximate solution $x^{(a)}$ for $r = 0.2$.

t	$x^{(e)}$	$x^{(a)}$
1.00	1.00000000	1.00000000
1.20	0.90956601	0.90893033
1.40	0.83610569	0.83559010
1.60	0.77621269	0.77599029
1.80	0.72689842	0.72695577
2.00	0.68580268	0.68608085
2.20	0.65112992	0.65156958
2.40	0.62152805	0.62208100
2.60	0.59597683	0.59660686
2.80	0.57370048	0.57438122
3.00	0.55410290	0.55481550
3.20	0.53672088	0.53745193
3.40	0.52119030	0.52193029
3.60	0.50722171	0.50796395
3.80	0.49458255	0.49532234
4.00	0.48308407	0.48381815
4.20	0.47257163	0.47329776
4.40	0.46291742	0.46363408
4.60	0.45401491	0.45472114
4.80	0.44577464	0.44646985
5.00	0.43812096	0.43880484

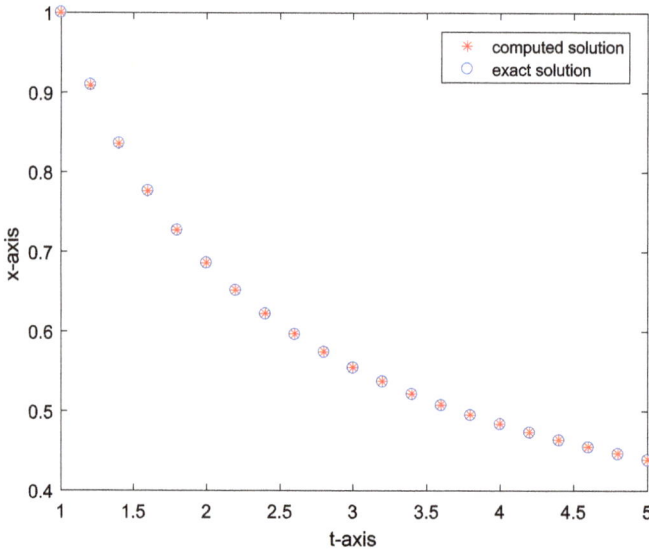

Figure 8.5: Approximate and exact solutions for $r = 0.2$.

Table 8.6: The values of the exact solution $x^{(e)}$ and approximate solution $x^{(a)}$ for $r = 0.4$.

t	$x^{(e)}$	$x^{(a)}$
1.00	1.00000000	1.00000000
1.40	0.83610569	0.83259459
1.80	0.72689842	0.72603732
2.20	0.65112992	0.65213300
2.60	0.59597683	0.59795040
3.00	0.55410290	0.55653392
3.40	0.52119030	0.52381116
3.80	0.49458255	0.49725594
4.20	0.47257163	0.47522753
4.60	0.45401491	0.45661836
5.00	0.43812096	0.44065556

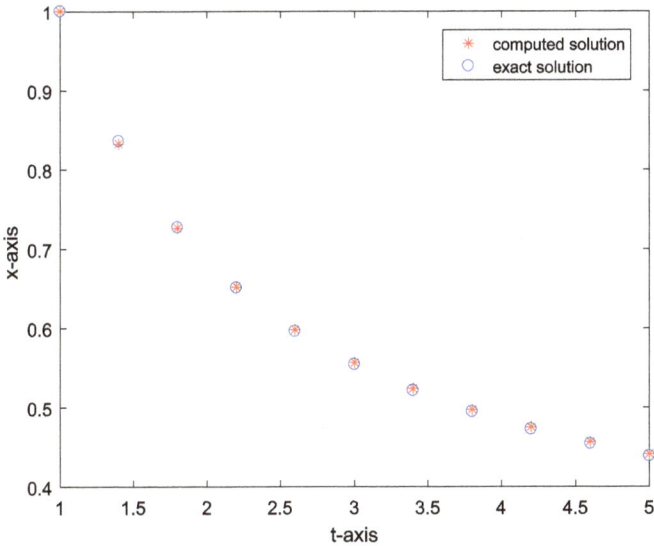

Figure 8.6: Approximate and exact solutions for $r = 0.4$.

Table 8.7: The values of the exact solution $x^{(e)}$ and approximate solution $x^{(a)}$ for $r = 0.8$.

t	$x^{(e)}$	$x^{(a)}$
1.00	1.00000000	1.00000000
1.80	0.72689842	0.71381132
2.60	0.59597683	0.59831702
3.40	0.52119030	0.52801758
4.20	0.47257163	0.48044577
5.00	0.43812096	0.44601078

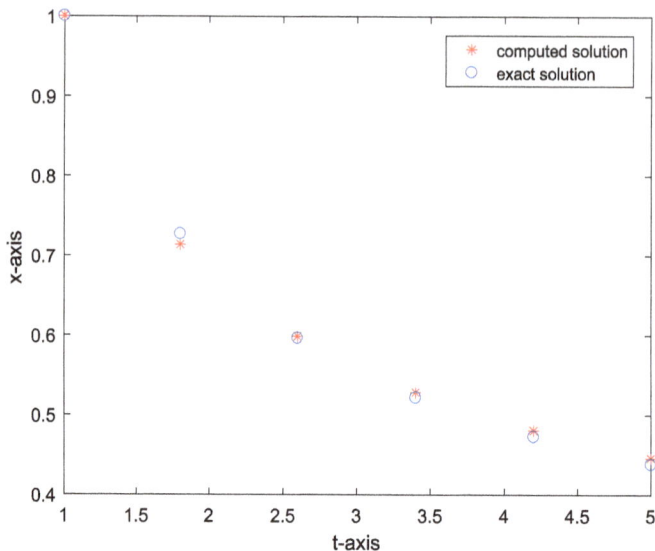

Figure 8.7: Approximate and exact solutions for $r = 0.8$.

In the next example, we will consider the treatment of the initial value problem from Example 5.7, but we consider the time scale $\mathbb{T} = a\mathbb{N}_0$, for a positive a, instead of $\mathbb{T} = \mathbb{N}_0$.

Example 8.3. Consider the IVP associated with the linear dynamic equation

$$x^{\Delta}(t) = \frac{1}{t+1}x(t) + \frac{1}{t^2+1}, \quad x(0) = \alpha, \tag{8.4}$$

where $t \in [0, 10]_{\mathbb{N}_0}$. We will apply a two-stage Runge–Kutta method with two different choices of the parameters, one resulting in a consistency order $p = 1$, and the other resulting in the order of consistency $p = 2$.

We take

$$t_0 = 0, \quad x_0 = x(0) = \alpha,$$

and

$$t^1_{n+1} = t_n + r^1_{n+1} = t_n + r,$$

where the step size r is constant. Note that on the given time scale $\mathbb{T} = a\mathbb{N}_0$ we have $\sigma(t) = t + a$, $\mu(t) = a$ for all $t \in \mathbb{T}$. For the given initial value problem, we have

$$f(t, x) = \frac{x}{t+1} + \frac{1}{t^2+1}, \quad f_x(t, x) = \frac{1}{t+1}, \quad t \in \mathbb{T}, \ x \in \mathbb{R}.$$

We will apply the Pötzsche chain rule to compute $g(t, x)$. First, note that

$$\Delta_1 f(t,x) = \frac{f(t+a,x) - f(t,x)}{a}$$

$$= \frac{\left(\frac{x}{t+a+1} + \frac{1}{(t+a)^2+1}\right) - \left(\frac{x}{t+1} + \frac{1}{t^2+1}\right)}{a}$$

$$= -\frac{x}{(t+1)(t+a+1)} - \frac{2t+a}{(t^2+1)((t+a)^2+1)}, \qquad t \in \mathbb{T}, \, x \in \mathbb{R},$$

and

$$f(\sigma(t), x + s\mu(t)f(t,x)) = f\left(t+a, x + sa\left(\frac{x}{t+1} + \frac{1}{t^2+1}\right)\right)$$

$$= \frac{x + sa\left(\frac{x}{t+1} + \frac{1}{t^2+1}\right)}{t+a+1} + \frac{1}{(t+a)^2+1}, \qquad t \in \mathbb{T}, \, s, x \in \mathbb{R},$$

so that we have

$$\frac{\partial}{\partial x} f(\sigma(t), x + s\mu(t)f(t,x)) = \frac{1}{t+a+1} + \frac{sa}{(t+1)(t+a+1)}, \qquad t \in \mathbb{T}, \, s, x \in \mathbb{R}.$$

Then, we compute

$$g(t,x) = \Delta_1 f(t,x) + \left(\int_0^1 \frac{\partial}{\partial x} f(\sigma(t), x + s\mu(t)f(t,x)) \Delta s\right) f(t,x)$$

$$= -\frac{x}{(t+1)(t+a+1)} - \frac{2t+a}{(t^2+1)((t+a)^2+1)}$$

$$+ \left(\int_0^1 \left(\frac{1}{t+a+1} + \frac{sa}{(t+1)(t+a+1)}\right) \Delta s\right) \left(\frac{x}{t+1} + \frac{1}{t^2+1}\right)$$

$$= -\frac{x}{(t+1)(t+a+1)} - \frac{2t+a}{(t^2+1)((t+a)^2+1)}$$

$$+ \left(\frac{x}{t+1} + \frac{1}{t^2+1}\right) \left(\frac{s}{t+a+1} + \frac{s^2 a}{2(t+1)(t+a+1)}\right)\Big|_0^1$$

$$= -\frac{x}{(t+1)(t+a+1)} - \frac{2t+a}{(t^2+1)((t+a)^2+1)}$$

$$+ \left(\frac{x}{t+1} + \frac{1}{t^2+1}\right) \left(\frac{1}{t+a+1} + \frac{a}{2(t+1)(t+a+1)}\right), \qquad t \in \mathbb{T}, \, s, x \in \mathbb{R}.$$

We also have

$$h_2(t_{n+1}^1, t_n) = \frac{(t_{n+1}^1 - t_n)(t_{n+1}^1 - t_n - a)}{2} = \frac{r(r-a)}{2}.$$

Let

$$Q_n = x_n + \frac{h_2(t_{n+1}^1, t_n)}{r_{n+1}^1} \cdot \frac{g(t_n, x_n)}{f_x(t_n, x_n)}.$$

Then

$$Q_n = x_n + \frac{r(r-a)}{2r} \frac{1}{\frac{1}{t_{n+1}}} \left[\left(-\frac{x_n}{(t_n+1)(t_n+a+1)} - \frac{2t_n+a}{(t_n^2+1)((t_n+a)^2+1)} \right) \right.$$

$$+ \left(\frac{x_n}{t_n+1} + \frac{1}{t_n^2+1} \right) \left(\frac{1}{n+a+1} + \frac{a}{2(n+1)(t_n+a+1)} \right) \right]$$

$$= x_n + (t_n+1)\frac{r-a}{2} \left[\left(-\frac{x_n}{(t_n+1)(t_n+a+1)} - \frac{2t_n+a}{(t_n^2+1)((t_n+a)^2+1)} \right) \right.$$

$$+ \left(\frac{x_n}{t_n+1} + \frac{1}{t_n^2+1} \right) \left(\frac{1}{t_n+a+1} + \frac{a}{2(t_n+1)(t_n+a+1)} \right) \right].$$

Then the two-stage Runge–Kutta method becomes

$$k_1 = \frac{x_n}{t_n+1} + \frac{1}{t_n^2+1}, \quad k_2 = a_1\left(\frac{x_n}{t_{n+1}^1+1} + \frac{1}{(t_{n+1}^1)^2+1} \right) + a_2\left(\frac{Q_n}{t_n+1} + \frac{1}{t_n^2+1} \right),$$

$$x_{n+1} = x_n + r(b_1 k_1 + b_2 k_2),$$

where $x_0 = a$, $t_0 = 0$ and $t_{n+1}^1 = t_n + r$ for $n = 0, \dots, \frac{10}{r}$.

We perform the computation with *MATLAB* for different values of the parameter a, the initial value a, the step size r and the parameters a_1, a_2, b_1, b_2.

First, we take a two-step Runge–Kutta method consistent of order $p = 1$ by choosing $a_1 = 0$, $a_2 = 0.5$, $b_1 = 0.5$, $b_2 = 1$, so that $b_1 + b_2(a_1 + a_2) = 1$ and $a_2 b_2 = 0.5 \neq 1$.

We choose $a = 0.4$, $x(0) = a = 1$ and $r = 0.8$. The approximate and exact solutions are compared in Table 8.8 and their graphs are shown in Figure 8.8.

Then we take $a = 0.2$, $x(0) = a = 1$, and $r = 0.4$. The approximate and exact solutions are compared in Table 8.9 and their graphs are shown in Figure 8.9.

Table 8.8: The values of the exact solution $x^{(e)}$ and approximate solution $x^{(a)}$ for $a = 0.4$, $a_1 = 0$, $a_2 = 0.5$, $b_1 = 0.5$, $b_2 = 1$, and $r = 0.8$.

t	$x^{(e)}$	$x^{(a)}$
0.00	1.00000000	1.00000000
0.80	2.65911330	2.65241379
1.60	4.29312420	4.30967564
2.40	5.82142631	5.85573599
3.20	7.30182274	7.35036839
4.00	8.75998114	8.82121628
4.80	10.20649681	10.27967998
5.60	11.64626723	11.73100179
6.40	13.08181407	13.17786764
7.20	14.51455340	14.62177969
8.00	15.94533686	16.06363841
8.80	17.37470556	17.50401440
9.60	18.80301900	18.94328598

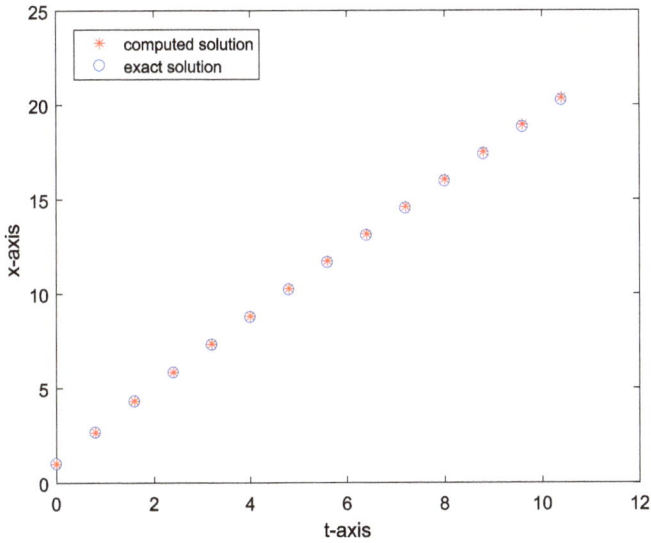

Figure 8.8: Approximate and exact solutions for $a = 0.4$, $a_1 = 0$, $a_2 = 0.5$, $b_1 = 0.5$, $b_2 = 1$, and $r = 0.8$.

Table 8.9: The values of the exact solution $x^{(e)}$ and approximate solution $x^{(a)}$ for $a = 0.2$, $a_1 = 0$, $a_2 = 0.5$, $b_1 = 0.5$, $b_2 = 1$, and $r = 0.4$.

t	$x^{(e)}$	$x^{(a)}$
0.00	1.00000000	1.00000000
0.40	1.82564103	1.81615385
0.80	2.68827709	2.67990887
1.20	3.51981834	3.51563446
1.60	4.31615069	4.31575117
2.00	5.08753630	5.09010707
2.40	5.84262105	5.84755645
2.80	6.58699088	6.59389658
3.20	7.32411400	7.33273643
3.60	8.05617463	8.06634743
4.00	8.78458851	8.79619940
4.40	9.51030240	9.52327331
4.80	10.23396853	10.24824368
5.20	10.95604855	10.97158705
5.60	11.67687725	11.69364847
6.00	12.39670271	12.41468320
6.40	13.11571227	13.13488380
6.80	13.83404989	13.85439804
7.20	14.55182789	14.57334107
7.60	15.26913509	15.29180388
8.00	15.98604260	16.00985928
8.40	16.70260803	16.72756617
8.80	17.41887848	17.44497269
9.20	18.13489285	18.16211857
9.60	18.85068352	18.87903686
10.00	19.56627768	19.59575530

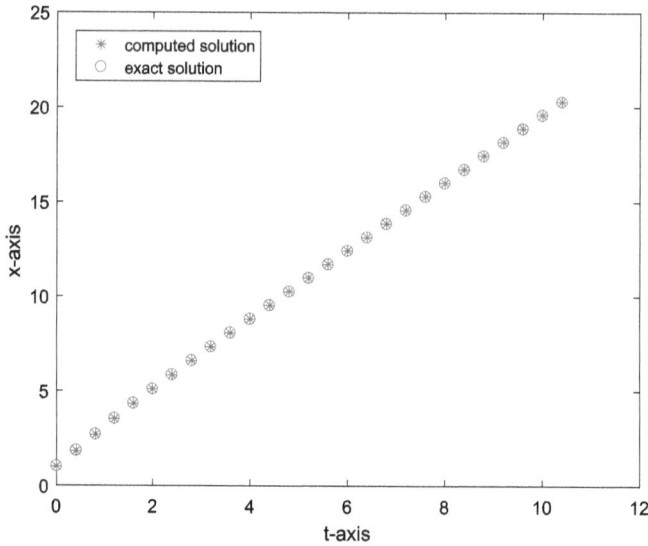

Figure 8.9: Approximate and exact solutions for $a = 0.2$, $a_1 = 0$, $a_2 = 0.5$, $b_1 = 0.5$, $b_2 = 1$, and $r = 0.4$.

Table 8.10: The values of the exact solution $x^{(e)}$ and approximate solution $x^{(a)}$ for $a = 0.4$, $a_1 = 0$, $a_2 = 1$, $b_1 = 0$, $b_2 = 1$, and $r = 0.8$.

t	$x^{(e)}$	$x^{(a)}$
0.00	1.00000000	1.00000000
0.80	2.65911330	2.70482759
1.60	4.29312420	4.37622615
2.40	5.82142631	5.93815579
3.20	7.30182274	7.45071010
4.00	8.75998114	8.94031362
4.80	10.20649681	10.41787140
5.60	11.64626723	11.88843233
6.40	13.08181407	13.35460156
7.20	14.51455340	14.81784515
8.00	15.94533686	16.27904663
8.80	17.37470556	17.73876842
9.60	18.80301900	19.19738481

Second, we take a two-step Runge–Kutta method consistent of order $p = 2$ by choosing $a_1 = 0$, $a_2 = 0.25$, $b_1 = 0$, $b_2 = 4$ so that the consistency condition is satisfied.

We choose $a = 0.4$, $x(0) = \alpha = 1$ and $r = 0.8$. The approximate and exact solutions are compared in Table 8.10 and their graphs are shown in Figure 8.10.

Finally, we take $a = 0.2$, $x(0) = \alpha = 1$ and $r = 0.4$. The approximate and exact solutions are compared in Table 8.11 and their graphs are shown in Figure 8.11.

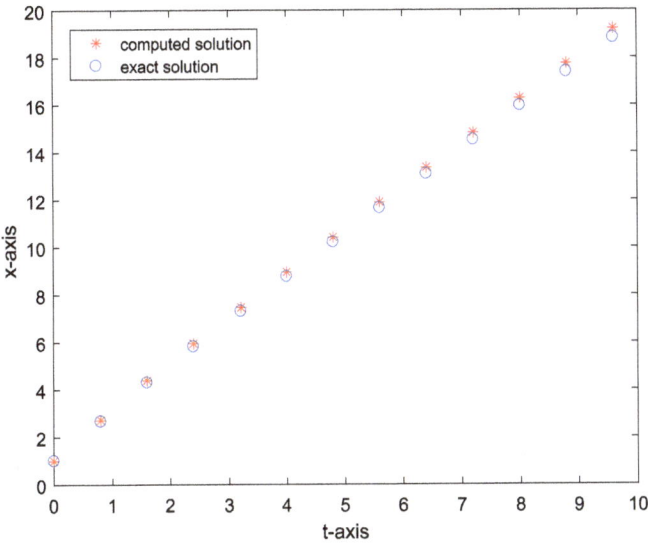

Figure 8.10: Approximate and exact solutions for $a = 0.4$, $a_1 = 0$, $a_2 = 1$, $b_1 = 0$, $b_2 = 1$, and $r = 0.8$.

Table 8.11: The values of the exact solution $x^{(e)}$ and approximate solution $x^{(a)}$ for $a = 0.2$, $a_1 = 0$, $a_2 = 1$, $b_1 = 0$, $b_2 = 1$, and $r = 0.4$.

t	$x^{(e)}$	$x^{(a)}$
0.00	1.00000000	1.00000000
0.40	1.82564103	1.83230769
0.80	2.68827709	2.70072505
1.20	3.51981834	3.53737714
1.60	4.31615069	4.33843059
2.00	5.08753630	5.11431504
2.40	5.84262105	5.87376394
2.80	6.58699088	6.62241056
3.20	7.32411400	7.36375045
3.60	8.05617463	8.09998473
4.00	8.78458851	8.83254016
4.40	9.51030240	9.56237100
4.80	10.23396853	10.29013479
5.20	10.95604855	11.01629706
5.60	11.67687725	11.74119547
6.00	12.39670271	12.46508030
6.40	13.11571227	13.18814063
6.80	13.83404989	13.91052175
7.20	14.55182789	14.63233706
7.60	15.26913509	15.35367626
8.00	15.98604260	16.07461118
8.40	16.70260803	16.79520002
8.80	17.41887848	17.51549037
9.20	18.13489285	18.23552156
9.60	18.85068352	18.95532631
10.00	19.56627768	19.67493210

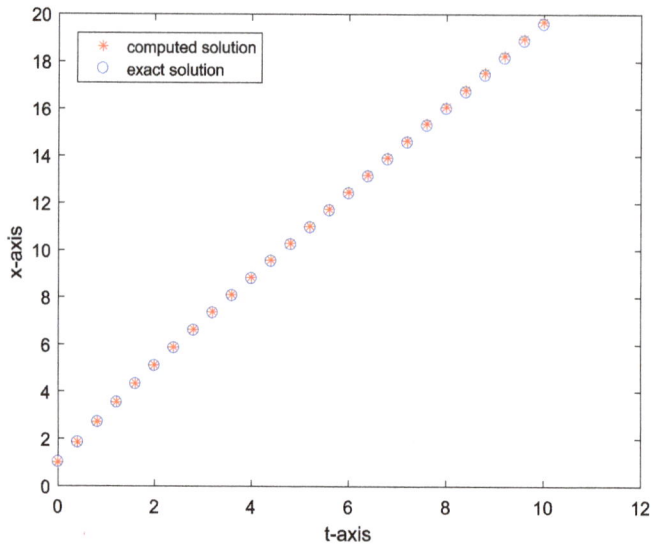

Figure 8.11: Approximate and exact solutions for $a = 0.2$, $a_1 = 0$, $a_2 = 1$, $b_1 = 0$, $b_2 = 1$, and $r = 0.4$.

8.6 Advanced practical problems

Problem 8.4. Let $\mathbb{T} = \frac{1}{3}\mathbb{N}_0$. Consider the IVP

$$\begin{cases} x^\Delta(t) = x(t)^2 + t^2, & t \in [0, 9], \\ x(0) = 1. \end{cases}$$

Write a two-stage Runge–Kutta method with $r = \frac{2}{3}$.

Problem 8.5. Let $\mathbb{T} = \mathbb{R}$. Consider the IVP

$$\begin{cases} x^\Delta(t) = e^{x(t)t^2} + 3t, & t \in [0, 5], \\ x(0) = -1. \end{cases}$$

Write a two-stage Runge–Kutta method with $r = 0.4$.

Problem 8.6. Let $\mathbb{T} = \mathbb{R}$. Consider the IVP

$$\begin{cases} x^\Delta(t) = \sin(x(t)) + \frac{t^2+1}{x(t)}, & t \in [0, \pi], \\ x(0) = 1. \end{cases}$$

Write a three-stage Runge–Kutta method with $r = 0.2$.

9 The series solution method – SSM

In this chapter, we develop the series solution method for Cauchy problems associated with dynamic equations and Caputo fractional dynamic equations. The method uses the Taylor series expansion of an unknown function and can be applied to first-order dynamic and fractional dynamic equations having a nonlinearity of polynomial type. It can also be extended to linear higher-order equations with nonconstant coefficients [12].

9.1 Preliminaries on series representations

We will start with some preliminary definitions and notations. Suppose that $\mathbb{T}$ is a time scale with forward jump operator σ, delta derivative operator Δ, and Δ-differentiable graininess function μ. Fix $t_0 \in \mathbb{T}$. Let $h_k(\cdot, t_0)$, $k \in \mathbb{N}_0$, denote the monomials on time scales and $S_n^{(k)}$ the set of all possible strings of length n containing exactly k times σ and $n - k$ times Δ operator. For example, $S_3^{(2)}$ has the form

$$S_3^{(2)} = \{\sigma\sigma\Delta, \sigma\Delta\sigma, \Delta\sigma\sigma\}.$$

First, we give the following result which plays a crucial role for the deduction of the series solution method. Its proof is based on the Taylor formula, Leibnitz formula, and some of the properties of the monomials $h_n(\cdot, t_0)$.

Theorem 9.1. *For every $m, n \in \mathbb{N}_0$, one has*

$$h_n(t, t_0) h_m(t, t_0) = \sum_{l=m}^{m+n} \left(\sum_{\Lambda_{l,m} \in S_m^{(l)}} h_n^{\Lambda_{l,m}}(t_0, t_0) \right) h_l(t, t_0)$$

for any $t, t_0 \in \mathbb{T}$.

Proof. If $m = 0$ or $n = 0$, the assertion is evident. Suppose that $m \neq 0$ and $n \neq 0$. By the Taylor formula, we have

$$h_n(t, t_0) h_m(t, t_0) = \sum_{l=0}^{\infty} (h_n(t, t_0) h_m(t, t_0))^{\Delta^l} \big|_{t=t_0} h_l(t, t_0), \quad t, t_0 \in \mathbb{T}.$$

By the Leibnitz formula, one has

$$(h_n(t, t_0) h_m(t, t_0))^{\Delta^l} = \sum_{k=0}^{l} \left(\sum_{\Lambda_{l,k} \in S_k^{(l)}} h_n^{\Lambda_{l,k}}(t, t_0) \right) h_m^{\Delta^k}(t, t_0), \quad t, t_0 \in \mathbb{T}.$$

If $l < m$, then

https://doi.org/10.1515/9783110787320-009

$$\left(h_n(t,t_0)h_m(t,t_0)\right)^{\Delta^l} = \sum_{k=0}^{l}\left(\sum_{\Lambda_{l,k}\in S_k^{(l)}} h_n^{\Lambda_{l,k}}(t,t_0)\right)h_{m-k}(t,t_0), \quad t,t_0 \in \mathbb{T}.$$

From here, for $l < m$, we have $h_{m-k}(t_0,t_0) = 0$ and

$$\left(h_n(t,t_0)h_m(t,t_0)\right)^{\Delta^l}\Big|_{t=t_0} = 0, \quad t,t_0 \in \mathbb{T}.$$

For $l \geq m$, using that $h_0(t,t_0) = 1$, we get

$$\left(h_n(t,t_0)h_m(t,t_0)\right)^{\Delta^l}\Big|_{t=t_0} = \sum_{k=0}^{m-1}\left(\sum_{\Lambda_{l,k}\in S_k^{(l)}} h_n^{\Lambda_{l,k}}(t,t_0)\right)h_{m-k}(t,\alpha)\Big|_{t=t_0}$$

$$+ \sum_{\Lambda_{l,m}\in S_m^{(l)}} h_n^{\Lambda_{l,m}}(t,t_0)\Big|_{t=t_0}$$

$$= \sum_{\Lambda_{l,m}\in S_m^{(l)}} h_n^{\Lambda_{l,m}}(t_0,t_0), \quad t,t_0 \in \mathbb{T}.$$

Hence, using the fact that $\Lambda_{l,m}$ consists of m times σ and $l - m$ times Δ, and

$$f^\sigma = f \quad \text{or} \quad f^\sigma = f + \mu f^\Delta,$$
$$f^{\sigma\sigma} = f \quad \text{or} \quad f^{\sigma\sigma} = f + \mu f^\Delta + \mu^\sigma(f^\Delta + \mu f^{\Delta^2}),$$

and so on, we obtain

$$h_n(t,t_0)h_m(t,t_0) = \sum_{l=m}^{\infty} \left(h_n(t,t_0)h_m(t,t_0)\right)^{\Delta^l}\Big|_{t=t_0} h_l(t,t_0)$$

$$= \sum_{l=m}^{\infty}\left(\sum_{\Lambda_{l,m}\in S_m^{(l)}} h_n^{\Lambda_{l,m}}(t_0,t_0)\right)h_l(t,t_0)$$

$$= \sum_{l=m}^{m+n}\left(\sum_{\Lambda_{l,m}\in S_m^{(l)}} h_n^{\Lambda_{l,m}}(t_0,t_0)\right)h_l(t,t_0), \quad t,t_0 \in \mathbb{T},$$

which completes the proof. $\qquad\square$

Consider an infinite series of the form

$$\sum_{i=0}^{\infty} Q_i h_i(t,t_0), \quad t,t_0 \in \mathbb{T}, \quad t > t_0, \tag{9.1}$$

for some constants $Q_r, r \in \mathbb{N}_0$. Define the constants $C_{r,k,l}$ as

$$C_{r,k,l} = \sum_{\Lambda_{r,k-l}\in S_{k-l}^{(r)}} h_l^{\Lambda_{r,k-l}}(t_0,t_0), \tag{9.2}$$

where $r \in \{k-l,\ldots,k\}$, $l \in \{0,\ldots,k\}$, $k \geq r$, and the constants $Q_{n,r}$ as

$$Q_{1,r} = Q_r,$$

$$Q_{n,r} = \sum_{k=r}^{\infty} \sum_{l=k-r}^{k} Q_{n-1,l} Q_{1,k-l} C_{r,k,l}, \tag{9.3}$$

for $r, k \in \mathbb{N}_0$, $k \geq r$, and $n \in \mathbb{N}$, $n > 1$. Using these notations and the result in Theorem 9.1, for $t, t_0 \in \mathbb{T}$, we compute the following:

$$\left(\sum_{i=0}^{\infty} Q_i h_i(t, t_0) \right)^2 = \sum_{i=0}^{\infty} \sum_{j=0}^{\infty} Q_i Q_j h_i(t, t_0) h_j(t, t_0)$$

$$= \sum_{k=0}^{\infty} \sum_{l=0}^{k} Q_l Q_{k-l} h_l(t, t_0) h_{k-l}(t, t_0)$$

$$= \sum_{k=0}^{\infty} \left(\sum_{l=0}^{k} Q_l Q_{k-l} \sum_{r=k-l}^{k} \left(\sum_{\Lambda_{r,k-l} \in S_{k-l}^{(r)}} h_l^{\Lambda_{r,k-l}}(t_0, t_0) \right) h_r(t, t_0) \right).$$

Now, we employ the constants $C_{r,k,l}$, defined in (9.2), and arrive at

$$\left(\sum_{i=0}^{\infty} Q_i h_i(t, t_0) \right)^2 = \sum_{k=0}^{\infty} \left(\sum_{l=0}^{k} Q_l Q_{k-l} \sum_{r=k-l}^{k} C_{r,k,l} h_r(t, t_0) \right)$$

$$= \sum_{k=0}^{\infty} \sum_{l=0}^{k} \sum_{r=k-l}^{k} Q_l Q_{k-l} C_{r,k,l} h_r(t, t_0)$$

$$= \sum_{r=0}^{\infty} \sum_{k=r}^{\infty} \sum_{l=k-r}^{k} Q_{1,l} Q_{1,k-l} C_{r,k,l} h_r(t, t_0)$$

$$= \sum_{r=0}^{\infty} Q_{2,r} h_r(t, t_0), \quad t, t_0 \in \mathbb{T}.$$

In a similar way, we compute

$$\left(\sum_{i=0}^{\infty} Q_i h_i(t, t_0) \right)^3 = \left(\sum_{i=0}^{\infty} Q_i h_i(t, t_0) \right)^2 \left(\sum_{j=0}^{\infty} Q_j h_j(t, t_0) \right)$$

$$= \left(\sum_{i=0}^{\infty} Q_{2,i} h_i(t, t_0) \right) \left(\sum_{j=0}^{\infty} Q_{1,j} h_j(t, t_0) \right)$$

$$= \sum_{k=0}^{\infty} \sum_{l=0}^{k} \sum_{r=k-l}^{k} Q_{2,l} Q_{1,k-l} C_{r,k,l} h_r(t, t_0)$$

$$= \sum_{r=0}^{\infty} \sum_{k=r}^{\infty} \left(\sum_{l=k-r}^{k} Q_{2,l} Q_{1,k-l} C_{r,k,l} \right) h_r(t, t_0)$$

$$= \sum_{r=0}^{\infty} Q_{3,r} h_r(t, t_0), \quad t, t_0 \in \mathbb{T}.$$

Generalizing this representation, we end up with the following formula:

$$\left(\sum_{i=0}^{\infty} Q_i h_i(t, t_0) \right)^n = \sum_{r=0}^{\infty} Q_{n,r} h_r(t, t_0), \quad t, t_0 \in \mathbb{T}, \, n \in \mathbb{N}. \tag{9.4}$$

9.2 The SSM for dynamic equations

Consider the Cauchy problem

$$\begin{cases} y^{\Delta}(t) = f(t, y(t)), & t > t_0, \\ y(t_0) = y_0, \end{cases} \tag{9.5}$$

where $f : \mathbb{T} \times \mathbb{R} \to \mathbb{R}$ is a given function and $y_0 \in \mathbb{R}$ is a given constant. Suppose that the nonlinear function f has the form

$$f(t, y(t)) = \sum_{p=1}^{n} a_p(t)(y(t))^p + a_0(t),$$

where

$$a_p(t) = \sum_{i=0}^{\infty} A_{i,p} h_i(t, t_0), \quad p \in \{0, \ldots, n\}, \quad t, t_0 \in \mathbb{T}, \, t > t_0, \tag{9.6}$$

and the coefficients $A_{i,p}$ are given real constants for $i \in \mathbb{N}_0, p \in \{0, \ldots, n\}$. The problem (9.5) is equivalent to the integral equation

$$y(t) = y_0 + \int_{t_0}^{t} f(u, y(u)) \Delta u, \quad t, t_0 \in \mathbb{T}, \, t > t_0,$$

or, equivalently,

$$y(t) = y_0 + \int_{t_0}^{t} \left(\sum_{p=1}^{n} a_p(u)(y(u))^p + a_0(u) \right) \Delta u, \quad t, t_0 \in \mathbb{T}, \, t > t_0. \tag{9.7}$$

We will search a solution of the equation (9.7) of the form

$$y(t) = \sum_{i=0}^{\infty} B_i h_i(t, t_0), \quad t, t_0 \in \mathbb{T}, \, t > t_0, \tag{9.8}$$

where $B_i, i \in \mathbb{N}_0$ are constants which will be determined below. Let

$$B_{1,r} = B_r,$$

$$B_{s,r} = \sum_{k=r}^{\infty} \sum_{l=k-r}^{k} B_{s-1,l} B_{1,k-l} C_{r,k,l}, \tag{9.9}$$

where $r, k, s \in \mathbb{N}_0$, $k \geq r$, $s \geq 2$. Then, using (9.4), we obtain

$$(y(t))^p = \sum_{r=0}^{\infty} B_{p,r} h_r(t, t_0), \quad p \in \{1, \ldots, n\}, \quad t, t_0 \in \mathbb{T}, \ t > t_0.$$

Consequently,

$$a_p(y)(y(t))^p = \sum_{i=0}^{\infty} A_{i,p} h_i(t, t_0) \sum_{j=0}^{\infty} B_{p,j} h_j(t, t_0)$$

$$= \sum_{k=0}^{\infty} \sum_{l=0}^{k} \sum_{r=k-l}^{k} A_{l,p} B_{p,k-l} C_{r,k,l} h_r(t, t_0)$$

$$= \sum_{r=0}^{\infty} \left(\sum_{k=r}^{\infty} \sum_{l=k-r}^{k} A_{l,p} B_{p,k-l} C_{r,k,l} \right) h_r(t, t_0), \quad t, t_0 \in \mathbb{T}, \ t > t_0,$$

where $p \in \{1, \ldots, n\}$. Let

$$D_{r,p} = \sum_{k=r}^{\infty} \sum_{l=k-r}^{k} A_{l,p} B_{p,k-l} C_{r,k,l}. \tag{9.10}$$

Then

$$a_p(y)(y(t))^p = \sum_{r=0}^{\infty} D_{r,p} h_r(t, t_0), \quad t, t_0 \in \mathbb{T}, \ t > t_0, \tag{9.11}$$

where $p \in \{1, \ldots, n\}$. Now, using (9.10) an (9.11), we get

$$\sum_{i=0}^{\infty} B_i h_i(t, t_0) = y_0 + \int_{t_0}^{t} \left(\sum_{p=1}^{n} \left(\sum_{r=0}^{\infty} D_{r,p} h_r(u, t_0) \right) + \sum_{r=0}^{\infty} A_{0,r} h_r(u, t_0) \right) \Delta u$$

$$= y_0 + \sum_{p=1}^{n} \sum_{r=0}^{\infty} D_{r,p} h_{r+1}(t, t_0) + \sum_{r=0}^{\infty} A_{0,r} h_{r+1}(t, t_0), \quad t, t_0 \in \mathbb{T}, \ t > t_0, \tag{9.12}$$

whereupon

$$B_0 = y_0,$$

$$B_i = \sum_{p=1}^{n} D_{i-1,p} + A_{0,i-1}, \quad i \in \mathbb{N}. \tag{9.13}$$

9.3 The SSM for Caputo fractional dynamic equation

Suppose that $\mathbb{T}$ is a time scale with forward jump operator σ, graininess function μ, and delta differential operator Δ, and that $\mathbb{T}$ has the form

$$\mathbb{T} = \{t_n : n \in \mathbb{N}_0\},$$

where

$$\lim_{n \to \infty} t_n = \infty, \quad \sigma(t_n) = t_{n+1}, \quad n \in \mathbb{N}_0, \quad w = \inf_{n \in \mathbb{N}_0} \mu(t_n) > 0.$$

Assume that the graininess function μ is delta differentiable. First, we will recall the Laplace transform on time scales.

Definition 9.2. Let $\mathbb{T}_0$ be a time scale such that $0 \in \mathbb{T}_0$ and $\sup \mathbb{T}_0 = \infty$. Let $f : \mathbb{T}_0 \to \mathbb{C}$ and define the set

$$\mathcal{D}(f) = \left\{ z \in \mathbb{C} : 1 + z\mu(t) \neq 0 \text{ for all } t \in \mathbb{T}_0 \right.$$

$$\left. \text{and the improper integral } \int_0^\infty f(y) e_{\ominus z}^\sigma(y, 0)\Delta y \text{ exists} \right\},$$

where $e_{\ominus z}^\sigma(y, 0) = (e_{\ominus z} \circ \sigma)(y, 0) = e_{\ominus z}(\sigma(y), 0)$.

The Laplace transform of the function f is defined as

$$\mathcal{L}(f)(z) = \int_0^\infty f(y) e_{\ominus z}^\sigma(y, 0)\Delta y, \tag{9.14}$$

for all $z \in \mathcal{D}(f)$.

Other concepts needed in the definition of the fractional Δ-derivative are the shift of a function and the convolution of two functions on a time scale.

Definition 9.3. For a given function $f : \mathbb{T} \to \mathbb{C}$, the shift (delay) of f is denoted by $\widehat{f}$ and defined as the solution of the shifting problem

$$\begin{cases} u^{\Delta_t}(t, \sigma(s)) = -u^{\Delta_s}(t, s), & t \in \mathbb{T}, \quad t \geq s \geq t_0, \\ u(t, t_0) = f(t), & t \in \mathbb{T}, \quad t \geq t_0. \end{cases} \tag{9.15}$$

Example 9.4.
1. Let $f : \mathbb{T} \to \mathbb{C}$ be any function where $\mathbb{T}$ is either $\mathbb{R}$ or $\mathbb{Z}$. Then the shift of f is

$$\widehat{f(\cdot)}(t, s) = f(t - s + t_0), \quad t \geq s \geq t_0.$$

2. The shift of $e_\lambda(t, t_0)$, where $t, t_0 \in \mathbb{T}$ and $t \geq t_0$, is

$$\widehat{e_\lambda(\cdot, t_0)}(t, s) = e_\lambda(t, s), \quad t, s \in \mathbb{T} \text{ and are independent of } t_0.$$

3. Let $f : [t_0, \infty] \to \mathbb{C}$ be a function of the form

$$f(t) = \sum_{k=0}^{\infty} a_k h_k(t, t_0),$$

where the coefficients a_k satisfy

$$|a_k| \le MR^k,$$

for some $M, R > 0$ and $k \in \mathbb{N}_0$. Then the shift of f has the form

$$\widehat{f(\cdot)}(t, s) = \sum_{k=0}^{\infty} a_k h_k(t, s), \quad t, s \in \mathbb{T}, \quad t \ge s \ge t_0.$$

In particular, we have

$$\widehat{h_k(\cdot, t_0)}(t, s) = h_k(t, s), \quad t, s \in \mathbb{T}, \quad t \ge s \ge t_0, \quad \text{and} \quad k \in \mathbb{N}_0.$$

Definition 9.5. For functions $f, g : \mathbb{T} \to \mathbb{C}$, the convolution $f * g$ is defined as

$$(f * g)(t) = \int_{t_0}^{t} \widehat{f}(t, \sigma(s)) g(s) \Delta s, \quad t \in \mathbb{T}, \ t \ge t_0. \tag{9.16}$$

The convolution is associative, that is, $(f * g) * h = f * (g * h)$.

Next, define the generalized Δ-power function, the Riemann–Liouville fractional Δ-integral and Δ-derivative, and the Caputo fractional Δ-derivative on the time scale $\mathbb{T}$ in the form given above. Take $\alpha \in \mathbb{R}$ arbitrarily.

Definition 9.6. The generalized Δ-power function $h_\alpha(t, t_0)$ on $\mathbb{T}$ is defined as

$$h_\alpha(t, t_0) = \mathcal{L}^{-1}\left(\frac{1}{z^{\alpha+1}}\right)(t), \quad t \ge t_0,$$

for all $z \in \mathbb{C}\backslash\{0\}$ such that $\mathcal{L}^{-1}$ exists. The fractional generalized Δ-power function $h_\alpha(t, s)$ on $\mathbb{T}$ is defined as the shift of $h_\alpha(t, t_0)$, that is,

$$h_\alpha(t, s) = \widehat{h_\alpha(\cdot, t_0)}(t, s), \quad t, s \in \mathbb{T}, \quad t \ge s \ge t_0.$$

The series solution method employs the following property of the generalized Δ-power functions.

Theorem 9.7. *Let $\alpha, \beta \in \mathbb{R}$. Then*

$$(h_\alpha(\cdot, t_0) * h_\beta(\cdot, t_0))(t) = h_{\alpha+\beta+1}(t, t_0), \quad t \in \mathbb{T}.$$

Definition 9.8. Let $\alpha \geq 0$ and let $\overline{[-\alpha]}$ denote the integer part of $-\alpha$. For a function $f : \mathbb{T} \to \mathbb{R}$, the Riemann–Liouville fractional Δ-integral of order α is defined as

$$(I^0_{\Delta,t_0} f)(t) = f(t),$$
$$(I^\alpha_{\Delta,t_0} f)(t) = (h_{\alpha-1}(\cdot, t_0) * f)(t)$$

$$= \int_{t_0}^{t} \widehat{h_{\alpha-1}(\cdot, t_0)}(t, \sigma(u)) f(u) \Delta u \tag{9.17}$$

$$= \int_{t_0}^{t} h_{\alpha-1}(t, \sigma(u)) f(u) \Delta u,$$

for $\alpha > 0$ and $t \geq t_0$.

Definition 9.9. Let $\alpha \geq 0$, $m = -\overline{[-\alpha]}$, and $f : \mathbb{T} \to \mathbb{R}$. For $s, t \in \mathbb{T}^{\kappa^m}$, $s < t$, the Riemann–Liouville fractional Δ-derivative of order α is defined by

$$D^\alpha_{\Delta,s} f(t) = D^m_\Delta I^{m-\alpha}_{\Delta,s} f(t), \quad t \in \mathbb{T}, \tag{9.18}$$

if it exists. For $\alpha < 0$, we define

$$D^\alpha_{\Delta,s} f(t) = I^{-\alpha}_{\Delta,s} f(t), \quad t, s \in \mathbb{T}, \quad t > s.$$
$$I^\alpha_{\Delta,s} f(t) = D^{-\alpha}_{\Delta,s} f(t), \quad t, s \in \mathbb{T}^{\kappa^r}, \quad t > s, \quad r = \overline{[-\alpha]} + 1. \tag{9.19}$$

Remark 9.10. Noting that the generalized monomials $h_\alpha(t, t_0)$ on the set of real numbers $\mathbb{R}$ are computed as

$$h_\alpha(t, t_0) = \mathcal{L}^{-1}\left(\frac{1}{z^{\alpha+1}}\right)(t) = \frac{(t - t_0)^\alpha}{\Gamma(\alpha)}, \quad t \geq t_0,$$

we observe that if $\mathbb{T} = \mathbb{R}$, that is, if the Δ derivative is replaced by the classical derivative, the Riemann–Liouville fractional Δ-derivative defined in (9.18) becomes the usual Riemann–Liouville fractional derivative.

Using these definitions, the Caputo fractional Δ-derivative is defined as follows.

Definition 9.11. For a function $f : \mathbb{T} \to \mathbb{R}$, the Caputo fractional Δ-derivative of order α is denoted by ${}^C D^\alpha_{\Delta,t_0}$ and defined via the Riemann–Liouville fractional Δ-derivative of order α as follows:

$$^C D^\alpha_{\Delta,t_0} = D^\alpha_{\Delta,t_0}\left(f(t) - \sum_{k=0}^{m-1} h_k(t, t_0) f^{\Delta^k}(t_0)\right), \quad t > t_0, \tag{9.20}$$

where $m = \overline{[\alpha]} + 1$ if $\alpha \notin \mathbb{N}$ and $m = \overline{[\alpha]}$ if $\alpha \in \mathbb{N}$.

Another representation of the Caputo fractional Δ-derivative is given in the following theorem.

Theorem 9.12. *Let $\alpha > 0$, $m = \overline{[\alpha]} + 1$ if $\alpha \notin \mathbb{N}$ and $m = \alpha$, if $\alpha \in \mathbb{N}$.*
1. *If $\alpha \notin \mathbb{N}$ then*

$$^{C}D_{\Delta,t_0}^{\alpha} f(t) = I_{\Delta,t_0}^{m-\alpha} D_{\Delta,t_0}^{m} f(t), \quad t \in \mathbb{T}, \quad t > t_0.$$

2. *If $\alpha \in \mathbb{N}$ then*

$$^{C}D_{\Delta,t_0}^{\alpha} f(t) = f^{\Delta^{m}}(t), \quad t \in \mathbb{T}, \quad t > t_0.$$

Remark 9.13. Regarding the result of Theorem 9.12, if $\mathbb{T} = \mathbb{R}$, the Caputo fractional Δ-derivative defined in (9.20) becomes the usual Caputo fractional derivative.

Let $^{C}D_{\Delta,t_0}^{\alpha}$ denote the Caputo fractional Δ-derivative. Suppose that $\alpha > 0$ and that $m = -\overline{[-\alpha]}$. We will consider the Cauchy problem associated with the Caputo fractional Δ-derivative given as

$$\begin{cases} ^{C}D_{\Delta,t_0}^{\alpha} y(t) = f(t, y(t)), & t > t_0, \\ ^{C}D_{\Delta,t_0}^{k} y(t) = b_k, & k \in \{0, \ldots, m-1\}, \end{cases} \tag{9.21}$$

where $f : \mathbb{T} \times \mathbb{R} \to \mathbb{R}$ is a given function and $b_k \in \mathbb{R}$ for $k \in \{0, \ldots, m-1\}$ are given constants.

We suppose that the nonlinear function f has the form

$$f(t, y(t)) = \left(h_{-\alpha-1}(\cdot, t_0) * \left(\sum_{p=1}^{n} a_p(\cdot)(y(\cdot))^p + a_0(\cdot) \right) \right)(t),$$

where

$$a_p(t) = \sum_{i=0}^{\infty} A_{i,p} h_i(t, t_0), \quad p \in \{0, \ldots, n\}, \tag{9.22}$$

and the coefficients $A_{i,p}$ are given real constants for $i \in \mathbb{N}_0$, $p \in \{0, \ldots, n\}$.

It can be shown that the Cauchy problem (9.21) is equivalent to an integral equation of the form

$$y(t) = \sum_{j=0}^{m-1} h_j(t, t_0) b_j + \left(h_{\alpha-1}(\cdot, t_0) * \left(h_{-\alpha-1}(\cdot, t_0) * \left(\sum_{p=1}^{n} a_p(\cdot)(y(\cdot))^p + a_0(\cdot) \right) \right) \right)(t)$$

$$= \sum_{j=0}^{m-1} h_j(t, t_0) b_j + \left(h_{-1}(\cdot, t_0) * \left(\sum_{p=1}^{n} a_p(\cdot)(y(\cdot))^p + a_0(\cdot) \right) \right)(t). \tag{9.23}$$

We will search for a solution of the equation (9.23) of the form

$$y(t) = \sum_{i=0}^{\infty} B_i h_i(t, t_0), \tag{9.24}$$

where B_i, $i \in \mathbb{N}_0$, are constants to be determined from the equation. As in the previous section, we let

$$B_{1,r} = B_r,$$
$$B_{s,r} = \sum_{k=r}^{\infty} \sum_{l=k-r}^{k} B_{s-1,l} B_{1,k-l} C_{r,k,l}, \tag{9.25}$$

where $r, k, s \in \mathbb{N}_0$, $k \geq r$, $s \geq 2$, and have

$$(y(t))^p = \sum_{r=0}^{\infty} B_{p,r} h_r(t, t_0), \quad p \in \{1, \ldots, n\}. \tag{9.26}$$

Also we use the fact that

$$a_p(y)(y(t))^p = \sum_{r=0}^{\infty} D_{r,p} h_r(t, t_0), \tag{9.27}$$

where $p \in \{1, \ldots, n\}$. Hence, we obtain

$$\sum_{i=0}^{\infty} B_i h_i(t, t_0) = \sum_{j=0}^{\infty} b_j h_j(t, t_0)$$
$$+ \left(h_{-1}(\cdot, t_0) * \left(\sum_{p=1}^{n} \left(\sum_{r=0}^{\infty} D_{r,p} h_r(\cdot, t_0) \right) + \sum_{r=0}^{\infty} A_{0,r} h_r(\cdot, t_0) \right) \right)(t)$$
$$= \sum_{j=0}^{\infty} b_j h_j(t, t_0) + \sum_{p=1}^{n} \sum_{r=0}^{\infty} D_{r,p} h_r(t, t_0) + \sum_{r=0}^{\infty} A_{0,r} h_r(t, t_0), \tag{9.28}$$

which implies the following relation for the computation of the coefficients B_i in the series representation of y:

$$B_i = b_i + \sum_{p=1}^{n} D_{i,p} + A_{0,i}, \quad i \in \{0, \ldots, m-1\},$$
$$B_i = \sum_{p=1}^{n} D_{i,p} + A_{0,i}, \quad i \in \{m, \ldots, \}. \tag{9.29}$$

Because of the nonlinear structure of the function f involved in the fractional dynamic equation, the recurrence relation (9.29) is also nonlinear.

9.4 Numerical examples

In this section, we consider some particular examples of Cauchy problems associated with dynamic equations and Caputo fractional dynamic equations on time scales.

Example 9.14. In the first example, we will apply the series solution method to a dynamic equation used in population growth models, known as the logistic model. The logistic model on an arbitrary time scale is described by the Cauchy problem

$$\begin{cases} N^\Delta(t) = \frac{\alpha N(t)}{\mu(t)}\left(1 - \frac{N(t)}{K}\right), & t \geq t_0, \\ N(t_0) = N_0. \end{cases} \tag{9.30}$$

Here $N(t)$ is the size of the population of a certain species at time t and $N(t_0) = N_0$ is the initial size of the population. The constant α represents the proportionality constant which is large for quickly growing species like bacteria and small for slowly growing populations like elephants. The constant K stands for the carrying capacity of the system, that is, the size of the population that the environment can sustain for the long term.

The logistic model discussed here is different from the model solved in Example 7.22. The logistic equation given in Example 7.22 was proposed by Bohner and Peterson [1] who argued that this model is more suitable as a generalization of the continuous logistic differential equation. On the other hand, the dynamic equation given in (9.30) appears quite often in the literature as the logistic model on time scales.

We will consider this model on the time scale $\mathbb{T} = a\mathbb{Z}$ for some positive constant a.

As noted above, the Cauchy problem (9.30) can be written as an integral equation of the form

$$N(t) = N_0 + \int_{t_0}^{t} \frac{\alpha N(u)}{a}\left(1 - \frac{N(u)}{K}\right)\Delta u, \quad t, t_0 \in a\mathbb{Z}, \tag{9.31}$$

which is a nonlinear Volterra integral equation of the second kind [10]. We will take the initial time as $t_0 = 0$ and the initial population as N_0 and apply the series solution method to solve this integral equation. Let

$$N(t) = \sum_{i=0}^{\infty} B_i h_i(t, 0) = \sum_{i=0}^{\infty} B_{1,i} h_i(t, 0), \quad t \in a\mathbb{Z}.$$

Then,

$$(N(t))^2 = \left(\sum_{i=0}^{\infty} B_{1,i} h_i(t, 0)\right)\left(\sum_{j=0}^{\infty} B_{1,j} h_j(t, 0)\right) = \sum_{r=0}^{\infty} B_{2,r} h_r(t, 0), \quad t \in a\mathbb{Z},$$

where

$$B_{2,r} = \sum_{k=r}^{\infty} \sum_{l=k-r}^{k} B_{1,l} B_{1,k-l} C_{r,k,l},$$

for $r \in \mathbb{N}_0$. We insert these series into equation (9.31) and get

$$\sum_{r=0}^{\infty} B_r h_r(t,0) = N_0 + \int_0^t \frac{\alpha}{a} \left(\sum_{r=0}^{\infty} B_r h_r(u,0) - \frac{1}{K} \sum_{r=0}^{\infty} B_{2,r} h_r(u,0) \right) \Delta u$$

$$= N_0 + \frac{\alpha}{a} \left(\sum_{r=0}^{\infty} B_r h_{r+1}(t,0) - \frac{1}{K} \sum_{r=0}^{\infty} B_{2,r} h_{r+1}(t,0) \right)$$

$$= N_0 + \sum_{r=0}^{\infty} \left(\frac{\alpha}{a} B_r - \frac{\alpha}{aK} B_{2,r} \right) h_{r+1}(t,0), \quad t \in a\mathbb{Z}.$$

Therefore, we have

$$B_0 = N_0,$$
$$B_{r+1} = \frac{\alpha}{a} B_r - \frac{\alpha}{aK} B_{2,r}, \quad \text{for all } r \in \mathbb{N}_0. \tag{9.32}$$

We recall that on $\mathbb{T} = a\mathbb{Z}$ the forward jump operator is $\sigma(t) = t + a$. The first five monomials $h_n(t,0)$, $n = 0, 1, 2, 3, 4$ are as follows:

$$h_0(t,0) = 1,$$
$$h_1(t,0) = t,$$
$$h_2(t,0) = \int_0^t x \Delta x = \frac{t(t-a)}{2},$$
$$h_3(t,0) = \int_0^t \frac{x(x-a)}{2} \Delta x = \frac{t(t-a)(t-2a)}{6},$$
$$h_4(t,0) = \int_0^t \frac{x(x-a)(x-2a)}{6} \Delta x = \frac{t(t-a)(t-2a)(t-3a)}{24}, \quad t \in a\mathbb{Z}.$$

To compute the first few coefficients $B_{2,r}$, we consider the series expansion of N^2 given as

$$(N(t))^2 = (B_0 h_0(t,0) + B_1 h_1(t,0) + B_2 h_2(t,0) + B_3 h_3(t,0) + \cdots)^2$$
$$= B_0 B_0 + (B_0 B_1 + B_1 B_0)t + B_1 B_1 t^2 + (B_0 B_2 + B_2 B_0)\frac{t(t-1)}{2}$$
$$+ (B_1 B_2 + B_2 B_1)\frac{t^2(t-1)}{2} + B_2 B_2 \frac{t^2(t-1)^2}{4} + \cdots, \quad t \in a\mathbb{Z},$$

and we note that

$$t^2 = ah_1(t,0) + 2h_2(t,0),$$
$$\frac{t^2(t-1)}{2} = 2ah_2(t,0) + 3h_3(t,0),$$
$$\frac{t^2(t-1)^2}{4} = a^2 h_2(t,0) + 6ah_3(t,0) + 6h_4(t,0), \quad t \in a\mathbb{Z}.$$

As a result, we obtain for $t \in a\mathbb{Z}$,

$$(N(t))^2 = \sum_{n=0}^{\infty} B_{2,n} h_n(t,0)$$

$$= B_0 B_0 h_0(t,0) + (B_0 B_1 + B_1 B_0 + aB_1 B_1) h_1(t,0)$$

$$+ (B_0 B_2 + B_2 B_0 + 2B_1 B_1 + 2aB_1 B_2 + 2aB_2 B_1 + a^2 B_2 B_2) h_2(t,0) + \cdots.$$

Then, the recurrence relation (9.32) yields

$$B_0 = N_0,$$

$$B_1 = \frac{\alpha}{a} B_0 - \frac{\alpha}{aK} B_{2,0} = \frac{\alpha}{a} B_0 - \frac{\alpha}{aK} B_0 B_0,$$

$$B_2 = \frac{\alpha}{a} B_1 - \frac{\alpha}{aK} B_{2,1} = \frac{\alpha}{a} B_1 - \frac{\alpha}{aK}(B_0 B_1 + B_1 B_0 + aB_1 B_1),$$

$$B_3 = \frac{\alpha}{a} B_2 - \frac{\alpha}{aK} B_{2,2}$$

$$= \frac{\alpha}{a} B_2 - \frac{\alpha}{aK}(B_0 B_2 + B_2 B_0 + 2B_1 B_1 + 2aB_1 B_2 + 2aB_2 B_1 + a^2 B_2 B_2).$$

Then, the solution $N(t)$ has the form

$$N(t) = B_0 + B_1 t + B_2 \frac{t(t-a)}{2} + B_3 \frac{t(t-a)(t-2a)}{6} + \cdots, \quad t \in a\mathbb{Z}.$$

On the time scale $\mathbb{T} = a\mathbb{Z}$, problem (9.30) can be written as

$$\begin{cases} \frac{N(t+a)-N(t)}{a} = \frac{\alpha N(t)}{a}\left(1 - \frac{N(t)}{K}\right), & t \geq t_0, \\ N(t_0) = N_0, \end{cases}$$

whose exact solution is computed as

$$N(t+a) = N(t) + \alpha N(t)\left(1 - \frac{N(t)}{K}\right), \quad \text{where } N(0) = N_0, \quad t \geq 0.$$

We compute the series and exact solutions of the problem for several values of the parameters N_0, α, K, and a on the interval $[0,5]$. Figures 9.1 and 9.2 show the comparison of the computed and exact solutions. It can be observed that the exact solution tends to the equilibrium K after approaching this value. The computed solution remains close to the exact solution at points close enough to the value 0 at which the Taylor series representation is considered. This behavior is typical for the series solutions, moreover, only 4 terms of the series solution are computed.

In Tables 9.1 and 9.2 the exact and approximate solutions are compared for the values $N_0 = 3$, $K = 20$, $a = 0.5$, and $\alpha = 0.4, 0.2$. The graphs of these solutions are also compared in Figures 9.1 and 9.2.

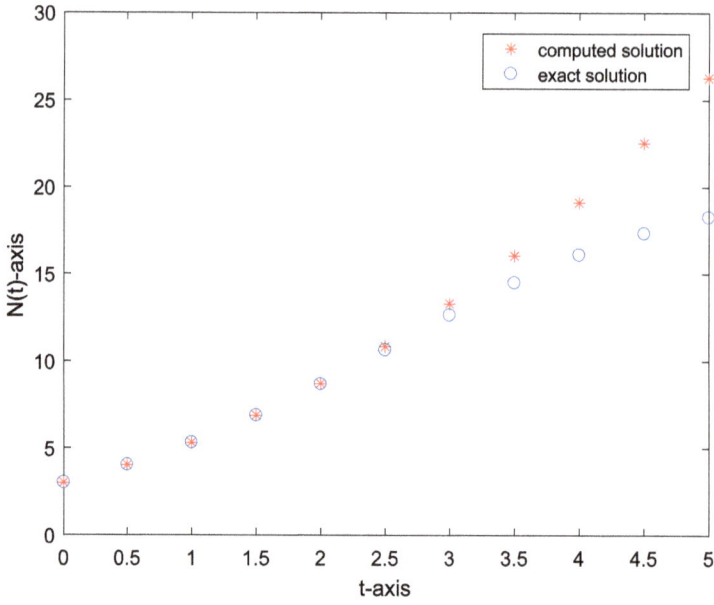

Figure 9.1: Computed and exact values of the solution with $N_0 = 3$, $K = 20$, $a = 0.5$, and $\alpha = 0.4$.

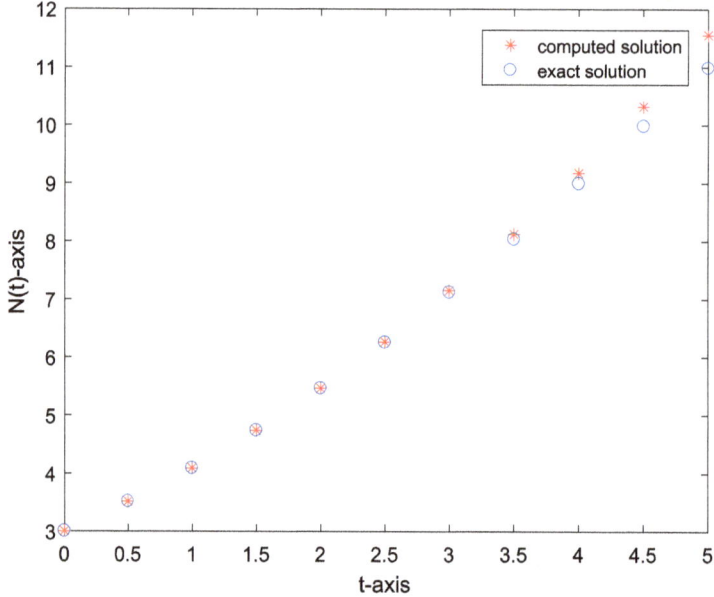

Figure 9.2: Computed and exact values of the solution with $N_0 = 3$, $K = 20$, $a = 0.5$, and $\alpha = 0.2$.

Table 9.1: The values of the approximate solution $N^{(a)}(t)$ and the exact solution $N^{(e)}(t)$ for $N_0 = 3$, $K = 20$, $a = 0.5$, and $\alpha = 0.4$.

t	$N^{(a)}(t)$	$N^{(e)}(t)$
0.00	3.00000000	3.00000000
0.50	4.02000000	4.02000000
1.00	5.30479200	5.30479200
1.50	6.86389244	6.86389244
2.00	8.70681775	8.66718902
2.50	10.84308437	10.63166132
3.00	13.28220873	12.62368140
3.50	16.03370729	14.48600732
4.00	19.10709646	16.08352209
4.50	22.51189269	17.34333727
5.00	26.25761241	18.26484522

Table 9.2: The values of the approximate solution $N^{(a)}(t)$ and the exact solution $N^{(e)}(t)$ for $N_0 = 3$, $K = 20$, $a = 0.5$, and $\alpha = 0.2$.

t	$N^{(a)}(t)$	$N^{(e)}(t)$
0.00	3.00000000	3.00000000
0.50	3.51000000	3.51000000
1.00	4.08879900	4.08879900
1.50	4.73937603	4.73937603
2.00	5.46471011	5.46263438
2.50	6.26778027	6.25675751
3.00	7.15156555	7.11663887
3.50	8.11904496	8.03350116
4.00	9.17319753	8.99482998
4.50	10.31700230	9.98472631
5.00	11.55343829	10.98472398

A larger value for the parameter K is also considered in Tables 9.3 and 9.4 where the exact and approximate solutions are compared for the values $N_0 = 10$, $K = 200$, $a = 0.25$, and $\alpha = 0.8$, 0.2. These solutions are also compared graphically in Figures 9.3 and 9.4. Finally, the results for the case $N_0 = 8$, $K = 40$, $a = 0.5$, and $\alpha = 0.1$ are presented in Table 9.5 and Figure 9.5, respectively.

Example 9.15. Consider the problem

$$
\begin{cases}
{}^C D_{\Delta,0}^{\frac{11}{4}} y(t) = \left(h_{-\frac{15}{4}}(\cdot,0) * \left(\left(\sum_{i=0}^{\infty} \frac{1}{i^2+3} h_i(\cdot,0)\right) y(\cdot)\right.\right. \\
\qquad\qquad\qquad + \left(\sum_{i=0}^{\infty} \frac{i}{i^2+i+1} h_i(\cdot,0)\right)(y(\cdot))^2\right)(t), \quad t > 0, \\
y(0) = 1, \quad y^{\Delta}(0) = -1, \quad y^{\Delta^2}(0) = 2.
\end{cases}
\tag{9.33}
$$

Table 9.3: The values of the approximate solution $N^{(a)}(t)$ and the exact solution $N^{(e)}(t)$ for $N_0 = 10$, $K = 200$, $a = 0.25$ and $\alpha = 0.8$.

t	$N^{(a)}(t)$	$N^{(e)}(t)$
0.00	10.00000000	10.00000000
0.25	17.60000000	17.60000000
0.50	30.44096000	30.44096000
0.75	51.08711982	51.08711982
1.00	82.10271927	81.51724043
1.50	185.49919634	158.52657334
1.75	263.00855360	184.82513419
2.00	361.14430976	196.04392063
2.25	482.47070464	199.14618187
2.50	629.55197805	199.82632035
2.75	804.95236982	199.96514341
3.00	1011.23611976	199.99302382
3.25	1250.96746769	199.99860457
3.50	1526.71065343	199.99972091
3.75	1841.02991679	199.99994418
4.00	2196.48949758	199.99998884
4.25	2595.65363564	199.99999777
4.50	3041.08657076	199.99999955
4.75	3535.35254278	199.99999991
5.00	4081.01579151	199.99999998

Table 9.4: The values of the approximate solution $N^{(a)}(t)$ and the exact solution $N^{(e)}(t)$ for $N_0 = 10$, $K = 200$, $a = 0.25$, and $\alpha = 0.2$.

t	$N^{(a)}(t)$	$N^{(e)}(t)$
0.00	10.00000000	10.00000000
0.25	11.90000000	11.90000000
0.50	14.13839000	14.13839000
0.75	16.76617393	16.76617393
1.00	19.83435571	19.83830413
1.25	23.39393928	23.41240664
1.50	27.49592856	27.54674718
1.75	32.19132749	32.29727334
2.00	37.53113998	37.71361414
2.25	43.56636997	43.83402028
2.50	50.34802138	50.67940300
2.75	57.92709815	58.24688171
3.00	66.35460421	66.50355883
3.25	75.68154347	75.38154726
3.50	85.95891987	84.77547904
3.75	97.23773733	94.54369300
4.00	109.56899980	104.51392172
4.25	123.00371118	114.49354623
4.50	137.59287542	124.28348334
4.75	153.38749643	133.69379578
5.00	170.43857816	142.55852391

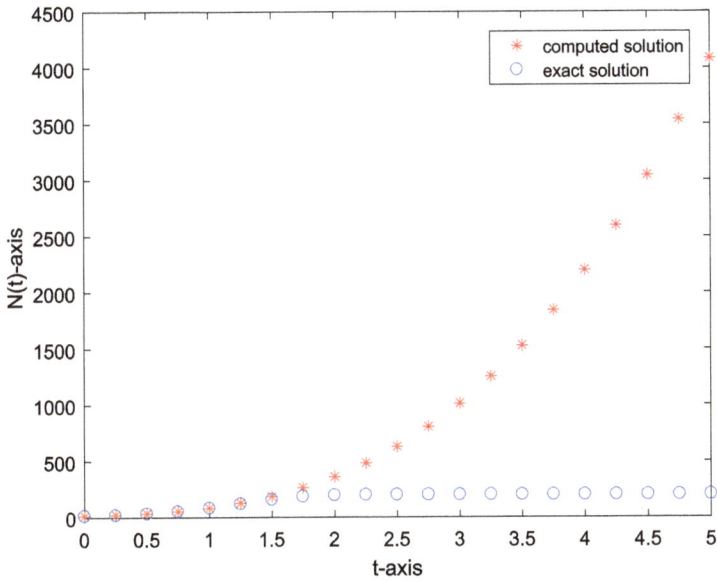

Figure 9.3: Computed and exact values of the solution with $N_0 = 10, K = 200, a = 0.25$, and $\alpha = 0.8$.

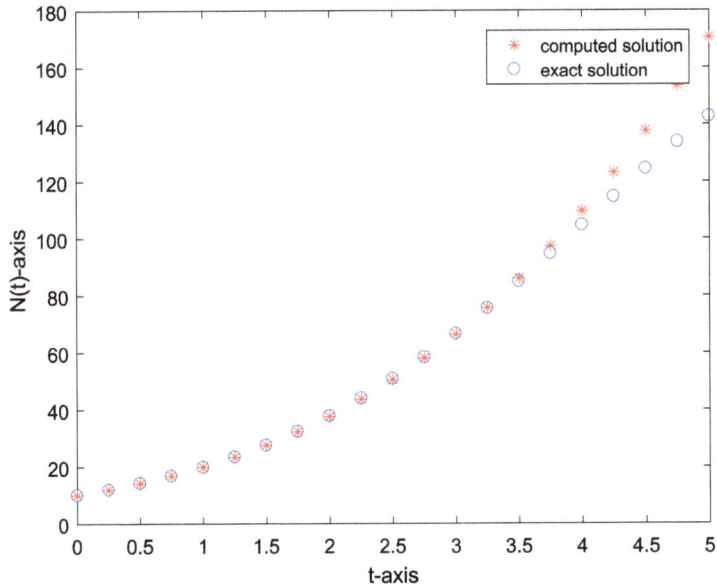

Figure 9.4: Computed and exact values of the solution with $N_0 = 10, K = 200, a = 0.25$, and $\alpha = 0.2$.

Table 9.5: The values of the approximate solution $N^{(a)}(t)$ and the exact solution $N^{(e)}(t)$ for $N_0 = 8$, $K = 40$, $a = 0.5$, and $\alpha = 0.1$.

t	$N^{(a)}(t)$	$N^{(e)}(t)$
0.00	8.00000000	8.00000000
0.50	8.64000000	8.64000000
1.00	9.31737600	9.31737600
1.50	10.03207986	10.03207986
2.00	10.78406344	10.78368128
2.50	11.57327861	11.57132995
3.00	12.39967722	12.39372376
3.50	13.26321114	13.24908516
4.00	14.16383223	14.13514803
4.50	15.10149234	15.04915681
5.00	16.07614334	15.98787969

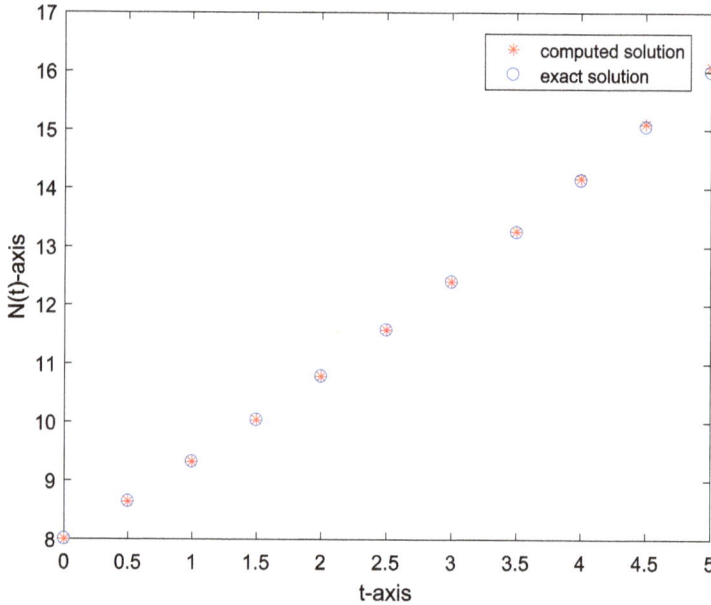

Figure 9.5: Computed and exact values of the solution with $N_0 = 8$, $K = 40$, $a = 0.5$, and $\alpha = 0.1$.

Here we have $\alpha = \frac{11}{4}$ and $m = -[-\frac{11}{4}] = 3$. Employing the integral equation form given in (9.23), we can rewrite problem (9.33) as

$$y(t) = h_0(t,0) - h_1(t,0) + 2h_2(t,0)$$

$$+ \left(h_{\frac{7}{4}}(\cdot,0) * h_{-\frac{15}{4}}(\cdot,0) * \left(\left(\sum_{i=0}^{\infty} \frac{1}{i^2 + 3} h_i(\cdot,0) \right) y(\cdot) \right) \right.$$

$$+ \left(\sum_{i=0}^{\infty} \frac{i}{i^2 + i + 1} h_i(\cdot, 0) \right) (y(\cdot))^2 \right) (t)$$

$$= h_0(t, 0) - h_1(t, 0) + 2h_2(t, 0)$$

$$+ \left(h_{-1}(\cdot, 0) * \left(\left(\sum_{i=0}^{\infty} \frac{1}{i^2 + 3} h_i(\cdot, 0) \right) y(\cdot) + \left(\sum_{i=0}^{\infty} \frac{i}{i^2 + i + 1} h_i(\cdot, 0) \right) (y(\cdot))^2 \right) \right) (t)$$

$$= h_0(t, 0) - h_1(t, 0) + 2h_2(t, 0) + \left(\left(\sum_{i=0}^{\infty} \frac{1}{i^2 + 3} h_{-1}(\cdot, 0) * h_i(\cdot, 0) \right) y(\cdot)$$

$$+ \left(\sum_{i=0}^{\infty} \frac{i}{i^2 + i + 1} h_{-1}(\cdot, 0) * h_i(\cdot, 0) \right) (y(\cdot))^2 \right) (t)$$

$$= h_0(t, 0) - h_1(t, 0) + 2h_2(t, 0)$$

$$+ \left(\sum_{i=0}^{\infty} \frac{1}{i^2 + 3} h_i(t, 0) \right) y(t) + \left(\sum_{i=0}^{\infty} \frac{i}{i^2 + i + 1} h_i(t, 0) \right) (y(t))^2. \tag{9.34}$$

Assume that

$$y(t) = \sum_{r=0}^{\infty} B_r h_r(t, 0), \quad t \in \mathbb{T}, t \geq 0, \tag{9.35}$$

where the coefficients B_r are going to be obtained. Then, by (9.9), we have

$$(y(t))^2 = \sum_{r=0}^{\infty} B_{2,r} h_r(t, 0), \quad t \in \mathbb{T}, t \geq 0, \tag{9.36}$$

where

$$B_{1,r} = B_r,$$

$$B_{2,r} = \sum_{k=r}^{\infty} \sum_{l=k-r}^{k} B_{1,l} B_{1,k-l} C_{r,k,l},$$

and $r \in \mathbb{N}_0$. On the other hand,

$$D_{r,1} = \sum_{k=r}^{\infty} \sum_{l=k-r}^{k} \frac{1}{l^2 + 3} B_{1,k-l} C_{r,k,l},$$

$$D_{r,2} = \sum_{k=r}^{\infty} \sum_{l=k-r}^{k} \frac{l}{l^2 + l + 1} B_{2,k-l} C_{r,k,l}. \tag{9.37}$$

Then, we get

$$\left(\sum_{i=0}^{\infty} \frac{1}{i^2 + 3} h_i(t, 0) \right) y(t) = \sum_{r=0}^{\infty} D_{r,1} h_r(t, 0)$$

$$\left(\sum_{i=0}^{\infty} \frac{i}{i^2 + i + 1} h_i(t,0) \right) (y(t))^2 = \sum_{r=0}^{\infty} D_{r,2} h_r(t,0), \quad t \in \mathbb{T}, \, t \geq 0.$$

We substitute these expressions and (9.35) into the equation (9.34) and get

$$\sum_{r=0}^{\infty} B_r h_r(t,0) = h_0(t,0) - h_1(t,0) + 2h_2(t,0) + \sum_{r=0}^{\infty} (D_{r,1} + D_{r,2}) h_r(t,0), \quad t \in \mathbb{T}, \, t \geq 0.$$

(9.38)

From this equation we conclude

$$
\begin{aligned}
B_0 &= 1 + D_{0,1} + D_{0,2}, \\
B_1 &= -1 + D_{1,1} + D_{1,2}, \\
B_2 &= 2 + D_{2,1} + D_{2,2}, \\
B_r &= D_{r,1} + D_{r,2}, \quad r \in \{3, 4, \ldots\}.
\end{aligned}
$$

(9.39)

Example 9.16. Consider the problem

$$
\begin{cases}
{}^C D_{\Delta,1}^{\frac{9}{5}} y(t) = \left(h_{-\frac{14}{5}}(\cdot,1) * \left(\sum_{i=0}^{\infty} \frac{1}{2+i} h_i(\cdot,1) + \left(\sum_{i=0}^{\infty} \frac{i-1}{i+4} h_i(\cdot,1) \right) y(\cdot) \right. \right. \\
\qquad\qquad \left. \left. + \left(\sum_{i=0}^{\infty} \frac{i+1}{2i+3} h_i(\cdot,1) \right) (y(\cdot))^3 \right) \right)(t), \quad t > 1, \\
y(1) = 0, \quad y^{\Delta}(1) = 1,
\end{cases}
$$

(9.40)

Here we have $\alpha = \frac{9}{5}$ and $m = -\overline{[-\frac{9}{5}]} = 2$. Employing the integral equation form (9.23), we can rewrite problem (9.40) as

$$
y(t) = h_1(t,1) + \left(h_{\frac{4}{5}}(\cdot,1) * h_{-\frac{14}{5}}(\cdot,1) * \left(\sum_{i=0}^{\infty} \frac{1}{2+i} h_i(\cdot,1) \right. \right.
$$
$$
\left. \left. + \left(\sum_{i=0}^{\infty} \frac{i-1}{i+4} h_i(\cdot,1) \right) y(\cdot) + \left(\sum_{i=0}^{\infty} \frac{i+1}{2i+3} h_i(\cdot,1) \right) (y(\cdot))^3 \right) \right)(t)
$$
$$
= h_1(t,1) + \left(h_{-1}(\cdot,1) * \left(\sum_{i=0}^{\infty} \frac{1}{2+i} h_i(\cdot,1) \right. \right.
$$
$$
\left. \left. + \left(\sum_{i=0}^{\infty} \frac{i-1}{i+4} h_i(\cdot,1) \right) y(\cdot) + \left(\sum_{i=0}^{\infty} \frac{i+1}{2i+3} h_i(\cdot,1) \right) (y(\cdot))^3 \right) \right)(t), \quad t > 1.
$$

Then we obtain

$$
y(t) = h_1(t,1) + \left(\left(\sum_{i=0}^{\infty} \frac{1}{2+i} h_{-1}(\cdot,1) * h_i(\cdot,1) + \left(\sum_{i=0}^{\infty} \frac{i-1}{i+4} h_{-1}(\cdot,1) * h_i(\cdot,1) \right) y(\cdot) \right. \right.
$$
$$
\left. \left. + \left(\sum_{i=0}^{\infty} \frac{i+1}{2i+3} h_{-1}(\cdot,1) * h_i(\cdot,1) \right) (y(\cdot))^3 \right) \right)(t)
$$

$$= h_1(t, 1) + \left(\left(\sum_{i=0}^{\infty} \frac{1}{2+i} h_i(t, 1) \right) \right.$$

$$+ \left(\sum_{i=0}^{\infty} \frac{i-1}{i+4} h_i(t, 1) \right) y(t) + \left(\sum_{i=0}^{\infty} \frac{i+1}{2i+3} h_i(t, 1) \right) (y(t))^3 \right), \quad t > 1. \tag{9.41}$$

Assume that

$$y(t) = \sum_{r=0}^{\infty} B_r h_r(t, 1), \quad t > 1, \tag{9.42}$$

where the coefficients B_r are going to be obtained. Then, by (9.9), we have

$$(y(t))^2 = \sum_{r=0}^{\infty} B_{2,r} h_r(t, 1),$$

$$(y(t))^3 = \sum_{r=0}^{\infty} B_{3,r} h_r(t, 1), \quad t > 1, \tag{9.43}$$

where

$$B_{1,r} = B_r,$$

$$B_{2,r} = \sum_{k=r}^{\infty} \sum_{l=k-r}^{k} B_{1,l} B_{1,k-l} C_{r,k,l},$$

$$B_{3,r} = \sum_{k=r}^{\infty} \sum_{l=k-r}^{k} B_{2,l} B_{1,k-l} C_{r,k,l},$$

and $r \in \mathbb{N}_0$. On the other hand, let

$$D_{r,1} = \sum_{k=r}^{\infty} \sum_{l=k-r}^{k} \frac{l-1}{l+4} B_{1,k-l} C_{r,k,l},$$

$$D_{r,3} = \sum_{k=r}^{\infty} \sum_{l=k-r}^{k} \frac{l+1}{2l+3} B_{3,k-l} C_{r,k,l}. \tag{9.44}$$

Then, we get

$$\left(\sum_{i=0}^{\infty} \frac{i-1}{i+4} h_i(t, 1) \right) y(t) = \sum_{r=0}^{\infty} D_{r,1} h_r(t, 1)$$

$$\left(\sum_{i=0}^{\infty} \frac{i+1}{2i+3} h_i(t, 1) \right) (y(t))^3 = \sum_{r=0}^{\infty} D_{r,3} h_r(t, 1), \quad t > 1.$$

We substitute these expressions and (9.42) into equation (9.41), which gives

$$\sum_{r=0}^{\infty} B_r h_r(t,1) = h_1(t,1) + \sum_{r=0}^{\infty} \left(\frac{1}{2+r} + D_{r,1} + D_{r,3} \right) h_r(t,1), \quad t > 1. \qquad (9.45)$$

From this equation we conclude

$$B_0 = \frac{1}{2} + D_{0,1} + D_{0,3},$$

$$B_1 = 1 + \frac{1}{3} + D_{1,1} + D_{1,3}, \qquad (9.46)$$

$$B_r = \frac{1}{2+r} + D_{r,1} + D_{r,3}, \quad r \in \{2,3,\ldots\}.$$

9.5 Advanced practical problems

Problem 9.17. Find a series solution for the Cauchy problem

$$y^\Delta(t) = y(t) - 2, \quad y(0) = 1,$$

on the time scale $\mathbb{T} = 2\mathbb{Z}$.

Problem 9.18. Find a series solution for the Cauchy problem

$$y^\Delta(t) = (y(t))^2 - y(t), \quad y(1) = 2,$$

on the time scale $\mathbb{T} = 2^{\mathbb{N}_0}$.

Problem 9.19. Find a series solution for the Cauchy problem associated with the fractional dynamic equation

$$\begin{cases} {}^C D_{\Delta,0}^{\frac{13}{4}} y(t) = (h_{-\frac{17}{4}}(\cdot,0) * ((\sum_{i=0}^{\infty} \frac{1}{i^2+1} h_i(\cdot,0)) y(\cdot) \\ \qquad + (\sum_{i=0}^{\infty} \frac{i-1}{i^2+1} h_i(\cdot,0))(y(\cdot))^2))(t), \quad t > 0, \\ y(0) = -1, \quad y^\Delta(0) = 0, \quad y^{\Delta^2}(0) = 1, \end{cases}$$

where $\mathbb{T}$ is any time scale.

10 The Adomian polynomials method

The Adomian polynomials and the Adomian decomposition method have been used to find solutions of nonlinear ordinary and partial differential equations by proposing a series representation for the solution. This method has also been combined with the Laplace transform method and used to approximately solve some nonlinear problems for which the series solution method is not suitable [13].

In this chapter, we derive the Adomian polynomial method on arbitrary time scale and present its application to a dynamic equation of arbitrary order with a nonlinear term.

Let $\mathbb{T}$ be a time scale with forward jump operator, delta differentiation operator, and graininess function, σ, Δ, and μ, respectively.

10.1 Analyzing the method

To derive the Adomian polynomials method, we employ the same notations and results as in Chapter 9 devoted to the series solution method. By Theorem 9.1, we have

$$h_n(t,s)h_m(t,s) = \sum_{l=m}^{m+n}\left(\sum_{\Lambda_{l,m}\in S_m^{(l)}} h_n^{\Lambda_{l,m}}(s,s)\right)h_l(t,s)$$

for every $t,s \in \mathbb{T}$. For $s \in \mathbb{T}$, $l,m,n \in \mathbb{N}_0$, set

$$C_{l,m,n} = \sum_{\Lambda_{l,m}\in S_m^{(l)}} h_n^{\Lambda_{l,m}}(s,s)$$

and then, for any $m,n \in \mathbb{N}_0$, we have

$$h_n(t,s)h_m(t,s) = \sum_{l=m}^{m+n} C_{l,m,n}h_l(t,s). \tag{10.1}$$

For $n \in \mathbb{N}_0$, $t,s \in \mathbb{T}$, define the polynomials

$$H_n^1(t,s) = \left(h_1(t,s)\right)^n, \quad t,s \in \mathbb{T}.$$

Note that

$$H_n^1(t,s)H_m^1(t,s) = H_{n+m}^1(t,s), \quad t,s \in \mathbb{T}.$$

Moreover,

$$H_1^1(t,s) = h_1(t,s), \quad t,s \in \mathbb{T}, \tag{10.2}$$

https://doi.org/10.1515/9783110787320-010

and, by (10.1), we arrive at

$$H_2^1(t,s) = h_1(t,s)h_1(t,s)$$

$$= \sum_{l=1}^{2} C_{l,1,1}h_l(t,s)$$

$$= C_{1,1,1}h_1(t,s) + C_{2,1,1}h_2(t,s)$$

$$= C_{1,1,1}H_1^1(t,s) + C_{2,1,1}h_2(t,s), \quad t,s \in \mathbb{T},$$

whereupon

$$h_2(t,s) = -\frac{C_{1,1,1}}{C_{2,1,1}}H_1^1(t,s) + \frac{1}{C_{2,1,1}}H_2^1(t,s), \quad t,s \in \mathbb{T},$$

and so on. Below we denote by B_i^j, $i,j \in \mathbb{N}$, the constants for which

$$H_n^1(t,s) = B_1^n h_1(t,s) + B_2^n h_2(t,s) + \cdots + B_n^n h_n(t,s), \quad t,s \in \mathbb{T}. \tag{10.3}$$

This notation provides an alternative series representation for a given function defined on a time scale, that is, a series in terms of H_n^1 instead of the usual Taylor series in terms of h_n.

Example 10.1. Let $\alpha \in \mathbb{R}$. Then

$$e_\alpha(t,s) = 1 + \alpha h_1(t,s) + \alpha^2 h_2(t,s) + \cdots$$

$$= 1 + \alpha H_1^1(t,s) + \alpha^2 \left(-\frac{C_{1,1,1}}{C_{2,1,1}}H_1^1(t,s) + \frac{1}{C_{2,1,1}}H_2^1(t,s) \right) + \cdots$$

$$= 1 + \left(\alpha - \alpha^2 \frac{C_{1,1,1}}{C_{2,1,1}} + \cdots \right)H_1^1(t,s) + \left(\frac{\alpha^2}{C_{2,1,1}} + \cdots \right)H_2^1(t,s) + \cdots, \quad t,s \in \mathbb{T}.$$

Suppose that $u : \mathbb{T} \to \mathbb{R}$ is a given function which has a convergent series expansion of the form

$$u = \sum_{j=0}^{\infty} u_j. \tag{10.4}$$

Suppose also that $f : \mathbb{R} \to \mathbb{R}$ is a given analytic function such that

$$f(u) = \sum_{n=0}^{\infty} A_n(u_0, u_1, \ldots, u_n), \tag{10.5}$$

where A_n, $n \in \mathbb{N}_0$, are given by

$$A_0 = f(u_0), \quad A_n = \sum_{v=1}^{n} c(v,n)f^{(v)}(u_0), \quad n \in \mathbb{N}.$$

Here the functions $c(v, n)$ denote the sum of products of v components u_j of u given in (10.4), whose subscripts sum up to n, divided by the factorial of the number of repeated subscripts, i. e.,

$$A_0 = f(u_0),$$
$$A_1 = c(1, 1)f'(u_0) = u_1 f'(u_0),$$
$$A_2 = c(1, 2)f'(u_0) + c(2, 2)f''(u_0) = u_2 f'(u_0) + \frac{u_1^2}{2!}f''(u_0),$$
$$A_3 = c(1, 3)f'(u_0) + c(2, 3)f''(u_0) + c(3, 3)f'''(u_0)$$
$$= u_3 f'(u_0) + u_1 u_2 f''(u_0) + \frac{u_1^3}{3!}f'''(u_0),$$
$$A_4 = c(1, 4)f'(u_0) + c(2, 4)f''(u_0) + c(3, 4)f'''(u_0) + c(4, 4)f^{(4)}(u_0)$$
$$= u_4 f'(u_0) + \left(u_1 u_3 + \frac{u_2^2}{2}\right)f''(u_0) + \frac{u_1^2 u_2}{2}f'''(u_0) + \frac{u_1^4}{4!}f^{(4)}(u_0),$$

and so on. Suppose now that u is given by the convergent series

$$u = \sum_{n=0}^{\infty} c_n H_n^1(x, x_0), \quad x, x_0 \in \mathbb{T}. \tag{10.6}$$

We wish to find the corresponding transformed series for $f(u)$. From (10.4), we have

$$u = \sum_{n=0}^{\infty} u_n = \sum_{n=0}^{\infty} c_n H_n^1(x, x_0), \quad x, x_0 \in \mathbb{T},$$

and hence,

$$u_n = c_n H_n^1(x, x_0), \quad x, x_0 \in \mathbb{T}, \quad n \in \mathbb{N}_0.$$

Thus,

$$f(u) = \sum_{n=0}^{\infty} A_n(u_0, u_1, \dots, u_n)$$
$$= f\left(\sum_{n=0}^{\infty} c_n H_n^1(x, x_0)\right)$$
$$= \sum_{n=0}^{\infty} A^n(c_0, c_1, \dots, c_n)H_n^1(x, x_0), \quad x, x_0 \in \mathbb{T}.$$

Therefore,

$$A_n(u_0, u_1, \dots, u_n) = A^n(c_0, c_1, \dots, c_n)H_n^1(x, x_0), \quad x, x_0 \in \mathbb{T}.$$

For $n = 0$, we have

$$u_0 = c_0 H_0^1(x, x_0) = c_0, \quad x, x_0 \in \mathbb{T}.$$

Thus,

$$A_0(u_0) = A^0(c_0)H_0^1(x, x_0) = A^0(c_0), \quad x, x_0 \in \mathbb{T}.$$

For $n = 1$, we find

$$A_1(u_0, u_1) = u_1 f'(u_0) = A^1(c_0, c_1)H_1^1(x, x_0), \quad x, x_0 \in \mathbb{T},$$

or

$$c_1 H_1^1(x, x_0)f'(u_0) = A^1(c_0, c_1)H_1^1(x, x_0), \quad x, x_0 \in \mathbb{T},$$

whereupon

$$A^1(c_0, c_1) = c_1 f'(u_0) = c_1 f'(c_0) = A_1(c_0, c_1).$$

For $n = 2$, we have

$$A_2(u_0, u_1, u_2) = A^2(c_0, c_1, c_2)H_2^1(x, x_0), \quad x, x_0 \in \mathbb{T},$$

or

$$u_2 f'(u_0) + \frac{u_1^2}{2}f''(u_0) = A^2(c_0, c_1, c_2)H_2^1(x, x_0), \quad x, x_0 \in \mathbb{T}.$$

Then

$$c_2 H_2^1(x, x_0)f'(c_0) + \frac{c_1^2(H_1^1(x, x_0))^2}{2}f''(c_0) = A^2(c_0, c_1, c_2)H_2^1(x, x_0), \quad x, x_0 \in \mathbb{T},$$

or

$$\left(c_2 f'(c_0) + \frac{c_1^2}{2}f''(c_0)\right)H_2^1(x, x_0) = A^2(c_0, c_1, c_2)H_2^1(x, x_0), \quad x, x_0 \in \mathbb{T},$$

from where

$$A^2(c_0, c_1, c_2) = c_2 f'(c_0) + \frac{c_1^2}{2}f''(c_0) = A_2(c_0, c_1, c_2).$$

For $n = 3$, we find

$$u_3 f'(u_0) + u_1 u_2 f''(u_0) + \frac{u_1^3}{3!}f'''(u_0) = A_3(u_0, u_1, u_2, u_3)$$

$$= A^3(c_0, c_1, c_2, c_3)H_3^1(x, x_0), \quad x, x_0 \in \mathbb{T},$$

or

$$c_3 H_3^1(x, x_0) f'(c_0) + c_1 c_2 H_3^1(x, x_0) f''(x_0)$$
$$+ \frac{c_1^3}{3!} f'''(c_0) H_3^1(x, x_0) = A^3(c_0, c_1, c_2, c_3) H_3^1(x, x_0), \quad x, x_0 \in \mathbb{T},$$

whereupon

$$c_3 f'(c_0) + c_1 c_2 f''(x_0) + \frac{c_1^3}{3!} f'''(c_0) = A^3(c_0, c_1, c_2, c_3) = A_3(c_0, c_1, c_2, c_3),$$

and so on. Therefore we get the following result.

Theorem 10.2. *Let* $u : \mathbb{T} \to \mathbb{R}$ *be a function with a convergent expansion given in* (10.6). *Let also* $f : \mathbb{R} \to \mathbb{R}$ *be an analytic function having the form* (10.5). *Then*

$$f(u) = f\left(\sum_{n=0}^{\infty} c_n H_n^1(x, x_0) \right) = \sum_{n=0}^{\infty} A_n(c_0, c_1, \dots, c_n) H_n^1(x, x_0), \quad x, x_0 \in \mathbb{T}.$$

The above representation of the function f will be called Adomian polynomial decomposition of f.

Example 10.3. For $\alpha = 1$, consider $u = e_\alpha(x, x_0)$ and $f(u) = u^2$. Using Example 10.1, we have

$$e_\alpha(x, x_0) = \sum_{m=0}^{\infty} c_m H_m^1(x, x_0), \quad x, x_0 \in \mathbb{T},$$

where

$$c_0 = 1, \quad c_1 = \alpha - \alpha^2 \frac{C_{1,1,1}}{C_{2,1,1}} + \cdots, \quad c_3 = \frac{\alpha^2}{C_{2,1,1}} + \cdots, \quad \cdots$$

Note that

$$(e_\alpha(x, x_0))^2 = c_0^2 + 2 c_0 c_1 H_1^1(x, x_0) + \cdots, \quad x, x_0 \in \mathbb{T}. \tag{10.7}$$

On the other hand, by Theorem 10.2, we obtain

$$(e_\alpha(x, x_0))^2 = \sum_{m=0}^{\infty} A_m H_m^1(x, x_0)$$

and

$$A_0(u_0) = A_0(c_0) = 1 = c_0^2,$$
$$A_1(u_0, u_1) = c_1 f'(c_0) = 2 c_0 c_1,$$

and so on, i. e., we get (10.7).

In what it follows, we present the Adomian polynomials method for a dynamic equation of arbitrary order on a general time scale $\mathbb{T}$. With $\mathcal{L}$ we will denote the Laplace transform on $\mathbb{T}$. Suppose that $t_0 \in \mathbb{T}$. Consider the initial value problem

$$
\begin{cases}
y^{\Delta^n} + a_1 y^{\Delta^{n-1}} + \cdots + a_n y = f(y), & t > t_0, \\
y(t_0) = y_0, \quad y^{\Delta}(t_0) = y_1, \quad \ldots, \quad y^{\Delta^{n-1}}(t_0) = y_{n-1},
\end{cases}
\tag{10.8}
$$

where $a_i \in \mathbb{R}, i \in \{1, \ldots, n\}, y_i \in \mathbb{R}, i \in \{0, \ldots, n-1\}$, are given constants, $f : \mathbb{R} \to \mathbb{R}$ is an analytic function. We will search for a solution of the IVP (10.8) of the form

$$
y(t) = \sum_{j=0}^{\infty} c_j H_j^1(t, t_0), \quad t \geq t_0.
$$

Assume that

$$
f(y) = \sum_{j=0}^{\infty} A_j(c_0, \ldots, c_j) H_j^1(t, t_0), \quad t \geq t_0.
$$

Using the fact that

$$
\mathcal{L}(h_k(t, t_0))(z) = \frac{1}{z^{k+1}}, \quad t \geq t_0, \quad k \in \mathbb{N}_0,
$$

we get

$$
\mathcal{L}(H_0^1(t, t_0))(z) = \frac{1}{z},
$$

$$
\mathcal{L}(H_j^1(t, t_0))(z) = \sum_{k=1}^{j} B_k^j \mathcal{L}(h_k(t, t_0))(z) = \sum_{k=1}^{j} B_k^j \frac{1}{z^{k+1}}, \quad t \geq t_0, \quad j \in \mathbb{N}.
$$

Let $Y(z) = \mathcal{L}(y(t))(z)$. We take the Laplace transform of both sides of the dynamic equation in (10.8) and, using the initial conditions, obtain

$$
z^n Y(z) - \sum_{l=0}^{n-1} z^l y_{n-1-l} + a_1 z^{n-1} Y(z) - a_1 \sum_{l=0}^{n-2} z^l y_{n-2-l} + \cdots + a_n Y(z)
$$

$$
= \sum_{j=0}^{\infty} \left(A_j(c_0, \ldots, c_j) \sum_{k=1}^{j} B_k^j \frac{1}{z^{k+1}} \right),
$$

or

$$
(z^n + a_1 z^{n-1} + \cdots + a_n) Y(z) = \sum_{l=0}^{n-1} z^l y_{n-1-l} + a_1 \sum_{l=0}^{n-2} z^l y_{n-2-l} + \cdots + a_{n-1} y_0
$$

$$
+ A_0(c_0) \frac{1}{z} + \sum_{j=1}^{\infty} \left(A_j(c_0, \ldots, c_j) \sum_{k=1}^{j} B_k^j \frac{1}{z^{k+1}} \right).
$$

From this equation, we get

$$Y(z) = \frac{1}{z^n + a_1 z^{n-1} + \cdots + a_n} \left(\sum_{l=0}^{n-1} z^l y_{n-1-l} + a_1 \sum_{l=0}^{n-2} z^l y_{n-2-l} \right.$$

$$\left. + \cdots + a_{n-1} y_0 + A_0(c_0) \frac{1}{z} + \sum_{j=1}^{\infty} \left(A_j(c_0, \ldots, c_j) \sum_{k=1}^{j} B_k^j \frac{1}{z^{k+1}} \right) \right).$$

Consequently,

$$y(t) = \mathcal{L}^{-1} \left(\frac{1}{z^n + a_1 z^{n-1} + \cdots + a_n} \left(\sum_{l=0}^{n-1} z^l y_{n-1-l} + a_1 \sum_{l=0}^{n-2} z^l y_{n-2-l} \right. \right.$$

$$\left. \left. + \cdots + a_{n-1} y_0 + A_0(c_0) \frac{1}{z} + \sum_{j=1}^{\infty} \left(A_j(c_0, \ldots, c_j) \sum_{k=1}^{j} B_k^j \frac{1}{z^{k+1}} \right) \right) \right)(t), \quad t \geq t_0,$$

or, by the linearity of the inverse Laplace transform, for $t \geq t_0$,

$$y(t) = \sum_{l=0}^{n-1} y_{n-1-l} \mathcal{L}^{-1} \left(\frac{z^l}{z^n + a_1 z^{n-1} + \cdots + a_n} \right)(t)$$

$$+ a_1 \sum_{l=0}^{n-2} y_{n-2-l} \mathcal{L}^{-1} \left(\frac{z^l}{z^n + a_1 z^{n-1} + \cdots + a_n} \right)(t)$$

$$+ \cdots + a_{n-1} y_0 \mathcal{L}^{-1} \left(\frac{1}{z^n + a_1 z^{n-1} + \cdots + a_n} \right)(t)$$

$$+ A_0(c_0) \mathcal{L}^{-1} \left(\frac{1}{z^{n+1} + a_1 z^n + \cdots + a_n z} \right)(t)$$

$$+ \sum_{j=1}^{\infty} \left(A_j(c_0, \ldots, c_j) \sum_{k=1}^{j} B_k^j \mathcal{L}^{-1} \left(\frac{1}{z^{n+k+1} + a_1 z^{n+k} + \cdots + a_n z^{k+1}} \right)(t) \right).$$

After computing the inverse Laplace transform of the right-hand-side, we equate the coefficients of the functions $h_k(t, t_0)$ on both sides. In general, this results in a nonlinear system for the constants c_k, $k \in \mathbb{N}_0$.

10.2 First-order nonlinear dynamic equations

As a particular case, we consider an IVP associated with a first-order dynamic equation of the form

$$y^\Delta = f(y), \quad t > t_0, \quad y(t_0) = 0, \tag{10.9}$$

where $f : \mathbb{R} \to \mathbb{R}$ is an analytic function. We propose a solution of the IVP (10.9) of the form

$$y(t) = \sum_{j=0}^{\infty} c_j H_j^1(t, t_0), \quad t \geq t_0.$$

Like in the general case, we suppose that

$$f(y) = \sum_{j=0}^{\infty} A_j(c_0, \ldots, c_j) H_j^1(t, t_0), \quad t \geq t_0.$$

On the other hand, by (10.3), we have

$$y(t) = c_0 + \sum_{j=1}^{\infty} \sum_{k=1}^{j} c_j B_k^j h_k(t, t_0), \quad t \geq t_0, \tag{10.10}$$

and

$$f(y) = A_0(c_0) + \sum_{j=1}^{\infty} \sum_{k=1}^{j} A_j(c_0, \ldots, c_j) B_k^j h_k(t, t_0), \quad t \geq t_0. \tag{10.11}$$

Let

$$\mathcal{L}(y(t))(z) = Y(z), \quad t \geq t_0.$$

Then, we have

$$\mathcal{L}(y^\Delta(t))(z) = zY(z) - y(t_0) = zY(z), \quad t \geq t_0.$$

Taking the Laplace transform of both sides of the dynamic equation (10.9), we obtain

$$zY(z) = \mathcal{L}\left(A_0(c_0) + \sum_{j=1}^{\infty} \sum_{k=1}^{j} A_j(c_0, \ldots, c_j) B_k^j h_k(t, t_0) \right)(z)$$

$$= A_0(c_0)\frac{1}{z} + \sum_{j=1}^{\infty} \sum_{k=1}^{j} A_j(c_0, \ldots, c_j) B_k^j \frac{1}{z^{k+1}}, \quad t \geq t_0.$$

This yields

$$Y(z) = A_0(c_0)\frac{1}{z^2} + \sum_{j=1}^{\infty} \sum_{k=1}^{j} A_j(c_0, \ldots, c_j) B_k^j \frac{1}{z^{k+2}}.$$

Now, by taking inverse Laplace transform of both sides, we get

$$y(t) = A_0(c_0) h_1(t, t_0) + \sum_{j=1}^{\infty} \sum_{k=1}^{j} A_j(c_0, \ldots, c_j) B_k^j h_{k+1}(t, t_0), \quad t \geq t_0.$$

Employing (10.10), we have

$$c_0 + \sum_{j=1}^{\infty}\sum_{k=1}^{j} c_j B_k^j h_k(t, t_0) = A_0(c_0) h_1(t, t_0) + \sum_{j=1}^{\infty}\sum_{k=1}^{j} A_j(c_0, \ldots, c_j) B_k^j h_{k+1}(t, t_0), \quad t \geq t_0.$$

In order to equate the coefficients of the time scale monomials $h_k(t, t_0)$ on both sides, we reorder the sums as follows:

$$c_0 + \left(\sum_{j=1}^{\infty} c_j B_1^j\right) h_1(t, t_0) + \sum_{k=2}^{\infty}\left(\sum_{j=k}^{\infty} c_j B_k^j\right) h_k(t, t_0)$$

$$= A_0(c_0) h_1(t, t_0) + \sum_{k=2}^{\infty}\sum_{j=k-1}^{\infty} A_j(c_0, \ldots, c_j) B_{k-1}^j h_k(t, t_0), \quad t \geq t_0.$$

This results in the following nonlinear system for the constants c_j, $j = 0, 1, \ldots$:

$$c_0 = 0,$$

$$\sum_{j=1}^{\infty} c_j B_1^j = A_0(c_0) = f(0),$$

$$\sum_{j=k}^{\infty} c_j B_k^j = \sum_{j=k-1}^{\infty} A_j(c_0, \ldots, c_j) B_{k-1}^j, \quad k \geq 2.$$

(10.12)

Notice that the system is infinite and nonlinear in its unknowns. However, the nonlinearity is of polynomial type.

Remark 10.4. If $\mathbb{T} = \mathbb{R}$, we have $H_1^k(t, t_0) = h_k(t, t_0) = \frac{(t-t_0)^k}{k!}$ for $k \in \mathbb{N}$, and hence $B_k^j = k! \delta_{k,j}$ for $k \in \mathbb{N}$ and $j = 1, \ldots, k$. In this case, system (10.12) becomes

$$c_0 = 0,$$

$$k! c_k = (k-1)! A_{k-1}(c_0, \ldots, c_{k-1}), \quad k = 1, 2, 3, \ldots,$$

(10.13)

or simply

$$c_k = \frac{1}{k} A_{k-1}(c_0, \ldots, c_{k-1}) \quad k = 1, 2, 3, \ldots$$

10.3 Numerical examples

Example 10.5. As a first example, we consider an IVP associated with a linear dynamic equation of the first order having the form

$$y^\Delta(t) = ay(t) + b, \quad y(0) = 0,$$

(10.14)

where a, b are real constants. Assume that

$$y(t) = \sum_{j=0}^{\infty} c_j H_j^1(t, 0), \quad t \geq 0,$$

where c_j, $j = 0, 1, \ldots$, are the coefficients to be determined. By Theorem 10.2, we have

$$f(y) = ay(t) + b = \sum_{j=0}^{\infty} A_j(c_0, \ldots, c_j) H_j^1(t, 0), \quad t \geq 0,$$

where

$$A_0 = f(c_0) = ac_0 + b,$$
$$A_1 = c_1 f'(c_0) = ac_1,$$
$$A_2 = c_2 f'(c_0) + \frac{c_1^2}{2!} f''(c_0) = ac_2,$$
$$A_3 = c_3 f'(c_0) + c_1 c_2 f''(c_0) + \frac{c_1^3}{3!} f'''(c_0) = ac_3,$$
$$A_4 = c_4 f'(c_0) + \left(c_1 c_3 + \frac{c_2^2}{2} \right) f''(c_0) + \frac{c_1^2 c_2}{2} f'''(c_0) + \frac{c_1^4}{4!} f^{(4)}(c_0) = ac_4,$$
$$\cdots$$
$$A_n = ac_n,$$

since $f'(c_0) = a$ and $f^{(k)}(c_0) = 0$ for $k \geq 2$. Therefore, the system (10.12) for this example takes the form

$$c_0 = 0,$$
$$\sum_{j=1}^{\infty} c_j B_1^j = c_0 \tag{10.15}$$
$$\sum_{j=k}^{\infty} c_j B_k^j = \sum_{j=k-1}^{\infty} ac_{j-1} B_{k-1}^j, \quad k \in \mathbb{N}, \ k \geq 2.$$

This is an infinite linear system having the following triangular form:

$$c_0 = 0,$$
$$c_1 B_1^1 + c_2 B_1^2 + c_3 B_1^3 + \cdots = b,$$
$$c_2 B_2^2 + c_3 B_2^3 + c_4 B_2^4 + \cdots = a(c_1 B_1^1 + c_2 B_1^2 + c_3 B_1^3 + \cdots) = ab,$$
$$c_3 B_3^3 + c_4 B_3^4 + c_5 B_3^5 + \cdots = a(c_2 B_2^2 + c_3 B_2^3 + c_4 B_2^4 + \cdots) = a^2 b, \tag{10.16}$$
$$\cdots$$
$$c_n B_n^n + c_{n+1} B_n^{n+1} + \cdots = a(c_{n-1} B_{n-1}^{n-1} + c_n B_{n-1}^n + \cdots) = a^{n-1} b,$$
$$\cdots.$$

The exact solution can be obtained is

$$y(t) = \frac{b}{a}(e_a(t, 0) - 1), \quad t \geq 0.$$

We consider the time scale $\mathbb{T} = a\mathbb{N}_0$. On this time scale, we have

$$H_1^1(t,0) = B_1^1 h_1(t,0) = h_1(t,0),$$
$$H_2^1(t,0) = B_1^2 h_1(t,0) + B_2^2 h_2(t,0) = \alpha h_1(t,0) + 2h_2(t,0),$$
$$H_3^1(t,0) = B_1^3 h_1(t,0) + B_2^3 h_2(t,0) + B_3^3 h_3(t,0),$$
$$= \alpha^2 h_1(t,0) + 6\alpha h_2(t,0) + 6h_3(t,0),$$
$$H_4^1(t,0) = B_1^4 h_1(t,0) + B_2^4 h_2(t,0) + B_3^4 h_3(t,0) + B_4^4 h_4(t,0),$$
$$= \alpha^3 h_1(t,0) + 14\alpha^2 h_2(t,0) + 27\alpha h_3(t,0) + 24h_t(t,0), \quad t \geq 0.$$

The infinite system (10.16) has been truncated to $n = 4$ and solved easily using back substitution to find c_1, c_2, c_3, and c_4. In Tables 10.1 and 10.2, the exact and approximate solutions are compared for the values $\alpha = 0.5$, $a = 1$, $b = 1$ and $\alpha = 0.5$, $a = 1$, $b = 2$. The graphs of these solutions are also compared in Figures 10.1 and 10.2.

Table 10.1: The values of the approximate solution $y^{(a)}(t)$ and the exact solution $y^{(e)}(t)$ for $\alpha = 0.5$, $a = 1$, $b = 1$.

t	$y^{(e)}(t)$	$y^{(a)}(t)$
0.00	0.00000000	0.00000000
0.50	0.50000000	0.50000000
1.00	1.25000000	1.25000000
1.50	2.37500000	2.39843750
2.00	4.06250000	4.15625000
2.50	6.59375000	6.79687500
3.00	10.39062500	10.65625000
3.50	16.08593750	16.13281250
4.00	24.62890625	23.68750000
4.50	37.44335938	33.84375000
5.00	56.66503906	47.18750000

Table 10.2: The values of the approximate solution $y^{(a)}(t)$ and the exact solution $y^{(e)}(t)$ for $\alpha = 0.5$, $a = 1$, $b = 2$.

t	$y^{(e)}(t)$	$y^{(a)}(t)$
0.00	0.00000000	0.00000000
0.50	1.00000000	1.00000000
1.00	2.50000000	2.50000000
1.50	4.75000000	4.79687500
2.00	8.12500000	8.31250000
2.50	13.18750000	13.59375000
3.00	20.78125000	21.31250000
3.50	32.17187500	32.26562500
4.00	49.25781250	47.37500000
4.50	74.88671875	67.68750000
5.00	113.33007813	94.37500000

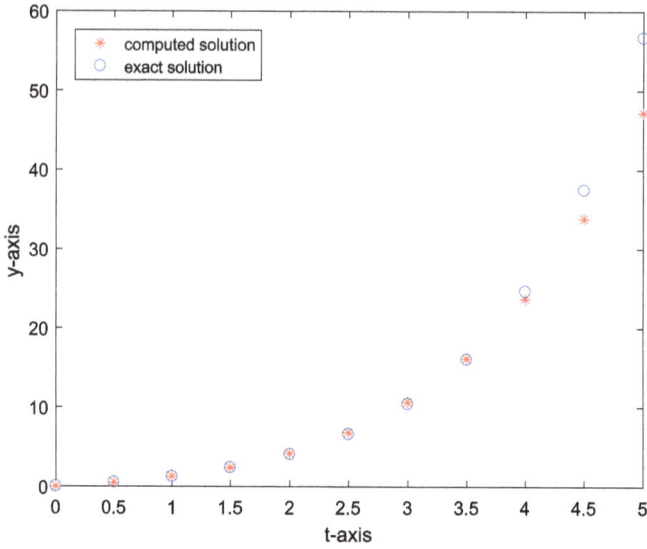

Figure 10.1: Computed and exact values of the solution with $\alpha = 0.5$, $a = 1$, and $b = 1$.

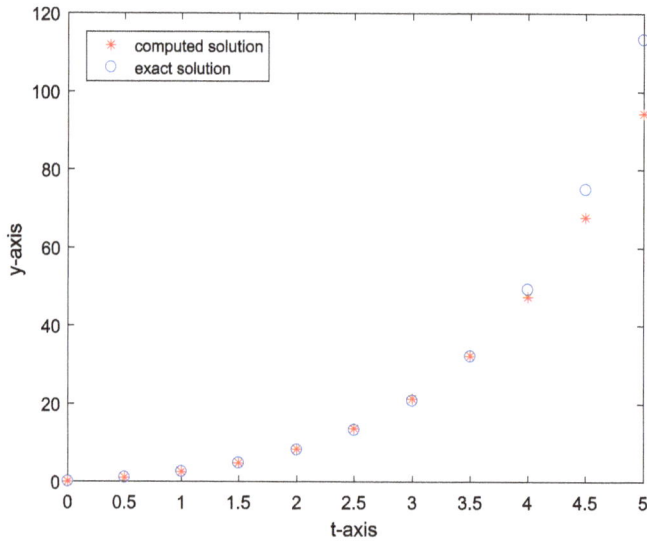

Figure 10.2: Computed and exact values of the solution with $\alpha = 0.5$, $a = 1$, and $b = 2$.

We also find the approximate and exact solutions of the problem on the time scale $\mathbb{T} = \mathbb{R}$. On this time scale, by Remark 10.4, we have $B_k^j = k! \delta_{k,j}$ for $k \in \mathbb{N}$ and $j = 1, \ldots, k$. As a result, the infinite system (10.16) becomes diagonal. It has been again truncated to $n = 4$ and the values of the constants c_1, c_2, c_3, and c_4 follow directly. The graphs of the exact and approximate solutions are compared in Figures 10.3 and 10.4 for the values $a = 1$, $b = -1$ and $a = \frac{1}{2}$, $b = 1$ on the interval $[0, 4]$. We observe good approximation near the point $t = 0$, which is the center of convergence in the series solution.

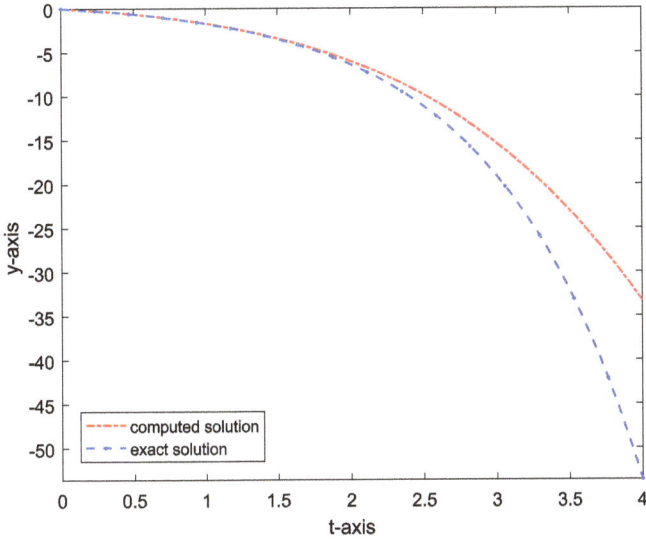

Figure 10.3: Computed and exact values of the solution with $a = 1$, $b = -1$ on $[0, 4]$.

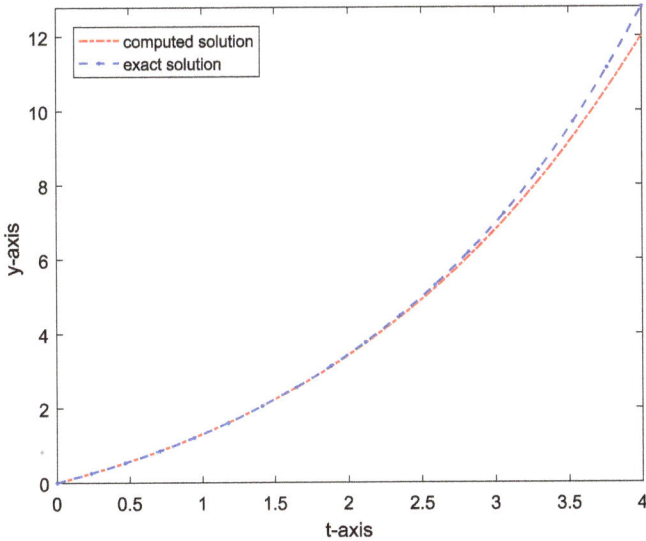

Figure 10.4: Computed and exact values of the solution with $a = \frac{1}{2}$, $b = 1$ on $[0, 4]$.

In the next two examples, we take f to be a nonlinear function.

Example 10.6. Consider the initial value problem associated with the first-order non-linear dynamic equation of the form

$$y^\Delta(t) = e^{y(t)}, \quad t \geq 0, \quad y(0) = 0, \tag{10.17}$$

where $e^{y(t)}$ is the exponential function on the set of real numbers. Assume that the solution has the series representation

$$y(t) = \sum_{j=0}^{\infty} c_j H_j^1(t, 0), \quad t \geq 0,$$

where $c_j, j \in \mathbb{N}_0$, are the coefficients to be determined. By Theorem 10.2, we have

$$f(y) = e^{y(t)} = \sum_{j=0}^{\infty} A_j(c_0, \ldots, c_j) H_j^1(t, 0), \quad t \geq 0,$$

where

$$A_0 = f(c_0) = e^{c_0},$$
$$A_1 = c_1 f'(c_0) = c_1 e^{c_0},$$
$$A_2 = c_2 f'(c_0) + \frac{c_1^2}{2!} f''(c_0) = \left(c_2 + \frac{c_1^2}{2!} \right) e^{c_0},$$
$$A_3 = c_3 f'(c_0) + c_1 c_2 f''(c_0) + \frac{c_1^3}{3!} f'''(c_0) = \left(c_3 + c_1 c_2 + \frac{c_1^3}{3!} \right) e^{c_0},$$
$$A_4 = c_4 f'(c_0) + \left(c_1 c_3 + \frac{c_2^2}{2} \right) f''(c_0) + \frac{c_1^2 c_2}{2} f'''(c_0) + \frac{c_1^4}{4!} f^{(4)}(c_0)$$
$$= \left(c_4 + c_1 c_3 + \frac{c_2^2}{2} + \frac{c_1^2 c_2}{2} + \frac{c_1^4}{4!} \right) e^{c_0},$$
$$\ldots.$$

The infinite nonlinear system (10.12) for this example has the form

$$c_0 = 0,$$

$$\sum_{j=1}^{\infty} c_j B_1^j = A_0(c_0),$$

$$\sum_{j=k}^{\infty} c_j B_k^j = \sum_{j=k-1}^{\infty} A_j B_{k-1}^j, \quad k \geq 2,$$

or, more explicitly,

$$c_0 = 0,$$

$$c_1 B_1^1 + c_2 B_1^2 + c_3 B_1^3 + \cdots = 1,$$

$$c_2 B_2^2 + c_3 B_2^3 + c_4 B_2^4 + \cdots = c_1 B_1^1 + \left(c_2 + \frac{c_1^2}{2!} \right) B_1^2 + \cdots,$$

$$c_3 B_3^3 + c_4 B_3^4 + c_5 B_3^5 + \cdots = \left(c_2 + \frac{c_1^2}{2!} \right) B_2^2 + \cdots,$$

$$\cdots.$$

Solving this nonlinear system, one can approximately obtain c_i, $i \in \mathbb{N}$, and hence, the approximate solution of the initial value problem is

$$y(t) = c_1 H_1^1(t,0) + c_2 H_2^1(t,0) + c_3 H_3^1(t,0) + \cdots, \quad t \geq 0. \tag{10.18}$$

Example 10.7. In the last example, we consider the initial value problem associated with the first-order nonlinear dynamic equation of the form

$$y^\Delta(t) = y^2 + 1, \quad y(0) = 0, \quad t \geq 0.$$

Assume that

$$y(t) = \sum_{j=0}^\infty c_j H_j^1(t,0), \quad t \geq 0,$$

where the coefficients c_j, $j \in \mathbb{N}$, will be determined from the nonlinear system (10.12). Let

$$f(y) = y^2 + 1 = \sum_{j=0}^\infty A_j(c_0, \ldots, c_j) H_j^1(t,0), \quad t \geq 0,$$

where

$$A_0 = f(c_0),$$
$$A_1 = c_1 f'(c_0),$$
$$A_2 = c_2 f'(c_0) + \frac{c_1^2}{2!} f''(c_0),$$
$$A_3 = c_3 f'(c_0) + c_1 c_2 f''(c_0) + \frac{c_1^3}{3!} f'''(c_0),$$
$$A_4 = c_4 f'(c_0) + \left(c_1 c_3 + \frac{c_2^2}{2} \right) f''(c_0) + \frac{c_1^2 c_2}{2} f'''(c_0) + \frac{c_1^4}{4!} f^{(4)}(c_0),$$

$$\cdots.$$

Since $f'(c_0) = 2c_0$, $f''(c_0) = 2$, and $f^{(m)}(c_0) = 0$ for $m \geq 3$, we obtain

$$A_0 = c_0^2 + 1,$$
$$A_1 = 2c_0 c_1,$$
$$A_2 = 2c_0 c_2 + c_1^2,$$
$$A_3 = 2c_0 c_3 + 2c_1 c_2,$$
$$A_4 = 2c_0 c_4 + 2c_1 c_3 + 2c_2^2,$$
$$\cdots.$$

The nonlinear infinite system (10.12) becomes

$$c_0 = 0,$$
$$\sum_{j=1}^{\infty} c_j B_1^j = A_0(c_0),$$
$$\sum_{j=k}^{\infty} c_j B_k^j = \sum_{j=k-1}^{\infty} A_j(c_0, \dots, c_j) B_{k-1}^j, \quad k \geq 2.$$

(10.19)

If, in particular, the time scale under consideration is $\mathbb{T} = \mathbb{Z}$, then

$$h_0(t,0) = 1, \quad h_1(t,0) = t, \quad h_k(t,0) = \frac{t(t-1)\dots(t-k+1)}{k!}, \quad t \geq 0, \quad k = 2,3,\dots,$$

and hence, we compute

$$H_1^1(t,0) = t = h_1(t,0),$$
$$H_2^1(t,0) = t^2 = 2h_2(t,0) + h_1(t,0),$$
$$H_3^1(t,0) = t^3 = 6h_3(t,0) + 6h_2(t,0) + h_1(t,0),$$
$$H_4^1(t,0) = t^4 = 24h_4(t,0) + 36h_3(t,0) + 14h_2(t,0) + h_1(t,0),$$
$$\cdots, \quad t \geq 0.$$

Then, system (10.19) turns into

$$c_0 = 0,$$
$$c_1 + c_2 + c_3 + c_4 + \cdots = 1,$$
$$2c_2 + 6c_3 + 14c_4 + \cdots = c_1^2 + 2c_1 c_2 + (2c_1 c_3 + 2c_2^2) + \cdots,$$
$$6c_3 + 36c_4 + \cdots = 2c_1^2 + 12c_1 c_2 + 14(2c_1 c_3 + 2c_2^2) + \cdots,$$
$$24c_4 + \cdots = 6c_1 c_2 + 36(2c_1 c_3 + 2c_2^2) + \cdots,$$
$$\cdots.$$

(10.20)

The nonlinear system (10.20) can be solved by any numerical method for solving a nonlinear system with the Newton method or the steepest descent methods in classical numerical analysis.

10.4 Advanced practical problems

Problem 10.8. Apply the Adomian polynomials method to the initial value problem

$$y^{\Delta}(t) = 3y^2 - 1, \quad y(0) = 0, \quad t \geq 0,$$

and derive the infinite system (10.12) on the time scale $\mathbb{T} = 2\mathbb{Z}$.

Problem 10.9. Apply the Adomian polynomials method to the initial value problem

$$y^{\Delta}(t) = y^2 + 2y, \quad y(0) = 2, \quad t \geq 0,$$

and derive the infinite system (10.12) on the time scale $\mathbb{T} = \mathbb{Z}$.

Problem 10.10. Apply the Adomian polynomials method to the initial value problem

$$y^{\Delta}(t) = e^y - 1, \quad y(0) = 1, \quad t \geq 0,$$

and derive the infinite system (10.12) on the time scale $\mathbb{T} = \mathbb{R}$.

11 Weak solutions and variational methods for some classes of linear first-order dynamic systems

Let $\mathbb{T}$ be a time scale with forward jump operator σ and delta differentiation operator Δ. Let also $0 \in \mathbb{T}$.

11.1 Variational methods for first-order linear dynamic systems – I

Let $T \in \mathbb{T}$ and U be a bounded, closed, and convex subset of $\mathbb{R}$. Denote

$$U_{ad} = \{u(t) : t \in [0, T], \ u \text{ is } \Delta\text{-measurable and } u(t) \in U\}.$$

Consider the following problem: find $u_0 \in U_{ad}$ such that

$$J(u_0(\cdot)) \leq J(u(\cdot)) \quad \text{for any } u \in U_{ad}, \tag{11.1}$$

where J is the cost functional given by

$$J(u(\cdot)) = \int_{[0,\sigma(T))} l(x(t,u), u(t)) \Delta t, \tag{11.2}$$

and $x(\cdot, u) \in AC([0, T])$ is a solution corresponding to the control $u \in U_{ad}$ of the following problem:

$$\begin{cases} x^\Delta(t) = p(t)x(t) + f(t) + u(t) & \text{for } \Delta\text{-a.e. } t \in [0, \rho(T)], \\ x(0) = x_0, \end{cases} \tag{11.3}$$

with a given $x_0 \in \mathbb{R}$. Here p is a regressive rd-continuous function, $f \in \mathbb{L}^1([0, T])$, the scalar function $l(x, u)$, along with its partial derivatives $\{l_x, l_u\}$, is continuous and uniformly bounded on $\mathbb{R} \times U$ for almost all $t \in [0, T]$. Let

$$M_1 = \sup_{t \in [0,T]} |e_p(t, 0)|, \quad M_2 = \sup_{t,\tau \in [0,T]} |e_p(t, \tau)|.$$

Define the Hamiltonian $H(x, \psi^\sigma, u)$ as follows:

$$H(x, \psi^\sigma, u) = l(x, u) + \psi^\sigma(px + f + u).$$

In the next result, we will give a necessary condition for the positive definiteness of the defined Hamiltonian.

https://doi.org/10.1515/9783110787320-011

Theorem 11.1. *Let u_0 be an optimal solution of the problem (11.1) and $x_0(\cdot, u_0)$ an optimal trajectory corresponding to u_0. Then it is necessary that there exists a function $\psi \in AC([\sigma(0), \sigma(T)])$ satisfying the following conditions:*

$$\int_{[0,\sigma(T))} \langle H_u(x_0(t), \psi^\sigma(t), u_0(t)), u(t) - u_0(t) \rangle \Delta t \geq 0$$

for any $u \in U_{ad}$, and

$$\psi^\Delta(t) = -H_x(x_0(t), \psi^\sigma(t), u_0(t)) = -p(t)\psi^\sigma(t) - l_x(x_0(t), u_0(t)), \quad t \in [\sigma(0), T],$$

$$\psi(\sigma(T)) = 0.$$

Proof. Let $\varepsilon \in [0, 1]$ be arbitrarily chosen. For $u \in U_{ad}$, define

$$u_\varepsilon = u_0 + \varepsilon(u - u_0).$$

Since U is a bounded, closed, convex subset of $\mathbb{R}$, we have that U_{ad} is a closed, convex subset of $L^\infty([0, T])$, and $u_0 \in U_{ad}$. Because u_0 is an optimal solution of (11.1), we have

$$J(u_0(\cdot)) \leq J(u(\cdot)), \quad \varepsilon \in [0, 1], \quad u \in U_{ad},$$

and

$$\lim_{\varepsilon \to 0} u_\varepsilon(t) = u_0(t), \quad t \in [0, T].$$

Observe that

$$|x_\varepsilon(t, u_\varepsilon)| = \left| e_p(t, 0)x_0 + \int_{[0,t)} e_p(t, \sigma(\tau))(f(\tau) + u_\varepsilon(\tau))\Delta\tau \right|$$

$$\leq |e_p(t, 0)||x_0| + \int_{[0,t)} |e_p(t, \sigma(\tau))||f(\tau) + u_\varepsilon(\tau)|\Delta\tau$$

$$\leq M_1|x_0| + M_2 \int_{[0,T)} |f(\tau) + u_\varepsilon(\tau)|\Delta\tau$$

$$\leq M_1|x_0| + M_2 \int_{[0,T)} |f(\tau)|\Delta\tau + M_2 \int_{[0,T)} |u_\varepsilon(\tau)|\Delta\tau$$

$$= M, \quad t \in [0, T].$$

Therefore, the sequence $\{x_\varepsilon(\cdot, u_\varepsilon)\}_{\varepsilon \in [0,1]}$ is uniformly bounded on $[0, T]$. Now we take $t_1, t_2 \in [0, T]$ arbitrarily. Without loss of generality, suppose that $t_1 < t_2$. Then

$$\left| x_\varepsilon(t_1, u_\varepsilon) - x_\varepsilon(t_2, u_\varepsilon) \right| = \left| e_p(t_1, 0)x_0 + \int_{[0,t_1)} e_p(t_1, \sigma(\tau))(f(\tau) + u_\varepsilon(\tau))\Delta\tau \right.$$

$$\left. - e_p(t_2, 0)x_0 - \int_{[0,t_2)} e_p(t_2, \sigma(\tau))(f(\tau) + u_\varepsilon(\tau))\Delta\tau \right|$$

$$= \left| (e_p(t_1, 0) - e_p(t_2, 0))x_0 \right.$$

$$+ \int_{[0,t_1)} (e_p(t_1, \sigma(\tau)) - e_p(t_2, \sigma(\tau)))(f(\tau) + u_\varepsilon(\tau))\Delta\tau$$

$$\left. - \int_{[t_1,t_2)} e_p(t_2, \sigma(\tau))(f(\tau) + u_\varepsilon(\tau))\Delta\tau \right|$$

$$= \left| e_p(t_1, 0)(1 - e_p(t_2, t_1))x_0 \right.$$

$$+ (1 - e_p(t_2, t_1)) \int_{[0,t_1)} e_p(t_1, \sigma(\tau))(f(\tau) + u_\varepsilon(\tau))\Delta\tau$$

$$\left. - \int_{[t_1,t_2)} e_p(t_2, \sigma(\tau))(f(\tau) + u_\varepsilon(\tau))\Delta\tau \right|$$

$$\leq |1 - e_p(t_2, t_1)| |e_p(t_1, 0)| |x_0|$$

$$+ |1 - e_p(t_2, t_1)| \int_{[0,t_1)} |e_p(t_1, \sigma(\tau))| |f(\tau) + u_\varepsilon(\tau)| \Delta\tau$$

$$+ \int_{[t_1,t_2)} |e_p(t_2, \sigma(\tau))| |f(\tau) + u_\varepsilon(\tau)| \Delta\tau$$

$$\to 0, \quad \text{as} \quad t_1 \to t_2.$$

Hence, $\{x_\varepsilon(\cdot, u_\varepsilon)\}_{\varepsilon \in [0,1]}$ is equicontinuous on $[0, T]$. Also,

$$\left| x_\varepsilon(t, u_\varepsilon) - x_0(t, u_0) \right| = \left| e_p(t, 0)x_0 + \int_{[0,t)} e_p(t, \sigma(\tau))(f(\tau) + u_\varepsilon(\tau))\Delta\tau \right.$$

$$\left. - e_p(t, 0)x_0 - \int_{[0,t)} e_p(t, \sigma(\tau))(f(\tau) + u_0(\tau))\Delta\tau \right|$$

$$= \left| \int_{[0,t)} e_p(t, \sigma(\tau))(u_\varepsilon(\tau) - u_0(\tau))\Delta\tau \right|$$

$$\leq \int_{[0,T)} |e_p(t, \sigma(\tau))| |u_\varepsilon(\tau) - u_0(\tau)| \Delta\tau$$

$$\leq M_2 \int_{[0,T)} |u_\varepsilon(\tau) - u_0(\tau)| \Delta\tau$$

$$\to 0, \quad \text{as} \quad \varepsilon \to 0.$$

Now, applying the Arzela–Ascoli theorem, we conclude that

$$x_\varepsilon \to x_0 \quad \text{in } C([0,T]), \quad \text{as} \quad \varepsilon \to 0, \, t \in [0,T].$$

Denote

$$y(t) = \lim_{\varepsilon \to 0} \frac{x_\varepsilon(t) - x_0(t)}{\varepsilon}, \quad t \in [0,T].$$

Then y satisfies the following initial value problem:

$$\begin{cases} y^\Delta(t) = p(t)y(t) + (u(t) - u_0(t)) & \text{for } \Delta\text{-a.e.} \quad t \in [0, \rho(T)], \\ y(0) = 0. \end{cases}$$

We will compute the Gateaux differential J at $u_0 \in U_{ad}$ in the direction $u - u_0$. Let $\{t_i\}_{i \in I}$, $I \subseteq \mathbb{N}$, be the set of all right-scattered points of $[0,T]$. Let also,

$$\tilde{u}_\varepsilon(t) = \begin{cases} u_\varepsilon(t) & \text{if } t \in [0,T], \\ u_\varepsilon(t_i) & \text{if } t \in (t_i, \sigma(t_i)) \text{ for some } i \in I. \end{cases}$$

Then

$$0 \le \lim_{\varepsilon \to 0} \frac{J(u_\varepsilon(\cdot)) - J(u_0(\cdot))}{\varepsilon}$$

$$= \lim_{\varepsilon \to 0} \int_{[0,\sigma(T))} \frac{l(x_\varepsilon(t,u_\varepsilon), u_\varepsilon(t)) - l(x_0(t,u_0), u_0(t))}{\varepsilon} \Delta t$$

$$= \lim_{\varepsilon \to 0} \int_{[0,\sigma(T))} \frac{l(\tilde{x}_\varepsilon(t,u_\varepsilon), \tilde{u}_\varepsilon(t)) - l(\tilde{x}_0(t,u_0), \tilde{u}_0(t))}{\varepsilon} dt$$

$$= \lim_{\varepsilon \to 0} \int_{[0,\sigma(T))} \left(\int_0^1 \left(\left\langle l_x(\tilde{x}_0 + \theta(\tilde{x}_\varepsilon - \tilde{x}_0), \tilde{u}_0 + \theta\varepsilon(\tilde{u} - \tilde{u}_0)), \frac{P\tilde{x}_\varepsilon - \tilde{x}_0}{\varepsilon} \right\rangle \right. \right.$$

$$\left. \left. + \langle l_u(\tilde{x}_0 + \theta(\tilde{x}_\varepsilon - \tilde{x}_0)), \tilde{u}_0 + \theta\varepsilon(\tilde{u} - \tilde{u}_\varepsilon)\rangle \right) d\theta \right) dt$$

$$= \int_{[0,\sigma(T)]} \left(\langle l_x(\tilde{x}_0(t), \tilde{u}_0), \tilde{y}(t)\rangle + \langle l_u(\tilde{x}_0(t), \tilde{u}_0), \tilde{u}(t) - \tilde{u}_0(t)\rangle \right) dt$$

$$= \int_{[0,\sigma(T))} \left(\langle l_x(x_0(t), u_0), y(t)\rangle + \langle l_u(x_0(t), u_0), u(t) - u_0(t)\rangle \right) \Delta t.$$

Define the operator

$$T_1 : \mathbb{L}^1([0,T)) \to C([0,T])$$

as follows:

$$y(t) = T_1(u - u_0)(t) = \int_{[0,t)} e_p(t, \sigma(\tau))(u(\tau) - u_0(\tau))\Delta\tau, \quad t \in [0, T].$$

Note that T_1 is a linear continuous operator. Next, define the functional

$$T_2 : \mathcal{C}([0, T]) \to \mathbb{R}$$

as follows:

$$T_2 y(t) = \int_{[0,\sigma(t))} \langle l_x(x_0(t), u_0(t)), y(t) \rangle \Delta t,$$

which is a linear continuous functional. Hence,

$$T_2 \cdot T_1 : \mathbb{L}^1([0, T]) \to \mathbb{R}$$

defined by

$$T_2 \cdot T_1(u - u_0) = \int_{[0,\sigma(t))} \langle l_x(x_0(t), u_0(t)), y(t) \rangle \Delta t, \quad t \in [0, T],$$

is a linear bounded functional. By the Riesz representation theorem, it follows that there is a $\psi^\sigma \in \mathbb{L}^\infty([0, T])$ such that

$$\int_{[0,\sigma(T))} \langle l_x(x_0(t), u_0(t)), y(t) \rangle \Delta t = \int_{[0,\sigma(T))} \langle u(t) - u_0(t), \psi^\sigma(t) \rangle \Delta t.$$

Then

$$0 \leq \int_{[0,\sigma(T))} (\langle l_x(x_0(t), u_0(t)), y(t) \rangle + \langle l_u(x_0(t), u_0(t)), u(t) - u_0(t) \rangle) \Delta t$$

$$= \int_{[0,\sigma(T))} \langle l_u(x_0(t), u_0(t)) + \psi^\sigma(t), u(t) - u_0(t) \rangle \Delta t$$

$$= \int_{[0,\sigma(T))} \langle H_u(x_0(t), \psi^\sigma(t), u_0(t)), u(t) - u_0(t) \rangle \Delta t, \quad u \in U_{ad}.$$

Next,

$$T_2 y(t) = \int_{[0,\sigma(T))} \langle l_x(x_0(t), u_0(t)), y(t) \rangle \Delta t$$

$$= \int_{[0,\sigma(T))} \langle y^\Delta(t) - p(t)y(t), \psi^\sigma(t) \rangle \Delta t$$

$$= \int_{[0,\sigma(T))} \langle y^\Delta(t), \psi^\sigma(t)\rangle \Delta t - \int_{[0,\sigma(T))} \langle p(t)y(t), \psi^\sigma(t)\rangle \Delta t$$

$$= y(\sigma(T))\psi(\sigma(T)) - \int_{[0,\sigma(T))} \langle y(t), \psi^\Delta(t)\rangle \Delta t - \int_{[0,\sigma(T))} \langle y(t), p(t)\psi^\sigma(t)\rangle \Delta t$$

$$= y(\sigma(T))\psi(\sigma(T)) - \int_{[0,\sigma(T))} \langle y(t), \psi^\Delta(t) + p(t)\psi^\sigma(t)\rangle \Delta t, \quad t \in [0,T],$$

whereupon

$$\int_{[0,\sigma(T))} \langle l_x(x_0(t), u_0(t)), y(t)\rangle \Delta t = y(\sigma(T))\psi(\sigma(T))$$

$$- \int_{[0,\sigma(T))} \langle y(t), \psi^\Delta(t) + p(t)\psi^\sigma(t)\rangle \Delta t,$$

or

$$y(\sigma(T))\psi(\sigma(T)) - \int_{[0,\sigma(T))} \langle y(t), \psi^\Delta(t) + p(t)\psi^\sigma(t) + l_x(x_0(t), u_0(t))\rangle \Delta t = 0.$$

Then ψ^σ can be chosen as the solution of the following backward problem:

$$\psi^\Delta(t) = -H_x(x_0(t), \psi^\sigma(t), u_0(t)) - p(t)\psi^\sigma(t) - l_x(x_0(t), u_0(t)), \quad t \in [\sigma(0), T],$$
$$\psi(\sigma(T)) = 0.$$

This completes the proof. □

Below, we will illustrate the previous result with the following examples.

Example 11.2. Let $\mathbb{T} = \mathbb{Z}$. Consider the following problem:

$$\text{minimize } J(u(\cdot)) = \int_{[0,11)} u(t)\Delta t, \quad u \in U_{ad},$$

subject to the dynamic equation

$$x^\Delta(t) = x(t) + t + u(t), \quad t \in [0,9],$$
$$x(0) = 1.$$

Here

$$\sigma(t) = t + 1, \quad t \in \mathbb{T},$$
$$T = 10, \quad \sigma(T) = 11, \quad \rho(T) = 9,$$
$$f(t) = t, \quad l(x(t,u), u(t)) = u(t),$$
$$p(t) = 1, \quad t \in \mathbb{T}.$$

Then the Hamiltonian is

$$H(x, \psi^\sigma, u) = l(x, u) + \psi^\sigma (px + f + u) = u + \psi^\sigma (x + t + u).$$

Next,

$$H_u(x, \psi^\sigma, u) = 1 + \psi^\sigma,$$
$$H_x(x, \psi^\sigma, u) = \psi^\sigma,$$
$$l_x(x, u) = 0.$$

The necessary conditions for the problem are as follows:

$$\int_{[0,11)} \langle 1 + \psi^\sigma(t), u(t) - u_0(t) \rangle \Delta t \geq 0$$

and

$$\psi^\Delta(t) = -\psi^\sigma(t), \quad t \in [1, 10],$$
$$\psi(11) = 0.$$

Example 11.3. Let $\mathbb{T} = 2^{\mathbb{N}_0} \cup \{0\}$. Consider the problem

$$\text{minimize } J(u(\cdot)) = \int_{[0,32)} (u(t))^2 \Delta t, \quad u \in U_{ad},$$

subject to the dynamic equation

$$x^\Delta(t) = e_1(t, 0)x(t) + t^2 + u(t), \quad t \in [0, 8],$$
$$x(0) = 1.$$

Here

$$\sigma(0) = 1, \quad \sigma(t) = 2t, \quad t \in 2^{\mathbb{N}_0}, \quad T = 16,$$
$$l(x, u) = u^2, \quad f(t) = t^2, \quad p(t) = e_1(t, 0), \quad t \in \mathbb{T}.$$

Then the Hamiltonian is given by

$$H(x, \psi^\sigma, u) = l(x, u) + \psi^\sigma (px + f + u) = u^2 + \psi^\sigma (e_1(t, 0)x + t^2 + u).$$

We have

$$H_u(x, \psi^\sigma, u) = 2u + \psi^\sigma,$$
$$H_x(x, \psi^\sigma, u) = \psi^\sigma e_1(t, 0),$$

$$l_x(x, u) = 0.$$

The necessary conditions for the problem are as follows:

$$\int_{[0,32)} \langle 2u_0(t) + \psi^\sigma(t), u(t) - u_0(t)\rangle\Delta t \geq 0,$$

and

$$\psi^\Delta(t) = -e_1(t, 0)\psi^\sigma(t), \quad t \in [1, 16],$$
$$\psi(32) = 0.$$

Exercise 11.4. Let $\mathbb{T} = 2\mathbb{Z}$. Write necessary conditions for the problem

$$\text{minimize } J(u(\cdot)) = \int_{[0,64)} (u(t))^4\Delta t, \quad u \in U_{ad},$$

subject to the dynamic equation

$$x^\Delta(t) = tx(t) + t^2 + t + u(t), \quad t \in [0, 60],$$
$$x(0) = 10.$$

11.2 Variational methods for first-order linear dynamic systems – II

In this section, we will use the notations from the previous section. Consider the following problem: Find $u_0 \in U_{ad}$ such that

$$J(u_0(\cdot)) = \inf_{u \in U_{ad}} J(u) = m, \tag{11.4}$$

where the cost functional J is given by

$$J(u(\cdot)) = \int_{[0,T)} l(t, x^\sigma(t), u(t))\Delta t, \quad u \in U_{ad},$$

and $x(\cdot, u) \in AC([0, T))$ is a solution corresponding to the control $u \in U_{ad}$ of the following initial value problem:

$$x^\Delta(t) = p(t)x(t) + f(t) + b(t)u(t) \quad \Delta\text{-a.e. } t \in [0, T],$$
$$x(0) = x_0 \in \mathbb{R}, \tag{11.5}$$

and

(H1) $p : [0, T) \to \mathbb{R}$ is a regressive rd-continuous function,

(H2) $f \in \mathbb{L}^1([0, T])$, $b \in \mathbb{L}^2([0, T])$,

(H3) $l : [0, T] \times \mathbb{R} \times \mathbb{R} \to \mathbb{R}$ is such that

1. $l(t, x, u)$ is measurable in t for x and u fixed, and continuous in x and u for t fixed, the continuity in x is uniform with respect to u,

2. $l(t, x, u)$ is convex in u for each t and x,

3. There exists a constant $L > 0$ such that

$$\left| l(t, x, u) \right| \leq L |u|^2, \quad t \in [0, T], \quad x, u \in \mathbb{R}.$$

We have that the problem (11.5) has a unique solution given by the expression

$$x(t) = e_p(t, 0)x_0 + \int_{[0,t)} e_p(t, \sigma(\tau))(f(\tau) + b(\tau)u(\tau))\Delta\tau, \quad t \in [0, T).$$

For $u : [0, T) \to \mathbb{R}$, define its extension to the real-valued interval $[0, T)$ as follows:

$$\tilde{u}(t) = \begin{cases} u(t) & \text{if } t \in [0, T), \\ u(t_i) & \text{if } t \in (t_i, \sigma(t_i)), \ i \in I, \end{cases}$$

where $\{t_i\}_{i \in I}$, $I \subseteq \mathbb{N}$, is the set of all right-scattered points of $[0, T)$. For $u \in \mathbb{L}^2([0, T))$, define the operator

$$Tu = \tilde{u}.$$

Note that T is a linear continuous operator. Now, we will give a criterion for the existence of an admissible control.

Theorem 11.5. *Suppose that* (H1)–(H3) *hold and* $-\infty < m < \infty$. *Then there exists an admissible control* $u_0 \in U_{ad}$ *such that*

$$J(u_0) = m.$$

Proof. Since $-\infty < m < \infty$, there exists a sequence $\{u_n(\cdot)\}_{n \in \mathbb{N}}$ such that

$$\inf_{u \in U_{ad}} J(u) = \lim_{n \to \infty} J(u_n) = m.$$

We have that $\{u_n(\cdot)\}_{n \in \mathbb{N}}$ is bounded in $\mathbb{L}^2([0, T))$, and hence, $\{Tu_n(\cdot)\}_{n \in \mathbb{N}}$ is bounded in $T(U_{ad})$. Therefore, there exists a subsequence, relabeled as $\{Tu_n(\cdot)\}_{n \in \mathbb{N}}$, such that

$$Tu_n(\cdot) \to z(\cdot) \quad \text{in } \mathbb{L}^2([0, T]).$$

Let $v \in \mathbb{L}^2([0, T))$. Then $Tv \in \mathbb{L}^2([0, T])$ and

$$\int\limits_{[0,T)} v(t)(u_n(t) - u_0(t))\Delta t = \int\limits_{[0,T]} Tv(t)(Tu_n(t) - Tu_0(t))dt$$

$$\to 0, \quad \text{as } n \to \infty.$$

Therefore,

$$u_n \to u_0 \quad \text{weakly in } \mathbb{L}^2([0,T]), \text{ as } n \to \infty.$$

We have

$$x(t, u_n) = e_p(t, 0)x_0 + \int\limits_{[0,t)} e_p(t, \sigma(\tau))(f(\tau) + b(\tau)u_n(\tau))\Delta\tau,$$

$$x(t, u_0) = e_p(t, 0)x_0 + \int\limits_{[0,t)} e_p(t, \sigma(\tau))(f(\tau) + b(\tau)u_0(\tau))\Delta\tau, \quad t \in [0, T).$$

Also,

$$|x(t, u_n)| = \left| e_p(t, 0)x_0 + \int\limits_{[0,t)} e_p(t, \sigma(\tau))(f(\tau) + b(\tau)u_n(\tau))\Delta\tau \right|$$

$$\leq |e_p(t, 0)x_0| + \left| \int\limits_{[0,t)} e_p(t, \sigma(\tau))(f(\tau) + b(\tau)u_n(\tau))\Delta\tau \right|$$

$$\leq M_1|x_0| + \int\limits_{[0,t)} |e_p(t, \sigma(\tau))|(|f(\tau)| + |b(\tau)||u_n(\tau)|)\Delta\tau$$

$$\leq M_1|x_0| + M_2 \int\limits_{[0,t)} |f(\tau)|\Delta\tau + M_2 \int\limits_{[0,t)} |b(\tau)||u_n(\tau)|\Delta\tau$$

$$\leq M_1|x_0| + M_2 \int\limits_{[0,t)} |f(\tau)|\Delta\tau + M_2 \left(\int\limits_{[0,t)} |b(\tau)|^2\Delta\tau \right)^{\frac{1}{2}} \left(\int\limits_{[0,t)} |u_n(\tau)|^2\Delta\tau \right)^{\frac{1}{2}}$$

$$\leq M_1|x_0| + M_2\|f\|_{\mathbb{L}^1([0,T))} + M_2\|b\|_{\mathbb{L}^2([0,T))}\|u_n\|_{\mathbb{L}^2([0,T))}, \quad t \in [0, T].$$

Consequently, the sequence $\{x(\cdot, u_n)\}_{n\in\mathbb{N}}$ is uniformly bounded on $[0, T]$. Now take $t_1, t_2 \in [0, T]$ arbitrarily. Without loss of generality, suppose that $t_1 < t_2$. Then

$$|x(t_1, u_n) - x(t_2, u_n)| = \left| e_p(t_1, 0)x_0 + \int\limits_{[0,t_1)} e_p(t_1, \sigma(\tau))(f(\tau) + b(\tau)u_n(\tau))\Delta\tau \right.$$

$$\left. - e_p(t_2, 0)x_0 - \int\limits_{[0,t_2)} e_p(t_2, \sigma(\tau))(f(\tau) + b(\tau)u_n(\tau))\Delta\tau \right|$$

$$\leq |e_p(t_1, 0)||1 - e_p(t_2, t_1)||x_0|$$

$$+\left|\int_{[0,t_1)}(e_p(t_1,\sigma(\tau))-e_p(t_2,\sigma(\tau)))(f(\tau)+b(\tau)u_n(\tau))\Delta\tau\right|$$

$$+\left|\int_{[t_1,t_2)}e_p(t_2,\sigma(\tau))(f(\tau)+b(\tau)u_n(\tau))\Delta\tau\right|$$

$$\le M_1|1-e_p(t_2,t_1)|\,|x_0|$$

$$+M_2|1-e_p(t_2,t_1)|\int_{[0,t_1)}(|f(\tau)|+|b(\tau)|\,|u_n(\tau)|)\Delta\tau$$

$$+\int_{[t_1,t_2)}|e_p(t_2,\sigma(\tau))|(|f(\tau)|+|b(\tau)|\,|u_n(\tau)|)\Delta\tau$$

$$\to 0,\quad\text{as}\quad t_1\to t_2.$$

Therefore $\{x(\cdot,u_n)\}_{n\in\mathbb{N}}$ is equicontinuous on $[0,T]$. For any fixed $t\in[0,T)$, define

$$\phi(\tau)=\{e_p(0,\sigma(\tau))b(\tau),\ \tau\in[0,t),\ \tau\in[t,T)\}.$$

We have that $\phi(\cdot)\in\mathbb{L}^2([0,T))$. Using that

$$u_n\to u_0\quad\text{weakly in }\mathbb{L}^2([0,T)),\text{ as }n\to\infty,$$

we get

$$\lim_{n\to\infty}e_p(t,0)\int_{[0,t)}e_p(0,\sigma(\tau))b(\tau)u_n(\tau)\Delta\tau$$

$$=e_p(t,0)\int_{[0,t)}e_p(0,\sigma(\tau))b(\tau)u_0(\tau)\Delta\tau,\quad t\in[0,T].$$

Hence, using the representations of $x(\cdot,u_n)$ and $x(\cdot,u_0)$, we conclude that

$$|x(t,u_n)-x(t,u_0)|\to 0,\quad\text{as}\quad n\to\infty,\ t\in[0,T].$$

Therefore, by the Arzela–Ascoli theorem, it follows that

$$\|x(\cdot,u_n)-x(\cdot,u_0)\|_{C([0,T])}\to 0,\quad\text{as}\quad n\to\infty.$$

By Mazur's theorem (see the Appendix E), it follows that for every positive integer k and positive integers $n(k),m(k)$ increasing with k and $n(k),m(k)\to\infty$, as $k\to\infty$, we can construct a suitable convex combination of $\{Tu_k(\cdot)\}_{k\in\mathbb{N}}$ such that

$$Tv_k(\cdot)=\sum_{j=1}^{m(k)}a_{kj}Tu_{n(k)+j}(\cdot)\to Tu_0(\cdot)\quad\text{in }\mathbb{L}^2([0,T)),$$

as $k \to \infty$, where

$$a_{kj} \geq 0, \quad \sum_{j=1}^{m(k)} a_{kj} = 1, \quad j \in \{1, \ldots, m(k)\}.$$

We have

$$v_k(\cdot) = \sum_{j=1}^{m(k)} a_{kj} v_{n(k)+j}(\cdot) \to u_0(\cdot) \quad \text{in } \mathbb{L}^2([0,T)), \text{ as } k \to \infty.$$

Therefore, there exists a subsequence, relabeled as $\{v_k(\cdot)\}_{k \in \mathbb{N}}$, such that

$$v_k(t) \to u_0(t) \quad \Delta\text{-a.e.} \quad t \in [0,T).$$

Then

$$J(u_0) = \int_{[0,T)} l(t, x(\sigma(t), u_0), u_0(t)) \Delta t$$

$$= \lim_{k \to \infty} \int_{[0,T)} l\left(t, x(\sigma(t), u_0), \sum_{j=1}^{m(k)} a_{kj} u_{n(k)+j}(t)\right) \Delta t$$

$$+ \lim_{k \to \infty} \int_{[0,T)} \left(l(t, x(\sigma(t), u_0), u_0(t)) - l\left(t, x(\sigma(t), u_0), \sum_{j=1}^{m(k)} a_{kj} u_{n(k)+j}(t)\right) \right) \Delta t$$

$$= \lim_{k \to \infty} \int_{[0,T)} l\left(t, x(\sigma(t), u_0), \sum_{j=1}^{m(k)} a_{kj} u_{n(k)+j}(t)\right) \Delta t$$

$$\leq \lim_{k \to \infty} \sum_{j=1}^{m(k)} a_{kj} \int_{[0,T)} l(t, x(\sigma(t), u_0), u_{n(k)+j}(t)) \Delta t$$

$$= \lim_{k \to \infty} \sum_{j=1}^{m(k)} a_{kj} \int_{[0,T)} l(t, x(\sigma(t), u_{n(k)+j}), u_{n(k)+j}(t)) \Delta t$$

$$+ \lim_{k \to \infty} \sum_{j=1}^{m(k)} a_{kj} \int_{[0,T)} (l(t, x(\sigma(t), u_0), u_{n(k)+j}(t)) - l(t, x(\sigma(t), u_{n(k)+j}), u_{n(k)+j}(t))) \Delta t$$

$$= \lim_{k \to \infty} \sum_{j=1}^{m(k)} a_{kj} \int_{[0,T)} l(t, x(\sigma(t), u_{n(k)+j})(t), u_{n(k)+j}(t)) \Delta t$$

$$= \lim_{k \to \infty} \sum_{j=1}^{m(k)+1} a_{kj} J(u_{n(k)+j}).$$

For any $\varepsilon > 0$, there exists a $K > 0$ such that $k > K$ implies

$$|J(u_{n(k)+j}) - m| < \varepsilon.$$

Hence,

$$\left|\sum_{j=1}^{m(k)} a_{kj} J(u_{n(k)+j}) - m\right| = \left|\sum_{j=1}^{m(k)} a_{kj}(J(u_{n(k)+j}) - m)\right|$$

$$\leq \sum_{j=1}^{m(k)} a_{kj}|J(u_{n(k)+j}) - m|$$

$$< \varepsilon \sum_{j=1}^{m(k)} a_{kj}$$

$$= \varepsilon.$$

Therefore, there exists an admissible control $u_0 \in U_{ad}$ such that

$$J(u_0) = m.$$

This completes the proof. □

Now, we will illustrate the above result with the following example.

Example 11.6. Let $\mathbb{T} = \mathbb{Z}$. Consider

$$x^{\Delta}(t) = au(t), \quad t \in [0, 10),$$
$$x(0) = 1, \quad x(10) = 0,$$
$$J(u(\cdot)) = \frac{1}{2} \int_{[0,10)} (u(t))^2 \Delta t,$$

where $a \in \mathbb{R}$, $a \neq 0$. Here

$$\sigma(t) = t + 1, \quad p(t) = 0, \quad f(t) = 0, t \in \mathbb{T}, \quad l(x, u) = \frac{1}{2}u^2,$$
$$H(x, \psi^{\sigma}, u) = l(x, u) + \psi^{\sigma}(px + f + u) = \frac{1}{2}u^2 + a\psi^{\sigma}u, \quad t \in [0, 10).$$

Then

$$H_u(x, \psi^{\sigma}, u) = u + a\psi^{\sigma},$$
$$H_x(x, \psi^{\sigma}, u) = 0, \quad t \in [0, 10),$$

and the necessary conditions are as follows:

$$u(t) + a\psi^{\sigma}(t) = 0, \quad \psi^{\Delta}(t) = 0,$$
$$x(0) = 1, \quad x(10) = 0, \quad t \in [0, 10).$$

We have

$$\psi(t) = C, \quad u(t) = -aC, \quad x^\Delta(t) = -a^2C,$$
$$x(t) = x(0) - a^2Ct = 1 - a^2Ct, \quad t \in [0, 10).$$

Now using that

$$x(10) = 0,$$

we get

$$0 = 1 - 10a^2C,$$

whereupon

$$C = \frac{1}{10a^2}$$

and the optimal trajectories are

$$u(t) = -\frac{1}{10a}, \quad x(t) = 1 - \frac{1}{10}t, \quad t \in [0, 10),$$

respectively.

Exercise 11.7. Let $\mathbb{T} = 3\mathbb{Z}$. Consider

$$x^\Delta(t) = x(t) + 3u(t), \quad t \in [0, 30),$$
$$x(0) = -1, \quad x(30) = 0,$$
$$J(u(\cdot)) = \frac{1}{2} \int_{[0,10)} (u(t))^2 \Delta t.$$

Find the optimal trajectories.

11.3 Advanced practical problems

Problem 11.8. Let $\mathbb{T} = 4\mathbb{Z}$. Write necessary conditions for the problem

$$\text{minimize } J(u(\cdot)) = \int_{[0,88)} (u(t) + (u(t))^3) \Delta t, \quad u \in U_{ad},$$

subject to the dynamic equation

$$x^\Delta(t) = \frac{1}{1+t^2}x(t) + t^4 + u(t), \quad t \in [0, 80],$$
$$x(0) = 15.$$

Problem 11.9. Let $\mathbb{T} = 3^{\mathbb{N}_0} \cup \{0\}$. Write necessary conditions for the problem

$$\text{minimize } J(u(\cdot)) = \int_{[0,243)} \left(2u(t) - 4(u(t))^5 \right) \Delta t, \quad u \in U_{ad},$$

subject to the dynamic equation

$$x^\Delta(t) = e_1(t,0)x(t) + \cos_3(t,0) + u(t), \quad t \in [0,27],$$
$$x(0) = 18.$$

Problem 11.10. Let $\mathbb{T} = 2^{\mathbb{N}_0} \cup \{0\}$. Consider

$$x^\Delta(t) = -tx(t) + u(t), \quad t \in [0,128),$$
$$x(0) = 1, \quad x(128) = 0,$$
$$J(u(\cdot)) = \int_{[0,128)} (u(t))^2 \Delta t.$$

Find the optimal trajectories.

Problem 11.11. Let $\mathbb{T} = 3^{\mathbb{N}_0} \cup \{0\}$. Consider

$$x^\Delta(t) = e_1(t,0)x(t) + u(t), \quad t \in [0,243),$$
$$x(0) = 1, \quad x(243) = 0,$$
$$J(u(\cdot)) = \int_{[0,243)} \frac{(u(t))^2}{1 + (u(t))^2} \Delta t.$$

Find the optimal trajectories.

Problem 11.12. Let $\mathbb{T} = 4^{\mathbb{N}_0} \cup \{0\}$. Consider

$$x^\Delta(t) = 3\cos_1(t,0)x(t) - 7u(t), \quad t \in [0,1024),$$
$$x(0) = 1, \quad x(1024) = 0,$$
$$J(u(\cdot)) = 7 \int_{[0,243)} \frac{(u(t))^2}{1 + (u(t))^6} \Delta t.$$

Find the optimal trajectories.

12 Variational methods for nonlinear dynamic equations

Let $\mathbb{T}$ be a time scale with forward jump operator σ and delta differentiation operator Δ. Let also $a, b \in \mathbb{T}, a < b$.

12.1 Existence of solutions

We will start with the following useful theorem.

Theorem 12.1. *Suppose that the function $f : \mathbb{R} \times \mathbb{R} \to \mathbb{R}$ satisfies the following conditions:*
1. *$f(\cdot, \cdot)$ is lower semicontinuous on $\mathbb{R} \times \mathbb{R}$,*
2. *$f \geq 0$ and $f(\xi, \cdot)$ is convex on $\mathbb{R}$ for every $\xi \in \mathbb{R}$. Set*

$$J(x, u) = \int_{[a,b)} f(x(t), u(t)) \Delta t.$$

If $\{x_n\}_{n \in \mathbb{N}}, \{u_n\}_{n \in \mathbb{N}} \subseteq \mathbb{L}^1(\mathbb{T})$ and $x_n \xrightarrow{s} x$, $u_n \xrightarrow{w} u$ in $\mathbb{L}^1(\mathbb{T})$, then

$$J(x, u) \leq \lim_{n \to \infty} J(x_n, u_n).$$

Proof. Let $\{x_n\}_{n \in \mathbb{N}}, \{u_n\}_{n \in \mathbb{N}} \subseteq \mathbb{L}^1(\mathbb{T})$ and $x_n \xrightarrow{s} x$, $u_n \xrightarrow{w} u$ in $\mathbb{L}^1(\mathbb{T})$. Then there exists a positive constant c such that

$$J(x_n, u_n) \leq c, \quad n \in \mathbb{N}.$$

Set

$$\alpha_n(t) = f(x_n(t), u_n(t)), \quad t \in \mathbb{T}.$$

By Mazur's theorem (see the Appendix), it follows that there are $\lambda_j^n \geq 0, j \in \{1, \ldots, k_n\}$, $n \in \mathbb{N}$, and $\alpha \in \mathbb{L}^1(\mathbb{T})$ such that

$$\sum_{j=1}^{k_n} \lambda_j^n = 1, \quad \alpha_n'(t) = \sum_{j=1}^{k_n} \lambda_j^n \alpha_{n+j}(t) \to \alpha(t) \quad \Delta\text{-a.e. on } \mathbb{T}, \text{ as } n \to \infty.$$

Since $x_n \xrightarrow{s} x$ in $\mathbb{L}^1(\mathbb{T})$, as $n \to \infty$, it follows that there exists a subsequence, relabeled as $\{x_n\}_{n \in \mathbb{N}}$, such that

$$x_n(t) \to x(t) \quad \Delta\text{-a.e. on } \mathbb{T}, \text{ as } n \to \infty.$$

https://doi.org/10.1515/9783110787320-012

Define

$$W(f, x, \alpha_n + \beta_n)(t) = \{v \in \mathbb{R} : f(x(t), v) \le \alpha_n(t) + \beta_n(t)\}, \quad t \in \mathbb{T},$$

and, for $y \in \mathbb{R}$, set

$$\beta_n(t) = \max\left\{0, \min_{v \in \mathbb{R}} f(x(t), v) - \alpha_n(t)\right\},$$

$$d_n^y(t) = \max\left\{0, yu_n(t) - \sup_{v \in W(f, x, \alpha_n + \beta_n)(t)} yv\right\},$$

$$d^y(v, W(f, x, \alpha_n + \beta_n)(t)) = \max\left\{0, yv - \sup_{z \in W(f, x, \alpha_n + \beta_n)(t)} yz\right\}, \quad t \in \mathbb{T}.$$

Note that

$$\beta_n \to 0, \quad d_n^y \to 0, \quad \text{as } n \to \infty,$$

in Lebesgue Δ-measure. Hence,

$$\beta_n(t) \to 0, \quad d_n^y(t) \to 0 \quad \Delta\text{-a.e. on } \mathbb{T}, \text{ as } n \to \infty.$$

Let

$$B^y(W(y, x, \alpha_n + \beta_n)(t), d_n^y(t)) = \{v \in \mathbb{R} : d^y(v, W(f, x, \alpha_n + \beta_n)(t)) \le d_n^y(t)\}.$$

Then

$$u_n(t) \in B^y(W(f, x, \alpha_n + \beta_n)(t), d_n^y(t)) \quad \Delta\text{-a.e. on } \mathbb{T}, \ n \in \mathbb{N}.$$

Set

$$u_n' = \sum_{j=1}^{k_n} \lambda_j^n u_{n+j}, \quad n \in \mathbb{N}, \quad \beta_n' = \sum_{j=1}^{k_n} \lambda_j^n \beta_{n+j}, \quad d_n'^y = \sum_{j=1}^{k_n} \lambda_j^n d_{n+j}^y.$$

Then

$$u_n'(t) \in B^y(W(f, x, \alpha_n' + \beta_n')(t), d_n'^y(t)) \quad \Delta\text{-a.e. on } \mathbb{T}, \ n \in \mathbb{N},$$
$$u_n' \to u, \quad \beta_n' \to 0 \quad \text{as } n \to \infty, \Delta\text{-a.e. on } \mathbb{T},$$

and

$$u(t) \in B^y(W(f, x, \alpha)(t), 0) \quad \Delta\text{-a.e. on } \mathbb{T}.$$

Assume that $\{y_n\}_{n \in \mathbb{N}}$ is dense in $\mathbb{R}$. Then

$$W(f,x,\alpha)(t) = \bigcap_{n\in\mathbb{N}} B^{y_n}(W(f,x,\alpha)(t),0) \quad \Delta\text{-a.e. on } \mathbb{T}.$$

Because $y \in \mathbb{R}$ was arbitrarily chosen, we get

$$u(t) \in W(f,x,\alpha)(t) \quad \Delta\text{-a.e. on } \mathbb{T}.$$

Thus,

$$f(x(t), u(t)) \le \alpha(t) \quad \Delta\text{-a.e. on } \mathbb{T}.$$

Hence,

$$J(x,u) \le \int_{[a,b)} \alpha(t)\Delta t$$

$$= \int_{[a,b)} \varliminf_{n\to\infty} \alpha'_n(t)\Delta t$$

$$\le \varliminf_{n\to\infty} \int_{[a,b)} \alpha'_n(t)\Delta t$$

$$= \varliminf_{n\to\infty} \int_{[a,b)} \alpha_n(t)\Delta t = \varliminf_{n\to\infty} J(x_n, u_n).$$

This completes the proof. □

Let U_{ad} be a nonempty, closed, convex subset of $\mathbb{L}^1(\mathbb{T})$, and $\Gamma_1(\mathbb{T})$ be the corresponding set. Consider the following problem: Find $u_0 \in U_{ad}$ such that

$$J(u_0) \le J(u) \quad \text{for all } u \in U_{ad},$$

where

$$J(u) = \int_{[a,b)} g(x(\tau), x^\sigma(\tau))\Delta\tau + \int_{[a,b)} h(u(\tau))\Delta\tau, \tag{12.1}$$

and x is a weak solution of the following dynamic system:

$$x^\Delta(t) + p(t)x^\sigma(t) = f(t, x(t), x^\sigma(t)) + u(t), \quad t > a,$$
$$x(a) = x_0 \in \mathbb{R}, \tag{12.2}$$
$$u \in U_{ad},$$

$p \in \Gamma_1(\mathbb{T})$, and f, g, and h satisfy the following conditions:
(H1)

1. $f : \mathbb{T} \times \mathbb{R} \times \mathbb{R} \to \mathbb{R}$ is Δ-measurable in $t \in \mathbb{T}$ and for all $x_1, x_2, y_1, y_2 \in \mathbb{R}$, $|x_1| \le \rho, |x_2| \le \rho, |y_1| \le \rho, |y_2| \le \rho$, we have

$$|f(t, x_1, y_1) - f(t, x_2, y_2)| \le L(|x_1 - x_2| + |y_1 - y_2|), \quad t \in \mathbb{T},$$

   for some constant $L = L(\rho) > 0$.
2. There exist a constant $\lambda \in (0, 1)$ and a function $q \in \mathbb{L}^1(\mathbb{T})$ such that

$$|f(t, x, y)| \le q(t)(1 + |x| + |y|^\lambda), \quad x, y \in \mathbb{R}, \quad t \in \mathbb{T}.$$

(H2)

1. $g : \mathbb{R} \times \mathbb{R} \to \mathbb{R}$ is lower semicontinuous.
2. There is a constant $c \in \mathbb{R}$ such that

$$g(x, y) \ge c, \quad x, y \in \mathbb{R}.$$

(H3)

1. $h : \mathbb{R} \to \mathbb{R}$ is convex.
2. $\lim_{|u| \to \infty} \frac{h(u)}{|u|} = \infty.$

Note that for any $u \in U_{ad}$, by Theorem D.58, it follows that the dynamic system (12.2) has a unique solution

$$x(t) = e_{\ominus p}(t, a)x_0 + \int_{[a,t)} e_{\ominus p}(t, \tau)(f(\tau, x(\tau), x^\sigma(\tau)) + u(\tau))\Delta\tau, \quad t \in [a, b).$$

Theorem 12.2. *Let $p \in \Gamma_1(\mathbb{T})$ and (H1)–(H3) hold. Then the problem (12.1), (12.2) has at least one solution.*

Proof. If

$$\inf_{u \in U_{ad}} J(u) = \infty,$$

then the theorem is proved. Assume that there is a constant $c < \infty$ such that

$$\inf_{u \in U_{ad}} J(u) \le c.$$

By (H2)(2) and (H3)(2), it follows that $c > -\infty$. Then there exists a minimizing sequence $\{u_n\}_{n \in \mathbb{N}} \subseteq U_{ad}$ such that

$$c \leq J(u_n)$$

$$= \int_{[a,b)} g(x_n(\tau), x_n^\sigma(\tau)) \Delta\tau + \int_{[a,b)} h(u_n(\tau)) \Delta\tau$$

$$\leq c + \frac{1}{n}$$

for any $n \geq N$ and for some $N \in \mathbb{N}$. Here x_n is the weak solution of the controlled system (12.2) corresponding to the control u_n. By (H3)(2), it follows that for any $\delta > 0$ there exists a $\theta = \theta(\delta)$ such that

$$h(u) \geq \theta(\delta)|u|$$

for all $|u| \geq \delta$, where

$$\lim_{\delta \to \infty} \theta(\delta) = \infty. \tag{12.3}$$

Hence, for any measurable subset $E \subseteq \mathbb{T}$, we have

$$\int_E |u_n(\tau)| \Delta\tau = \int_{E \cap \{s \in \mathbb{T}: |u_n(s)| < \delta\}} |u_n(\tau)| \Delta\tau + \int_{E \cap \{s \in \mathbb{T}: |u_n(s)| \geq \delta\}} |u_n(\tau)| \Delta\tau$$

$$\leq \delta \mu_\Delta(E) + \frac{1}{\theta(\delta)} \int_E |h(u_n(\tau))| \Delta\tau$$

$$\leq \delta \mu_\Delta(E) + \frac{C}{\theta(\delta)},$$

where $C > 0$ is independent of δ. Because of (12.3), we get

$$\limsup_{\delta \to \infty} \int_{E \cap \{s \in \mathbb{T}: |u_n(s)| \geq \delta\}} |u_n(\tau)| \Delta\tau \leq \lim_{\delta \to \infty} \frac{C}{\theta(\delta)} = 0.$$

Therefore $\{u_n\}_{n \in \mathbb{N}} \subset \mathbb{L}^1(\mathbb{T})$ is uniformly integrable, and hence $\{u_n\}_{n \in \mathbb{N}}$ is weakly compact in $\mathbb{L}^1(\mathbb{T})$. Because $U_{ad} \subseteq \mathbb{L}^1(\mathbb{T})$ is closed and convex, by Mazur's theorem, there are a subsequence, relabeled as $\{u_n\}_{n \in \mathbb{N}}$, and $\bar{u} \in U_{ad}$ for which we have

$$u_n \overset{w}{\to} \bar{u} \quad \text{in } \mathbb{L}^1(\mathbb{T}).$$

Then there is a constant $r > 0$ such that

$$\|x_n\|_{C_{rd}(\mathbb{T})} \leq r.$$

Let

$$F_n(t) = f(t, x_n(t), x_n^\sigma(t)), \quad t \in \mathbb{T}.$$

Then

$$|F_n(t)| = |f(t, x_n(t), x_n^\sigma(t))|$$
$$\leq q(t)(1 + |x_n(t)| + |x_n^\sigma(t)|)$$
$$\leq q(t)(1 + 2r), \quad t \in \mathbb{T}.$$

Therefore there are a subsequence, relabeled as $\{F_n\}_{n\in\mathbb{N}}$, and $\overline{F} \in \mathbb{L}^1(\mathbb{T})$ such that

$$F_n \xrightarrow{w} F \quad \text{in } \mathbb{L}^1(\mathbb{T}).$$

Define

$$v_n(t) = \int_{[a,t)} e_{\ominus p}(t, \tau)(F_n(\tau) + u_n(\tau))\Delta\tau,$$

$$\overline{v}(t) = \int_{[a,t)} e_{\ominus p}(t, \tau)(\overline{F}(\tau) + \overline{u}(\tau))\Delta\tau, \quad t \in \mathbb{T}.$$

By the Ascoli–Arzela theorem, it follows that

$$\|v_n - \overline{v}\|_{C_{rd}(\mathbb{T})} \to 0, \quad \text{as } n \to \infty.$$

Now, we consider the following dynamic equation:

$$y^\Delta(t) + p(t)y^\sigma(t) = \overline{F}(t) + \overline{u}(t), \quad t \in [a, b),$$
$$y(a) = x_0. \tag{12.4}$$

By Theorem D.55, we have that the dynamic equation (12.4) has a unique weak solution

$$\overline{x}(t) = e_{\ominus p}(t, a)x_0 + \int_{[a,t)} e_{\ominus p}(t, \tau)(\overline{F}(\tau) + \overline{u}(\tau))\Delta\tau, \quad t \in [a, b),$$

and $\overline{x} \in C_{rd}(\mathbb{T})$. Also,

$$x_n \to \overline{x}, \quad x_n^\sigma \to \overline{x}^\sigma \quad \text{in } C_{rd}(\mathbb{T}), \quad \text{as } n \to \infty.$$

Because

$$|f(t, x_n(t), x_n^\sigma(t)) - f(t, \overline{x}(t), \overline{x}^\sigma(t))| \leq L(\rho)(|x_n(t) - \overline{x}(t)| + |x_n^\sigma(t) - \overline{x}^\sigma(t)|), \quad t \in \mathbb{T},$$

for some $\rho > 0$, we get

$$F_n \to f(\cdot, \overline{x}(\cdot), \overline{x}^\sigma(\cdot)) \quad \text{in } \mathbb{L}^1(\mathbb{T}), \quad \text{as } n \to \infty.$$

Hence,

$$\overline{F}(\cdot) = f(\cdot, \overline{x}(\cdot), \overline{x}^{\sigma}(\cdot))$$

and

$$\overline{x}(t) = e_{\ominus p}(t, a)x_0 + \int_{[a,t)} e_{\ominus p}(t, \tau)(f(\tau, \overline{x}(\tau), \overline{x}^{\sigma}(\tau)) + \overline{u}(\tau))\Delta\tau, \quad t \in [a, b).$$

Therefore, $\overline{x}$ is a weak solution of the controlled problem (12.2) corresponding to the control $\overline{u}$. Hence, by Theorem 12.1, we obtain

$$c \leq J(\overline{u}) = \lim_{n \to \infty} J(u_n) = c,$$

i. e.,

$$J(\overline{u}) = c.$$

This completes the proof. □

We will illustrate the above result with the following example.

Example 12.3. Let $\mathbb{T} = \mathbb{Z}$,

$$J(u) = \int_{[0,10)} \frac{(x(\tau))^2}{1 + (x(\tau))^2} \Delta\tau + \int_{[0,10)} (u(\tau))^2 \Delta\tau,$$

and

$$x^{\Delta}(t) + x^{\sigma}(t) = \frac{(x(t))^2}{1 + (x(t))^2} + u(t), \quad t > 0,$$

$$x(0) = 15, \quad u \in U_{ad}.$$

Here

$$a = 0, \quad b = 15, \quad \sigma(t) = t + 1,$$

$$f(t, x(t), x^{\sigma}(t)) = \frac{(x(t))^2}{1 + (x(t))^2},$$

$$g(x(t), x^{\sigma}(t)) = \frac{(x(t))^2}{1 + (x(t))^2},$$

$$h(u(t)) = (u(t))^2, \quad t \in \mathbb{T}.$$

If $x_1, x_2, y_1, y_2 \in \mathbb{R}$, $|x_1| \leq \rho$, $|x_2| \leq \rho$, $|y_1| \leq \rho$, $|y_2| \leq \rho$, for some $\rho > 0$, we have

$$|f(t, x_1, y_1) - f(t, x_2, y_2)| = \left| \frac{x_1^2}{1 + x_1^2} - \frac{x_2^2}{1 + x_2^2} \right| \leq 2\rho|x_1 - x_2|,$$

because

$$\left(\frac{x^2}{1+x^2}\right)' = \frac{2x(1+x^2) - 2x^3}{(1+x^2)^2} = \frac{2x}{(1+x^2)^2}, \quad x \in \mathbb{R}.$$

Also, for any $\lambda \in [0,1]$, we have

$$\frac{x^2}{1+x^2} \le 1 + |x|, \quad x \in \mathbb{R},$$

and

$$|f(t,x,y)| \le 1 + |x| + |y|^\lambda, \quad x, y \in \mathbb{R}, \quad t \in \mathbb{T}.$$

Next, $h : \mathbb{R} \to \mathbb{R}$ is convex and

$$\lim_{|u| \to \infty} \frac{h(u)}{|u|} = \lim_{|u| \to \infty} \frac{u^2}{|u|} = \infty.$$

Hence, by Theorem 12.2, it follows that the considered problem has at least one solution.

Exercise 12.4. Let $\mathbb{T} = 2^{\mathbb{N}_0} \cup \{0\}$. Prove that the problem

$$J(u) = 2 \int_{[0,16)} \frac{(x(\tau))^4}{1 + (x(\tau))^4} \Delta\tau + 3 \int_{[0,16)} (u(\tau))^3 \Delta\tau,$$

and

$$x^\Delta(t) + tx^\sigma(t) = \frac{(x(t))^6}{1 + (x(t))^6} + u(t), \quad t > 0,$$

$$x(0) = -1, \quad u \in U_{ad},$$

has at least one solution.

12.2 Necessary conditions for the existence of solutions

In this section, we will derive some necessary conditions of optimality for a system, involving an adjoint equation and optimal inequality. We will start with the following useful result.

Theorem 12.5. *Suppose that $\phi \in C_{rd}(\mathbb{T})$, $\phi \ge 0$ on $\mathbb{T}$, $q,g \in L^1(\mathbb{T})$, $q \ge 0$, $g \ge 0$ on $\mathbb{T}$, $\alpha \ge 0$, $\lambda \in (0,1)$, and*

$$\phi(t) \le \alpha + \int_{[t,b)} q(\tau)(\phi(\tau))^\lambda \Delta\tau + \int_{[t,b)} g(\tau)\phi^\sigma(\tau)\Delta\tau, \quad t \in [a,b). \tag{12.5}$$

Then there exists a positive constant M such that

$$\phi(t) \le M, \quad t \in [a, b).$$

Proof. Define

$$f(t) = \alpha + \int_{[t,b)} q(\tau)(\phi(\tau))^\Lambda \Delta\tau,$$

$$\psi(t) = \int_{[t,b)} g(\tau)\phi^\sigma(\tau)\Delta\tau, \quad t \in [a, b).$$

Then

$$\psi(b) = 0,$$
$$\phi(t) \le f(t) + \psi(t), \quad t \in [a, b),$$
$$\psi^\Delta(t) = -g(t)\phi^\sigma(t)$$
$$\ge -g(t)(f^\sigma(t) + \psi^\sigma(t))$$
$$= -g(t)f^\sigma(t) - g(t)\psi^\sigma(t) \quad \Delta\text{-a.e. } t \in [a, b).$$

Hence,

$$\psi^\Delta(t)e_g(t, b) \ge -g(t)e_g(t, b)f^\sigma(t) - g(t)\psi^\sigma(t)e_g(t, b) \quad \Delta\text{-a.e. } t \in [a, b),$$

and

$$(\psi(\cdot)e_g(\cdot, b))^\Delta(t) \ge -g(t)e_g(t, b)f^\sigma(t) \quad \Delta\text{-a.e.} \quad t \in [a, b).$$

Then

$$-\psi(t)e_g(t, b) \ge -\int_{[t,b)} g(\tau)e_g(\tau, b)f^\sigma(\tau)\Delta\tau, \quad t \in [a, b),$$

and

$$\psi(t) \le \int_{[t,b)} g(\tau)e_g(b, t)e_g(\tau, b)f^\sigma(\tau)\Delta\tau$$

$$= \int_{[t,b)} g(\tau)e_g(\tau, t)f^\sigma(\tau)\Delta\tau, \quad t \in [a, b).$$

Hence, by (12.5), we obtain

$$\phi(t) \le \alpha + \int_{[t,b)} q(\tau)(\phi(\tau))^\lambda \Delta\tau + \int_{[t,b)} g(\tau)e_g(\tau,t)f^\sigma(\tau)\Delta\tau$$

$$= \alpha + \int_{[t,b)} q(\tau)(\phi(\tau))^\lambda \Delta\tau$$

$$+ \int_{[t,b)} g(\tau)e_g(\tau,t)\left(\alpha + \int_{[\sigma(\tau),b)} q(s)(\phi(s))^\lambda \Delta s \right)\Delta\tau$$

$$= \alpha + \int_{[t,b)} q(\tau)(\phi(\tau))^\lambda \Delta\tau + \alpha \int_{[t,b)} g(\tau)e_g(\tau,t)\Delta\tau$$

$$+ \int_{[t,b)} g(\tau)e_g(\tau,t)\left(\int_{[\sigma(\tau),b)} q(s)(\phi(s))^\lambda \Delta s \right)\Delta\tau$$

$$\le \alpha + \int_{[t,b)} q(\tau)(\phi(\tau))^\lambda \Delta\tau + \alpha \int_{[t,b)} g(\tau)e_g(\tau,t)\Delta\tau$$

$$+ \int_{[t,b)} g(\tau)e_g(\tau,t)\left(\int_{[\tau,b)} q(s)(\phi(s))^\lambda \Delta s \right)\Delta\tau$$

$$\le \alpha + \int_{[t,b)} g(\tau)(\phi(\tau))^\lambda \Delta\tau + \alpha \int_{[t,b)} g(\tau)e_g(\tau,t)\Delta\tau$$

$$+ \left(\int_{[t,b)} q(s)(\phi(s))^\lambda \Delta s \right) \int_{[t,b)} g(\tau)e_g(\tau,t)\Delta\tau$$

$$\le (\alpha+1)\left(1 + \int_{[a,b)} e_g(\tau,a)g(\tau)\Delta\tau \right)\left(1 + \int_{[t,b)} q(\tau)(\phi(\tau))^\lambda \Delta\tau \right), \quad t \in [a,b).$$

Define

$$\beta = \int_{[a,b)} e_g(\tau,a)g(\tau)\Delta\tau, \quad \gamma = (\alpha+1)(\beta+1),$$

$$h(t) = \gamma + \gamma \int_{[t,b)} q(\tau)(\phi(\tau))^\lambda \Delta\tau + \gamma \int_{[a,b)} q(\tau)(\phi(\tau))^\lambda \Delta\tau, \quad t \in [a,b).$$

Then

$$\phi(t) \le \gamma\left(1 + \int_{[t,b)} q(\tau)(\phi(\tau))^\lambda \Delta\tau \right) \le h(t), \quad t \in [a,b).$$

We have that h is a monotone decreasing function on $[a,b)$ and

$$h^\Delta(t) = -\gamma q(t)(\phi(t))^\lambda \ge -hq(t)(h(t))^\lambda, \quad t \in [a,b).$$

Observe that

$$((h(\cdot))^{1-\lambda})^{\Delta}(t) \ge (1-\lambda)(h^{\sigma}(t))^{-\lambda}h^{\Delta}(t)$$
$$\ge (1-\lambda)(h(t))^{-\lambda}h^{\Delta}(t), \quad t \in [a,b).$$

Then

$$((h(\cdot))^{1-\lambda})^{\Delta}(t) \ge -\gamma(1-\lambda)q(t), \quad t \in [a,b),$$

and

$$-(h(t))^{1-\lambda} + h(b))^{1-\lambda} \ge -\gamma(1-\lambda) \int_{[t,b)} q(\tau)\Delta\tau, \quad t \in [a,b),$$

or

$$(h(t))^{1-\lambda} - (h(b))^{1-\lambda} \le \gamma(1-\lambda) \int_{[t,b)} q(\tau)\Delta\tau, \quad t \in [a,b).$$

In particular, we have

$$(h(a))^{1-\lambda} - (h(b))^{1-\lambda} \le \gamma(1-\lambda) \int_{[a,b)} q(\tau)\Delta\tau.$$

Observe that

$$h(b) = \gamma + \gamma \int_{[a,b)} q(\tau)(\phi(\tau))^{\lambda}\Delta\tau,$$

whereupon

$$\frac{h(b) - \gamma}{\gamma} = \int_{[a,b)} q(\tau)(\phi(\tau))^{\lambda}\Delta\tau.$$

Then

$$h(a) = \gamma + 2\gamma \int_{[a,b)} q(\tau)(\phi(\tau))^{\lambda}\Delta\tau$$
$$= \gamma + 2\gamma\frac{h(b) - \gamma}{\gamma}$$
$$= \gamma + 2h(b) - 2\gamma = 2h(b) - \gamma.$$

Therefore,

$$(2h(b) - y)^{1-\lambda} - (h(b))^{1-\lambda} \le \gamma(1 - \lambda) \int_{[a,b)} q(\tau)\Delta\tau.$$

Let

$$r(z) = (2z - y)^{1-\lambda} - z^{1-\lambda}, \quad z \in \mathbb{R}.$$

Then

$$\lim_{z\to\infty} r(z) = \lim_{z\to\infty} \frac{r(z)}{z^{1-\lambda}} z^{1-\lambda} = \lim_{z\to\infty} \left(\left(2 - \frac{y}{z} \right)^{1-\lambda} - 1 \right) z^{1-\lambda} = \infty.$$

Consequently, there is a positive constant M such that

$$h(b) \le M.$$

Thus

$$\phi(t) \le h(b) \le M, \quad t \in [a, b).$$

This completes the proof. □

Now consider the backward problem

$$\phi^\Delta(t) + p(t)\phi^\sigma(t) = w(t, \phi(t), \phi^\sigma(t)), \quad a \le t < b, \\ \phi(b) = \phi_1, \tag{12.6}$$

where $\phi_1 \in \mathbb{R}$, $p \in \Gamma_1(\mathbb{T})$, and

(H4) $w : \mathbb{T} \times \mathbb{R} \times \mathbb{R} \to \mathbb{R}$ is Δ-measurable in $t \in \mathbb{R}$ and locally Lipschitz continuous, i. e., for all $x_1, x_2, y_1, y_2 \in \mathbb{R}$, $|x_1| \le \rho$, $|x_2| \le \rho$, $|y_1| \le \rho$, $|y_2| \le \rho$, for some positive constant ρ, there exists a constant $L = L(\rho)$ such that

$$\left| w(t, x_1, y_1) - w(t, x_2, y_2) \right| \le L(\rho)(|x_1 - x_2| + |y_1 - y_2|), \quad t \in \mathbb{T},$$

(H5) There exist a constant $\lambda \in (0, 1)$ and a function $q \in \mathbb{L}^1(\mathbb{T})$, $q \ge 0$ on $\mathbb{T}$, such that

$$\left| w(t, \phi, \psi) \right| \le q(t)(1 + |\phi|^\lambda + |\psi|), \quad \phi, \psi \in \mathbb{R}.$$

Using Theorem D.58, one can prove that the equation (12.6) has a unique weak solution $\phi \in \mathcal{C}_{rd}(\mathbb{T})$ given by

$$\phi(t) = e_{\ominus p}(b, t)\phi_1 - \int_{[t,b)} e_{\ominus p}(\sigma(\tau), t)w(\tau, \phi(\tau), \phi^\sigma(\tau))\Delta\tau, \quad t \in \mathbb{T}.$$

Let $(\overline{x}, \overline{u})$ be an optimal pair of (12.1), (12.2). Assume that

(H6) $f(t,\cdot,\cdot) : \mathbb{R} \times \mathbb{R} \to \mathbb{R}$ is partially differentiable and

$$f_{\bar{x}}(\cdot) = f_{\bar{x}}(\cdot, \bar{x}(\cdot), \bar{x}^\sigma(\cdot)) \in \mathbb{L}^1(\mathbb{T}),$$
$$f_{\bar{x}^\sigma}(\cdot) = f_{\bar{x}^\sigma}(\cdot, \bar{x}(\cdot), \bar{x}^\sigma(\cdot)) \in \mathbb{L}^1(\mathbb{T}),$$

(H7) $P(\cdot) = p(\cdot) - f_{\bar{x}^\sigma}(\cdot) \in \Gamma_1(\mathbb{T}),$

(H8) $g : \mathbb{R} \times \mathbb{R} \to \mathbb{R}$ is convex.

Theorem 12.6. *Let $p \in \Gamma_1(\mathbb{T})$ and (H1)–(H8) hold. Then, in order for $(\bar{x}, \bar{u})$ to be an optimal pair of the problem (12.1), (12.2), it is necessary that there is a function $\phi \in C_{rd}(\mathbb{T})$ such that*

$$\bar{x}^\Delta(t) + p(t)\bar{x}^\sigma(t) = f(t, \bar{x}(t), \bar{x}^\sigma(t)) + \bar{u}(t), \quad a \le t < b, \tag{12.7}$$
$$\bar{x}(a) = x_0,$$

$$\phi^\Delta(t) = P(t)\phi(t) - f_{\bar{x}}(t)\phi^\sigma(t) - (1 + \mu(t)P(t))\eta(t), \quad a \le t < b, \tag{12.8}$$
$$\eta \in \partial G(\bar{x}^\sigma), \quad \phi(b) = 0,$$

$$\int_{[a,b)} \left(\frac{\phi^\sigma(t)}{1 + \mu(t)P(t)} + \xi(t) \right)(u(t) - \bar{u}(t))\Delta t \ge 0, \quad \xi \in \partial H(\bar{u}), \tag{12.9}$$

where

$$\partial G(\bar{x}) = \left\{ \xi \in \mathbb{L}^1(\mathbb{T}) : \int_{[a,b)} \xi(t)(x(t) - \bar{x}(t))\Delta t \right.$$
$$\left. \le \int_{[a,b)} (g(x(t), x^\sigma(t)) - g(\bar{x}(t), \bar{x}^\sigma(t)))\Delta t \right\},$$

$$\partial H(\bar{u}) = \left\{ \xi \in \mathbb{L}^\infty(\mathbb{T}) : \int_{[a,b)} \xi(t)(u(t) - \bar{u}(t))\Delta t \le \int_{[a,b)} (h(u(t)) - h(\bar{u}(t)))\Delta t \right\}.$$

Proof. Let $(\bar{x}, \bar{u}) \in C_{rd}(\mathbb{T}) \times U_{ad}$ be an optimal pair. Then it satisfies (12.7). Since U_{ad} is convex, we have that

$$u_\varepsilon = \bar{u} + \varepsilon(u - \bar{u}) \in U_{ad}$$

for any $\varepsilon \in [0,1]$ and $u \in U_{ad}$. Let x_ε be the weak solution of the following dynamic equation:

$$x_\varepsilon^\Delta(t) + p(t)x_\varepsilon^\sigma(t) = f(t, x_\varepsilon(t), x_\varepsilon^\sigma(t)) + u_\varepsilon(t), \quad t \ge a,$$
$$x_\varepsilon(a) = x_0.$$

Then x_ε can be represented in the form

$$x_\varepsilon(t) = e_{\ominus p}(t,a)x_0 + \int_{[a,t)} e_{\ominus p}(t,\tau)(f(\tau,x_\varepsilon(\tau),x_\varepsilon^\sigma(\tau)) + u_\varepsilon(\tau))\Delta\tau, \quad t \in \mathbb{T}.$$

Consider

$$x_\varepsilon(t) - \overline{x}(t) = \int_{[a,t)} e_{\ominus p}(t,\tau)(f(\tau,x_\varepsilon(\tau),x_\varepsilon^\sigma(\tau)) - f(\tau,\overline{x}(\tau),\overline{x}^\sigma(\tau)))\Delta\tau$$

$$+ \varepsilon \int_{[a,t)} e_{\ominus p}(t,\tau)(u(\tau) - \overline{u}(\tau))\Delta\tau.$$

Set

$$y = \lim_{\varepsilon \to 0} \frac{x_\varepsilon - \overline{x}}{\varepsilon}.$$

By (H6), it follows that $u \to x(u)$ is continuously Gateaux differentiable at u in the direction $u - \overline{u}$. Its Gateaux derivative y^Δ satisfies

$$y^\Delta(t) + p(t)y^\sigma(t) = f_{\overline{x}}(t)y(t) + f_{\overline{x}^\sigma}(t)y^\sigma(t) + u(t) - \overline{u}(t), \quad a \le t < b,$$

$$y(a) = 0.$$

By Theorem D.58, it follows that the latter equation has a unique weak solution $y \in \mathcal{C}_{rd}(\mathbb{T})$ given by the expression

$$y(t) = \int_{[a,t)} e_{\ominus(p-f_{\overline{x}^\sigma})}(t,\tau)(f_{\overline{x}}(\tau)y(\tau) + u(\tau) - \overline{u}(\tau))\Delta\tau.$$

Define

$$G(x) = \int_{[a,b)} g(x(t),x^\sigma(t))\Delta t, \quad x \in \mathbb{L}^1(\mathbb{T}).$$

Since $g : \mathbb{R} \times \mathbb{R} \to \mathbb{R}$ is convex, we have that $g : \mathbb{R} \times \mathbb{R} \to \mathbb{R}$ is continuous. By Theorem 12.1, it follows that G is a lower semicontinuous functional on $\mathcal{C}_{rd}(\mathbb{T})$. For any $x_1, x_2 \in \mathcal{C}_{rd}(\mathbb{T})$ and $\lambda \in [0,1]$, we have

$$G(\lambda x_1 + (1-\lambda)x_2) = \int_{[a,b)} g(\lambda x_1(t) + (1-\lambda)x_2(t), \lambda x_1^\sigma(t) + (1-\lambda)x_2^\sigma(t))\Delta t$$

$$\le \lambda \int_{[a,b)} g(x_1(t),x_1^\sigma(t))\Delta t + (1-\lambda) \int_{[a,b)} g(x_2(t),x_2^\sigma(t))\Delta t$$

$$= \lambda G(x_1) + (1-\lambda)G(x_2).$$

Then G is convex on $\mathcal{C}_{rd}(\mathbb{T})$. Also, G is finite and continuous at $\overline{x}$, G is subdifferentiable at $\overline{x} \in \mathcal{C}_{rd}(\mathbb{T})$ and the subdifferential $\partial G(\overline{x})$ of G at $\overline{x}^\sigma$ is given by

$$\partial G(\overline{x}) = \left\{ \xi \in \mathbb{L}^1(\mathbb{T}) : \int_{[a,b)} \xi(t)(x(t) - \overline{x}(t))\Delta t \right.$$

$$\left. \leq \int_{[a,b)} (g(x(t), x^{\sigma}(t)) - g(\overline{x}(t), \overline{x}^{\sigma}(t)))\Delta t \right\}.$$

Note that $\partial G(\overline{x})$ is nonempty. Define

$$H(u) = \int_{[a,b)} H(u(t))\Delta t, \quad u \in \mathbb{L}^1(\mathbb{T}).$$

We have that H is subdifferentiable at $\overline{u} \in U_{ad}$ and

$$\partial H(\overline{u}) = \left\{ \xi \in \mathbb{L}^{\infty}(\mathbb{T}) : \int_{[a,b)} \xi(t)(u(t) - \overline{u}(t))\Delta t \leq \int_{[a,b)} (h(u(t)) - h(\overline{u}(t)))\Delta t \right\}.$$

Also, $\partial H(\overline{u})$ is nonempty. Since

$$J = G + H,$$

we conclude that J is subdifferentiable at $\overline{u} \in U_{ad}$. The subdifferential of J at $\overline{u}$ in the direction $u - \overline{u}$ is

$$\partial J(\overline{u}, u - \overline{u}) = \int_{[a,b)} y(t)\eta(t)\Delta t + \int_{[a,b)} \xi(t)(u(t) - \overline{u}(t))\Delta t$$

for any $\eta \in \partial G(\overline{x})$, $\xi \in \partial H(\overline{u})$. We have

$$J(x_{\varepsilon}, u_{\varepsilon}) - J(\overline{x}, \overline{u}) \geq 0, \quad \varepsilon \in [0, 1], \quad u \in U_{ad}.$$

Hence, for $\overline{u}$ to be optimal, it is necessary that

$$\int_{[a,b)} y(t)\eta(t)\Delta t + \int_{[a,b)} \xi(t)(u(t) - \overline{u}(t))\Delta t \geq 0 \tag{12.10}$$

for any $\eta \in \partial G(\overline{x})$, $\xi \in \partial H(\overline{u})$. For $\eta \in \partial G(\overline{x})$, consider the following adjoint equation:

$$\phi^{\Delta}(t) = P(t)\phi(t) - f_{\overline{x}}(t)\phi^{\sigma}(t) - (1 + \mu(t)P(t))\eta(t), \quad a \leq t < b,$$
$$\phi(b) = 0.$$

Note that $P \in \Gamma_1(\mathbb{T})$ and

$$\phi(t) = \int_{[t,b)} e_{\ominus p}(\sigma(\tau), t)(f_{\overline{x}}(\tau)\phi^{\sigma}(\tau) + (1 + \mu(\tau)P(\tau))\eta(\tau))\Delta\tau.$$

Also,

$$\eta(t) = \frac{P(t)\phi(t) - f_{\bar{x}}(t)\phi^\sigma(t) - \phi^\Delta(t)}{1 + \mu(t)P(t)}$$

and

$$\int_{[a,b)} y(t)\eta(t)\Delta t = \int_{[a,b)} y(t)\frac{P(t)(\phi^\sigma(t) - \mu(t)\phi^\Delta(t)) - \phi^\Delta(t) - f_{\bar{x}}(t)\phi^\sigma(t)}{1 + \mu(t)P(t)}\Delta t$$

$$= \int_{[a,b)} y(t)\left(-\phi^\Delta(t) + \frac{P(t) - f_{\bar{x}}(t)}{1 + \mu(t)P(t)}\phi^\sigma(t)\right)\Delta t$$

$$= \int_{[a,b)} \phi^\sigma(t)\left(y^\Delta(t) + \frac{P(t) - f_{\bar{x}}(t)}{1 + \mu(t)P(t)}y(t)\right)\Delta t$$

$$= \int_{[a,b)} \phi^\sigma(t)\frac{y^\Delta(t) + \mu(t)y(t)P(t) + P(t)y(t) - f_{\bar{x}}(t)y(t)}{1 + \mu(t)P(t)}\Delta t$$

$$= \int_{[a,b)} \phi^\sigma(t)\frac{y^\Delta(t) + y^\sigma(t)P(t) - y(t)P(t) + y(t)P(t) - f_{\bar{x}}(t)y(t)}{1 + \mu(t)P(t)}\Delta t$$

$$= \int_{[a,b)} \frac{\phi^\sigma(t)}{1 + \mu(t)P(t)}\left(-p(t)y^\sigma(t) + f_{\bar{x}}(t)y(t) + f_{\bar{x}^\sigma}(t)y^\sigma(t) + u(t)\right.$$

$$- \bar{u}(t) + y^\sigma(t)p(t) - y^\sigma(t)f_{\bar{x}^\sigma}(t) - f_{\bar{x}}(t)y(t)\Big)\Delta t$$

$$= \int_{[a,b)} \frac{\phi^\sigma(t)}{1 + \mu(t)(p(t) - f_{\bar{x}}(t))}(u(t) - \bar{u}(t))\Delta t.$$

Now, applying (12.10), we get

$$\int_{[a,b)} \frac{\phi^\sigma(t)}{1 + \mu(t)(p(t) - f_{\bar{x}}(t))}(u(t) - \bar{u}(t))\Delta t \geq 0$$

for $u \in U_{ad}$ and $\xi \in \partial H(\bar{u})$. This completes the proof. □

We will illustrate the above result with the following example.

Example 12.7. Let $\mathbb{T} = \mathbb{Z}$. Consider

$$J(u) = 2\int_{[0,20)} \frac{x^\sigma(t)}{(1 + (x(t))^2)(1 + (x^\sigma(t))^2)}\Delta t + \int_{[0,20)} (u(t))^2\Delta t,$$

and

$$x^\Delta(t) + 2x^\sigma(t) = \frac{x(t)}{1 + (x(t))^2} + x^\sigma(t) + u(t), \quad 0 \leq t < 20,$$

$$x(0) = 1, \quad u \in U_{ad}.$$

Here

$$\sigma(t) = t + 1, \quad g(x(t), x^\sigma(t)) = 2\frac{x^\sigma(t)}{(1 + (x(t))^2)(1 + (x^\sigma(t))^2)}, \quad h(u(t)) = (u(t))^2,$$

$$p(t) = 2, \quad f(t, x(t), x^\sigma(t)) = \frac{x(t)}{1 + (x(t))^2} + x^\sigma(t), \quad t \in \mathbb{T}.$$

Let

$$l(x) = \frac{x}{1 + x^2}, \quad x \in \mathbb{R}.$$

Then

$$l'(x) = \frac{1 + x^2 - 2x^2}{(1 + x^2)^2} = \frac{1 - x^2}{(1 + x^2)^2},$$

$$|l'(x)| = \left|\frac{1 - x^2}{(1 + x^2)^2}\right| \le \frac{1 + x^2}{(1 + x^2)^2} = \frac{1}{1 + x^2} \le 1, \quad x \in \mathbb{R},$$

and

$$|f(t, x_1, y_1) - f(t, x_2, y_2)| = \left|\frac{x_1}{1 + x_1^2} + y_1 - \frac{x_2}{1 + x_2^2} - y_2\right|$$

$$\le \left|\frac{x_1}{1 + x_1^2} - \frac{x_2}{1 + x_2^2}\right| + |y_1 - y_2|$$

$$\le |x_1 - x_2| + |y_1 - y_2|, \quad t \in \mathbb{T}, \quad x_1, x_2, y_1, y_2 \in \mathbb{R}.$$

Also,

$$f_x(t, x(t), x^\sigma(t)) = \frac{1 - (x(t))^2}{(1 + (x(t))^2)^2} \in \mathbb{L}^1([0, 20)),$$

$$f_{x^\sigma}(t, x(t), x^\sigma(t)) = 1 \in \mathbb{L}^1([0, 20)).$$

Next, $g : \mathbb{R} \times \mathbb{R} \to \mathbb{R}$ is lower semicontinuous. Since

$$\lim_{\substack{|x| \to \infty \\ |y| \to \infty}} |g(x, y)| = \lim_{\substack{|x| \to \infty \\ |y| \to \infty}} \left|2\frac{y}{(1 + x^2)(1 + y^2)}\right| = 0,$$

there is a constant $c \in \mathbb{R}$ such that

$$g(x, y) \ge c, \quad x, y \in \mathbb{R}.$$

Next,

$$P(t) = 2 - 1 = 1 \in \Gamma_1([0, 20)).$$

Then, for the pair $(\overline{x}, \overline{u})$ to be an optimal pair of the considered problem, it is necessary that there is a function $\phi \in C_{rd}(\mathbb{T})$ such that

$$\overline{x}^\Delta(t) + 2\overline{x}^\sigma(t) = \frac{\overline{x}(t)}{1 + (\overline{x}(t))^2} + \overline{x}^\sigma(t) + \overline{u}(t), \quad 0 \le t < 20, \quad \overline{x}(0) = 1,$$

$$\phi^\Delta(t) = \phi(t) - \frac{1 - (\overline{x}(t))^2}{(1 + (\overline{x}(t))^2)^2}\phi^\sigma(t) + 2\eta(t), \quad 0 \le t < 20,$$

$$\eta \in \partial G(\overline{x}^\sigma), \quad \phi(20) = 0,$$

and

$$\int_{[0,20)} \left(\frac{1}{2}\phi^\sigma(t) + \xi(t)\right)(u(t) - \overline{u}(t))\Delta t \ge 0, \quad u \in U_{ad}, \quad \xi \in \partial H(\overline{u}),$$

where

$$\partial G(\overline{x}) = \left\{ \xi \in \mathbb{L}^1([0, 20)) : \int_{[0,20)} \xi(t)(x(t) - \overline{x}(t))\Delta t \right.$$

$$\left. \le \int_{[0,20)} 2\left(\frac{x^\sigma(t)}{(1 + (x(t))^2)(1 + (x^\sigma(t))^2)} - \frac{\overline{x}^\sigma(t)}{(1 + (\overline{x}(t))^2)(1 + (\overline{x}^\sigma(t))^2)}\right)\Delta t \right\},$$

$$\partial H(\overline{u}) = \left\{ \xi \in \mathbb{L}^\infty([0, 20)) : \int_{[0,20)} \xi(t)(u(t) - \overline{u}(t))\Delta t \le \int_{[0,20)} ((u(t))^2 - (\overline{u}(t))^2)\Delta t \right\}.$$

Exercise 12.8. Let $\mathbb{T} = 3\mathbb{N}_0$. Find the necessary conditions of optimality for the following problem:

$$J(u) = -\int_{[0,100)} \frac{x(\tau)x^\sigma(\tau)}{(1 + (x(\tau))^2)(1 + (x^\sigma(\tau))^2)}\Delta\tau + 4\int_{[0,100)} (u(\tau))^2\Delta\tau,$$

and

$$x^\Delta(t) + 4\sin_2(t, 3)x^\sigma(t) = -\frac{(x(t))^2}{1 + (x(t))^4} + u(t), \quad t > 0,$$

$$x(0) = 1, \quad u \in U_{ad}.$$

Remark 12.9. Consider the problem (12.1) and

$$x^\Delta(t) = p(t)x(t) + f(t, x(t), x^\sigma(t)) + u(t), \quad t > a,$$
$$x(a) = x_0, \quad u \in U_{ad}, \tag{12.11}$$

where p, f, and u are as above. Using

$$x(t) = x^\sigma(t) - \mu(t)x^\Delta(t),$$

we rewrite the dynamic equation (12.11) as follows:

$$x^\Delta(t) = p(t)x^\sigma(t) - p(t)\mu(t)x^\Delta(t) + f(t, x(t), x^\sigma(t)) + u(t), \quad t > a,$$
$$x(a) = x_0, \quad u \in U_{ad},$$

or

$$(1 + \mu(t)p(t))x^\Delta(t) = p(t)x^\sigma(t) + f(t, x(t), x^\sigma(t)) + u(t), \quad t > a,$$
$$x(a) = x_0, \quad u \in U_{ad},$$

or

$$x^\Delta(t) = \frac{p(t)}{1 + \mu(t)p(t)}x^\sigma(t) + \frac{1}{1 + \mu(t)p(t)}f(t, x(t), x^\sigma(t)) + \frac{1}{1 + \mu(t)p(t)}u(t), \quad t > a,$$
$$x(a) = x_0, \quad u \in U_{ad}.$$

Exercise 12.10. Write the necessary conditions of optimality for the problem (12.1), (12.11).

12.3 Advanced practical problems

Problem 12.11. Let $\mathbb{T} = 3^{\mathbb{N}_0} \cup \{0\}$. Prove that the problem

$$J(u) = -\int_{[0,81)} \frac{(x(\tau))^8}{1 + (x(\tau))^8}\Delta\tau + 4\int_{[0,81)} (u(\tau))^5\Delta\tau,$$

and

$$x^\Delta(t) + e_2(t, 3)x^\sigma(t) = -\frac{(x(t))^{12}}{1 + (x(t))^{12}} + u(t), \quad t > 0,$$
$$x(0) = 1, \quad u \in U_{ad},$$

has at least one solution.

Problem 12.12. Let $\mathbb{T} = 7\mathbb{N}_0$. Find the necessary conditions of optimality for the following problem:

$$J(u) = -\int_{[0,49)} \frac{(x(\tau))^2 x^\sigma(\tau)}{(1 + (x(\tau))^4)(1 + (x^\sigma(\tau))^6)}\Delta\tau + \int_{[0,49)} (u(\tau))^4\Delta\tau,$$

and

$$x^\Delta(t) + \cos_2(t,7)x^\sigma(t) = \frac{(x(t))^4}{1 + (x(t))^4} + u(t), \quad t > 0,$$

$$x(0) = 1, \quad u \in U_{ad}.$$

Problem 12.13. Let $\mathbb{T} = 2^{\mathbb{N}_0} \cup \{0\}$. Find the necessary conditions of optimality for the following problem:

$$J(u) = -\int_{[0,64)} \frac{(x(\tau))^4(x^\sigma(\tau))^2}{(1 + 3(x(\tau))^4)(1 + 7(x^\sigma(\tau))^6)}\Delta\tau + \int_{[0,64)} (u(\tau))^4 \Delta\tau,$$

and

$$x^\Delta(t) + tx^\sigma(t) = \frac{(x(t))^4}{1 + (x(t))^4} + u(t), \quad t > 0,$$

$$x(0) = 1, \quad u \in U_{ad}.$$

Problem 12.14. Let $\mathbb{T} = 3^{\mathbb{N}_0} \cup \{0\}$. Find the necessary conditions of optimality for the following problem:

$$J(u) = -\int_{[0,243)} \frac{(x(\tau))^2 + 2(x^\sigma(\tau))^4}{(1 + (x(\tau))^4) + (1 + (x^\sigma(\tau))^6)}\Delta\tau + \int_{[0,243)} (u(\tau))^4 \Delta\tau,$$

and

$$x^\Delta(t) + (t^2 + t + 3)x^\sigma(t) = 12\frac{(x(t))^2}{1 + (x(t))^4} + 4x^\sigma(t) + u(t), \quad t > 0,$$

$$x(0) = 1, \quad u \in U_{ad}.$$

Problem 12.15. Let $\mathbb{T} = 4^{\mathbb{N}_0} \cup \{0\}$. Find the necessary conditions of optimality for the following problem:

$$J(u) = \int_{[0,256)} \frac{(x(\tau))^4 + (x^\sigma(\tau))^8}{(1 + (x(\tau))^{14})(1 + (x^\sigma(\tau))^{64})}\Delta\tau + \int_{[0,256)} (u(\tau))^4 \Delta\tau,$$

and

$$x^\Delta(t) + \frac{t^2 + t + 3}{t^8 + t^4 + 10}x^\sigma(t) = \frac{(x(t))^4}{1 + (x(t))^6} + (x^\sigma(t))^7 + u(t), \quad t > 0,$$

$$x(0) = 1, \quad u \in U_{ad}.$$

Problem 12.16. Suppose that $f(t, x(t), x^\sigma(t)) = f_1(t, x(t))$ in (12.2). Write the necessary conditions of optimality.

Problem 12.17. Suppose that $f(t, x(t), x^\sigma(t)) = f_1(t, x^\sigma(t))$ in (12.2). Write the necessary conditions of optimality.

Problem 12.18. Suppose that $f(t, x(t), x^\sigma(t)) = f_1(t, x(t)) + f_2(t, x^\sigma(t))$ in (12.2). Write the necessary conditions of optimality.

Problem 12.19. Suppose that $f(t, x(t), x^\sigma(t)) = f_1(t, x(t))f_2(t, x^\sigma(t))$ in (12.2). Write the necessary conditions of optimality.

Problem 12.20. Suppose that $f(t, x(t), x^\sigma(t)) = (f_1(t, x(t)))^2 + (f_2(t, x^\sigma(t)))^4$ in (12.2). Write the necessary conditions of optimality.

A Rolle's theorem

Suppose that $\mathbb{T}$ is a time scale with forward jump operator σ and delta differentiation operator Δ.

Definition A.1. Let $y : \mathbb{T} \to \mathbb{R}$ be $(k-1)$-times differentiable, $k \in \mathbb{N}$. We say that y has a generalized zero (GZ) of order greater than or equal to k at $t \in \mathbb{T}^{\kappa^{k-1}}$ provided

$$y^{\Delta^i}(t) = 0, \quad i \in \{0, \ldots, k-1\}, \tag{A.1}$$

or

$$y^{\Delta^i}(t) = 0 \quad \text{for } i \in \{0, \ldots, k-2\} \quad \text{and} \quad y^{\Delta^{k-1}}(\rho(t))y^{\Delta^{k-1}}(t) < 0 \tag{A.2}$$

holds.

Remark A.2. Note that in the Case (A.2) t must be left-scattered. Otherwise, $\rho(t) = t$ and

$$0 > y^{\Delta^{k-1}}(\rho(t))y^{\Delta^{k-1}}(t) = \left(y^{\Delta^{k-1}}(t)\right)^2 \geq 0,$$

which is a contradiction.

Theorem A.3. *Condition (A.2) holds if and only if*

$$y^{\Delta^j}(t) = 0, \quad j \in \{0, \ldots, k-2\}, \quad \text{and} \quad (-1)^{k-1}y(\rho(t))y^{\Delta^{k-1}}(t) < 0. \tag{A.3}$$

Proof.
1. Let (A.2) hold. Then t is left-scattered, $\sigma(\rho(t)) = t$, and

$$\begin{aligned}
y^{\Delta^{k-1}}(\rho(t)) &= \frac{y^{\Delta^{k-2}}(\sigma(\rho(t))) - y^{\Delta^{k-2}}(\rho(t))}{\mu(\rho(t))} \\
&= -\frac{y^{\Delta^{k-2}}(\rho(t))}{\mu(\rho(t))} \\
&\;\;\vdots \\
&= (-1)^{k-1}\frac{y(\rho(t))}{(\mu(\rho(t)))^{k-1}}.
\end{aligned} \tag{A.4}$$

Hence,

$$0 > y^{\Delta^{k-1}}(\rho(t))y^{\Delta^{k-1}}(t) = (-1)^{k-1}\frac{y(\rho(t))}{(\mu(\rho(t)))^{k-1}}y^{\Delta^{k-1}}(t)$$

and (A.3) holds.

https://doi.org/10.1515/9783110787320-013

2. Assume (A.3). Let t be left-dense. Then $\rho(t) = t$, and we have the following cases:

   (a) Let $k = 1$. Then $(y(t))^2 < 0$, which is a contradiction.

   (b) Let $k \geq 2$. Then $y(t) = 0$ and

$$(-1)^{k-1}y(t)y^{\Delta^{k-1}}(t) < 0.$$

This is a contradiction.

Consequently, t is left-scattered. Hence, using (A.4), we obtain (A.2).

This completes the proof. $\qquad\square$

Theorem A.4. *Let $j \in \mathbb{N}_0$ and $t \in \mathbb{T}^{\kappa^j}$. Then*

$$y^{\Delta^i}(t) = 0, \quad 0 \leq i \leq j, \tag{A.5}$$

if and only if

$$y^{\Delta^i}(\sigma^l(t)) = 0, \quad 0 \leq i \leq j - l, \quad 0 \leq l \leq j. \tag{A.6}$$

In this case,

$$y^{\Delta^{j+1-l}}(\sigma^l(t)) = \prod_{s=0}^{l-1} \mu(\sigma^s(t))y^{\Delta^{j+1}}(t). \tag{A.7}$$

Proof.

1. Let (A.5) hold.

   (a) Suppose that $j = 0$. Then $l = i = 0$, $y(t) = 0$, and

$$y^{\Delta^i}(\sigma^l(t)) = y(t) = 0.$$

   (b) Suppose that $j > 0$. Then

$$y(t) = y^{\Delta}(t) = \cdots = y^{\Delta^j}(t) = 0.$$

   i. Let $j = 1$. Then $i \in \{0, 1\}$, $l \in \{0, 1\}$.

      A. Let $l = 0$. Then $i \in \{0, 1\}$ and (A.6) holds.

      B. Let $l = 1$. Then $i = 0$ and

$$y(\sigma(t)) = y(t) + \mu(t)y^{\Delta}(t) = 0.$$

ii.   Assume that (A.6) is true for some $j \in \mathbb{N}$, i. e., assume

$$y(t) = y^{\Delta}(t) = \cdots = y^{\Delta^j}(t) = 0,$$
$$y(\sigma(t)) = y^{\Delta}(\sigma(t)) = \cdots = y^{\Delta^{j-1}}(\sigma(t)) = 0,$$
$$y(\sigma^2(t)) = y^{\Delta}(\sigma^2(t)) = \cdots = y^{\Delta^{j-2}}(\sigma^2(t)) = 0,$$
$$\vdots \qquad\qquad\qquad\qquad (A.8)$$
$$y(\sigma^l(t)) = y^{\Delta}(\sigma^l(t)) = \cdots = y^{\Delta^{j-l}}(\sigma^l(t)) = 0,$$
$$\vdots$$
$$y(\sigma^j(t)) = 0.$$

iii. We will prove (A.6) for $j + 1$, i. e., we will prove that

$$y(t) = y^{\Delta}(t) = \cdots = y^{\Delta^{j+1}}(t) = 0,$$
$$y(\sigma(t)) = y^{\Delta}(\sigma(t)) = \cdots = y^{\Delta^j}(\sigma(t)) = 0,$$
$$y(\sigma^2(t)) = y^{\Delta}(\sigma^2(t)) = \cdots = y^{\Delta^{j-1}}(\sigma^2(t)) = 0,$$
$$\vdots$$
$$y(\sigma^l(t)) = y^{\Delta}(\sigma^l(t)) = \cdots = y^{\Delta^{j-l+1}}(\sigma^l(t)) = 0,$$
$$\vdots$$
$$y(\sigma^{j+1}(t)) = 0.$$

By (A.8), it follows that we have to prove

$$y^{\Delta^{j+1}}(t) = y^{\Delta^j}(\sigma(t)) = \cdots = y^{\Delta^{j-l+1}}(\sigma^l(t)) = \cdots = y(\sigma^{j+1}(t)) = 0.$$

By (A.5), we have $y^{\Delta^{j+1}}(t) = 0$. Then

$$y^{\Delta^j}(\sigma(t)) = y^{\Delta^j}(t) + \mu(t)y^{\Delta^{j+1}}(t) = 0,$$
$$y^{\Delta^{j-1}}(\sigma^2(t)) = y^{\Delta^{j-1}}(\sigma(t)) + \mu(t)y^{\Delta^j}(\sigma(t)) = 0,$$
$$\vdots$$
$$y^{\Delta}(\sigma^j(t)) = 0,$$
$$y(\sigma^{j+1}(t)) = y(\sigma^j(t)) + \mu(t)y^{\Delta}(\sigma^j(t)) = 0.$$

Hence, by the principle of the mathematical induction, we conclude that (A.6) holds for any $j \in \mathbb{N}_0$.

2. Suppose that (A.6) holds.
   (a) Let $j = 0$. Then $i = l = 0$ and

$$y^{\Delta^i}(\sigma^l(t)) = y(t) = 0.$$

   (b) Let $j > 0$.
      i. Assume that $j = 1$. Then, by (A.6), we have

$$y(t) = y^\Delta(t) = y(\sigma(t)) = 0.$$

      ii. Assume that (A.5) holds for some $j \in \mathbb{N}_0$.
      iii. We will prove that (A.5) holds for $j + 1$. Since (A.6) holds for $j + 1$, by (A.8) we obtain

$$y(t) = y^\Delta(t) = \cdots = y^{\Delta^{j+1}}(t) = 0.$$

Hence, by the principle of the mathematical induction, it follows that (A.5) holds for any $j \in \mathbb{N}_0$.

Suppose (A.6) is true, i. e., we assume (A.8). Then we will prove (A.7). We have

$$y^{\Delta^j}(\sigma(t)) = y^{\Delta^j}(t) + \mu(t)y^{\Delta^{j+1}}(t)$$

$$= \mu(t)y^{\Delta^{j+1}}(t),$$

$$y^{\Delta^{j-1}}(\sigma^2(t)) = y^{\Delta^{j-1}}(\sigma(t)) + \mu(\sigma(t))y^{\Delta^j}(\sigma(t))$$

$$= \mu(\sigma(t))y^{\Delta^j}(\sigma(t))$$

$$= \mu(\sigma(t))\mu(t)y^{\Delta^{j+1}}(t),$$

$$\vdots$$

$$y^{\Delta^{j+1-l}}(\sigma^l(t)) = \prod_{s=0}^{l-1} \mu(\sigma^s(t))y^{\Delta^{j+1}}(t).$$

This completes the proof. □

Definition A.5. If y has a GZ of order greater than or equal to k at t, we will say that y has at least k GZs, counting multiplicities. By Theorem A.4, it follows that if y has a GZ of order greater than or equal to k at t, then y has a GZ of order greater than or equal to $k - 1$ at $\sigma(t)$. Therefore, if y has a GZ of order greater than or equal to k at t_1 and y has a GZ of order greater than or equal to k_2 at t_2 and $\sigma^{k-1}(t_1) < t_2$, then we will say that y has at least $k_1 + k_2$ GZs, counting multiplicities.

Theorem A.6 (Rolle's theorem). *If y has at least $k \in \mathbb{N}$ GZs on $[a, b]$, counting multiplicities, then y^Δ has at least $k - 1$ GZs on $[a, b]$, counting multiplicities.*

Proof.

1. Firstly, we will prove that if y has a GZ of order $\leq m$ at t, then y^Δ has a GZ of order $\leq m - 1$ at t.

   Since y has a GZ of order $\leq m$ at t, we have

   $$y^{\Delta^i}(t) = 0, \quad i \in \{0, \ldots, m - 1\},$$

   or

   $$y^{\Delta^i}(t) = 0, \quad i \in \{0, \ldots, m - 2\}, \quad \text{and} \quad y^{\Delta^{m-1}}(\rho(t))y^{\Delta^{m-1}}(t) < 0.$$

   If $m = 1$, then $y(t) = 0$ or $y(\rho(t))y(t) < 0$. If $y^\Delta(t) = 0$ or $y^\Delta(\rho(t))y^\Delta(t) < 0$, then $m > 1$, which is a contradiction. Therefore, y^Δ has no GZ at t. Let $m \geq 2$. Then

   $$(y^\Delta)^{\Delta^{i-1}}(t) = 0, \quad i \in \{0, \ldots, m - 2\},$$

   or

   $$(y^\Delta)^{\Delta^i}(t) = 0, \quad i \in \{0, \ldots, m - 3\}, \quad \text{and} \quad (y^\Delta)^{\Delta^{m-2}}(\rho(t))(y^\Delta)^{\Delta^{m-2}}(t) < 0.$$

   Thus, y^Δ has a GZ of order $\leq m - 1$ at t.

2. Now, we will prove that if y has a GZ of order $\leq m \in \mathbb{N}$ at t and y has a GZ of order ≤ 1 at s with $\sigma^{m-1}(t) < s$, then y^Δ has at least m GZs in $[t, s)$, which is equivalent to
   (a) If $y(r) = 0$ and y^Δ has no GZ in $[r, s)$, where $r < \rho(s)$, then y has no GZ at s.
   (b) If $y(\rho(r))y(r) < 0$ and y^Δ has no GZ in $[r, s)$, where $r < \rho(s)$, then y has no GZ at s.

   If the assumptions of (a) hold, then $y^\Delta(\tau) > 0$, $\tau \in [r, s)$, or $y^\Delta(\tau) < 0$, $\tau \in [r, s)$. Thus,

   $$y(\rho(s))y(s) = \left(\int_r^{\rho(s)} y^\Delta(\tau)\Delta\tau \right)\left(\int_r^s y^\Delta(\tau)\Delta\tau \right) > 0.$$

   Therefore, y has no GZ at s. If the assumptions of (b) hold, then $\rho(r) < r$ and

   $$y(\rho(r))y^\Delta(\rho(r)) = y(\rho(r))\frac{y(r) - y(\rho(r))}{\mu(\rho(r))} < 0.$$

   Since y^Δ has a constant sign on $[\rho(r), s)$, we get

   $$y(\rho(r))y^\Delta(\tau) < 0, \quad \tau[\rho(r), s).$$

   Therefore,

$$y(\rho(r))y(t) = y(\rho(r))\left(y(r) + \int_r^t y^\Delta(\tau)\Delta\tau \right) < 0, \quad t \in \{\rho(s), s\}.$$

Hence,

$$y(\rho(r))y(\rho(s)) < 0, \quad y(\rho(r))y(s) < 0,$$

and

$$(y(\rho(r)))^2 y(\rho(s))y(s) > 0.$$

Consequently, $y(\rho(s))y(s) > 0$ and y has no GZ at s.

This completes the proof. □

B Fréchet and Gâteaux derivatives

B.1 Remainders

Let X and Y be normed spaces. By $o(X, Y)$ we will denote the set of all maps $r : X \to Y$ for which there is some map $\alpha : X \to Y$ such that

1. $r(x) = \alpha(x)\|x\|$ for all $x \in X$,
2. $\alpha(0) = 0$,
3. α is continuous at 0.

Definition B.1. The elements of $o(X, Y)$ will be called remainders.

Exercise B.2. Prove that $o(X, Y)$ is a vector space.

Definition B.3. Let $f : X \to Y$ be a function and $x_0 \in X$. We say that f is stable at x_0 if there are some $\varepsilon > 0$ and $c > 0$ such that $\|x - x_0\| \le \varepsilon$ implies

$$\|f(x - x_0)\| \le c\|x - x_0\|.$$

Example B.4. Let $T : X \to Y$ be a linear bounded operator. Then

$$\|T(x - 0)\| = \|T(x)\| \le \|T\|\|x\|, \quad x \in X.$$

Hence, T is stable at 0.

Theorem B.5. *Let X, Y, Z, and W be normed spaces, $r \in o(X, Y)$, and assume $f : W \to X$ is stable at 0, while $g : Y \to Z$ is stable at 0. Then $r \circ f \in o(W, Y)$ and $g \circ r \in o(X, Z)$.*

Proof. Since $r \in o(X, Y)$, there is a map $\alpha : X \to Y$ such that

$$r(x) = \alpha(x)\|x\|, \quad x \in X,$$

$\alpha(0) = 0$, and α is continuous at 0. Define $\beta : W \to Y$ by

$$\beta(w) = \begin{cases} \frac{\|f(w)\|}{\|w\|}\alpha(f(w)) & \text{if } w \ne 0, \\ 0 & \text{if } w = 0, \end{cases}$$

for $w \in W$. Since $f : W \to Z$ is stable at 0, there are constants $\varepsilon > 0$ and $c > 0$ such that $\|w\| \le \varepsilon$ implies

$$\|f(w)\| \le c\|w\|.$$

Hence,

$$\|f(0)\| = 0 \quad \text{and} \quad f(0) = 0.$$

https://doi.org/10.1515/9783110787320-014

Next, $\beta(0) = 0$ and, if $w \neq 0$, $\|w\| \leq \varepsilon$, we get

$$\|\beta(w)\| = \frac{\|f(w)\|}{\|w\|}\|\alpha(f(w))\| \leq c\|\alpha(f(w))\|.$$

Now using

$$f(w) \to 0, \quad \text{as } w \to 0,$$

and

$$\alpha(f(w)) \to 0, \quad \text{as } w \to 0,$$

we get

$$\beta(w) \to 0, \quad \text{as } w \to 0.$$

Therefore, $\beta : W \to Y$ is continuous at 0. Also, we have
- If $w = 0$, then

$$\beta(0) = 0,$$
$$r \circ f(0) = \alpha(f(0))\|f(0)\| = 0.$$

- If $w \neq 0$, then

$$r \circ f(w) = \alpha(f(w))\|f(w)\|$$
$$= \frac{\|w\|\beta(w)}{\|f(w)\|}\|f(w)\|$$
$$= \|w\|\beta(w).$$

Therefore, $r \circ f \in o(W, Y)$.

Since $g : Y \to Z$ is stable at 0, there are constants $\varepsilon_1 > 0$ and $c_1 > 0$ such that $\|w\| \leq \varepsilon_1$ implies

$$\|g(w)\| \leq c_1\|w\|.$$

Define $y : X \to Y$ by

$$y(x) = \begin{cases} \frac{g(\|x\|\alpha(x))}{\|x\|} & \text{if } x \neq 0, \\ 0 & \text{if } x = 0. \end{cases}$$

Then

$$g(\|x\|\alpha(x)) = \|x\|y(x), \quad x \in X.$$

For $x \neq 0$, $x \in X$, we have

$$\|\gamma(x)\| = \frac{\|g(\|x\|\alpha(x))\|}{\|x\|} \leq \frac{c_1 \|x\|\alpha(x)}{\|x\|} = c_1 \alpha(x).$$

Then

$$\gamma(x) \to 0, \quad \text{as } x \to 0, \quad x \in X.$$

Also,

$$g \circ r(x) = g(r(x)) = g(\alpha(x)\|x\|) = \gamma(x)\|x\|, \quad x \in X.$$

This completes the proof. □

B.2 Definition and uniqueness of the Fréchet derivative

Suppose that X and Y are normed spaces, U is an open subset of X, and $x_0 \in U$. By $\mathcal{L}(X, Y)$ we will denote the vector space of all linear bounded operators from X to Y.

Definition B.6. We say that a function $f : X \to Y$ is Fréchet differentiable at x_0 if there are some $L \in \mathcal{L}(X, Y)$ and $r \in o(X, Y)$ such that

$$f(x) = f(x_0) + L(x - x_0) + r(x - x_0), \quad x \in U.$$

The operator L will be called the Fréchet derivative of the function f at x_0. We will write $Df(x_0) = L$.

Suppose that $L_1, L_2 \in \mathcal{L}(X, Y)$ and $r_1, r_2 \in o(X, Y)$ are such that

$$f(x) = f(x_0) + L_1(x - x_0) + r_1(x - x_0),$$
$$f(x) = f(x_0) + L_2(x - x_0) + r_2(x - x_0), \quad x \in U.$$

Then

$$f(x_0) + L_1(x - x_0) + r_1(x - x_0) = f(x_0) + L_2(x - x_0) + r_2(x - x_0), \quad x \in U,$$

or

$$L_1(x - x_0) - L_2(x - x_0) = r_2(x - x_0) - r_1(x - x_0), \quad x \in U.$$

Also, let $\alpha_1, \alpha_2 : X \to Y$ be such that

$$r_1(x) = \|x\|\alpha_2(x), \quad r_2(x) = \|x\|\alpha_1(x), \quad \alpha_1(0) = \alpha_2(0) = 0,$$

where α_1 and α_2 are continuous at 0. Then

$$L_1(x - x_0) - L_2(x - x_0) = \|x - x_0\|\alpha_1(x - x_0) - \|x - x_0\|\alpha_2(x - x_0)$$
$$= \|x - x_0\|(\alpha_1(x - x_0) - \alpha_2(x - x_0)), \quad x \in U.$$

Let $x \in X$ be arbitrarily chosen. Then there is some $h > 0$ such that for all $|t| \le h$ we have $x_0 + tx \in U$. Hence,

$$L_1(tx) - L_2(tx) = \|tx\|(\alpha_1(tx) - \alpha_2(tx))$$

or

$$t(L_1(x) - L_2(x)) = |t|\|x\|(\alpha_1(tx) - \alpha_2(tx)),$$

or

$$L_1(x) - L_2(x) = \text{sign}(t)\|x\|(\alpha_1(tx) - \alpha_2(tx)) \to 0, \quad \text{as } t \to 0.$$

Because $x \in X$ was arbitrarily chosen, we conclude that $L_1 = L_2$ and $r_1 = r_2$.

Definition B.7. We denote by $C^1(U, Y)$ the set of all functions $f : U \to Y$ that are Fréchet differentiable at each point of U and $Df : U \to \mathcal{L}(X, Y)$ is continuous. We denote by $C^2(U, Y)$ the set of all functions $f \in C^1(U, Y)$ such that $Df : U \to \mathcal{L}(X, Y)$ is Fréchet differentiable at each point of U and

$$D(Df) : U \to \mathcal{L}(X, \mathcal{L}(X, Y))$$

is continuous .

Theorem B.8. Let $f_1, f_2 : U \to Y$ be Fréchet differentiable at x_0 and $a, b \in \mathbb{R}$. Then $af_1 + bf_2$ is Fréchet differentiable at x_0.

Proof. Let $r_1, r_2 \in o(X, Y)$ be such that

$$f_1(x) = f_1(x_0) + Df_1(x_0)(x - x_0) + r_1(x - x_0),$$
$$f_2(x) = f_2(x_0) + Df_2(x_0)(x - x_0) + r_2(x - x_0), \quad x \in U.$$

Hence,

$$(af_1 + bf_2)(x) = a(f_1(x_0) + Df_1(x_0)(x - x_0) + r_1(x - x_0))$$
$$+ b(f_2(x_0) + Df_2(x_0)(x - x_0) + r_2(x - x_0))$$
$$= af_1(x_0) + bf_2(x_0) + (aDf_1(x_0) + bDf_2(x_0))(x - x_0)$$
$$+ (ar_1(x - x_0) + br_2(x - x_0)), \quad x \in U.$$

Note that $ar_1 + br_2 \in o(X, Y)$. This completes the proof. □

Theorem B.9. *A function $f : U \to Y$ is Fréchet differentiable at x_0 if and only if there is some function $F : U \to \mathcal{L}(X, Y)$ that is continuous at x_0 and for which*

$$f(x) - f(x_0) = F(x)(x - x_0), \quad x \in U.$$

Proof.

1. Suppose that there is a function $F : U \to \mathcal{L}(X, Y)$ that is continuous at x_0 and

$$f(x) - f(x_0) = F(x)(x - x_0), \quad x \in U.$$

Then

$$f(x) - f(x_0) = F(x)(x - x_0) - F(x_0)(x - x_0) + F(x_0)(x - x_0)$$
$$= F(x_0)(x - x_0) + r(x - x_0),$$

where

$$r(x) = \begin{cases} (F(x + x_0) - F(x_0))(x) & \text{for } x + x_0 \in U, \\ 0 & \text{for } x + x_0 \notin U. \end{cases}$$

Define

$$\alpha(x) = \begin{cases} \frac{(F(x+x_0)-F(x_0))(x)}{\|x\|} & \text{for } x + x_0 \in U, \, x \neq 0, \\ 0 & \text{for } x + x_0 \notin U, \\ 0 & \text{for } x = 0. \end{cases}$$

Then

$$r(x) = \alpha(x)\|x\|, \quad x \in X.$$

Let $\varepsilon > 0$ be arbitrarily chosen. Since $F : U \to \mathcal{L}(X, Y)$ is continuous at x_0, there exists some $\delta > 0$ for which $\|x\| < \delta$ implies

$$\|(F(x + x_0) - F(x_0))(x)\| \leq \|F(x + x_0) - F(x_0)\|\|x\| < \varepsilon\|x\|.$$

Therefore,

$$|\alpha(x)| < \varepsilon$$

for $\|x\| < \delta$, i. e., α is continuous at 0. From here, we conclude that $r \in o(X, Y)$ and $F(x_0) = Df(x_0)$.

2. Suppose that f is Fréchet differentiable at x_0. Then there is some $r \in o(X, Y)$ such that

$$f(x) = f(x_0) + Df(x_0)(x - x_0) + r(x - x_0), \quad x \in U,$$

where $Df(x_0) \in \mathcal{L}(X, Y)$. Since $r \in o(X, Y)$, there is some $\alpha : X \to Y$ such that

$$r(x) = \alpha(x)\|x\|,$$
$$\alpha(0) = 0,$$
$$\alpha(x) \to 0, \quad \text{as } x \to 0.$$

By the Hahn–Banach extension theorem, it follows that there is some $\lambda_x \in X^*$ such that

$$\lambda_x x = \|x\|$$

and

$$|\lambda_x v| \leq \|v\|, \quad v \in X.$$

Then

$$r(x) = (\lambda_x x)\alpha(x), \quad x \in X,$$

and

$$f(x) = f(x_0) + Df(x_0)(x - x_0) + (\lambda_{x-x_0}(x - x_0))\alpha(x - x_0), \quad x \in U.$$

Let $F : U \to \mathcal{L}(X, Y)$ be defined as follows:

$$F(x)(v) = Df(x_0)(v) + (\lambda_{x-x_0} v)\alpha(x - x_0), \quad x \in U, \quad v \in X.$$

We have

$$f(x) = f(x_0) + F(x)(x - x_0),$$
$$r(x - x_0) = (\lambda_{x-x_0}(x - x_0))\alpha(x - x_0)$$
$$= f(x) - f(x_0) - Df(x_0)(x - x_0)$$
$$= F(x)(x - x_0) - Df(x_0)(x - x_0), \quad x \in U.$$

Note that

$$\|F(x)(v) - F(x_0)(v)\| = \|Df(x_0)(v) + (\lambda_{x-x_0} v)\alpha(x - x_0) - Df(x_0)(v)\|$$
$$= \|(\lambda_{x-x_0} v)\alpha(x - x_0)\|$$
$$= |\lambda_{x-x_0} v|\|\alpha(x - x_0)\|$$
$$\leq \|v\|\|\alpha(x - x_0)\|, \quad x \in U, v \in X.$$

Then

$$\|F(x) - F(x_0)\| \le \|a(x - x_0)\|, \quad x \in U.$$

Consequently, F is continuous at x_0. This completes the proof. $\qquad\square$

Theorem B.10. *Let Z be a normed space, assume $f : U \to Z$ is Fréchet differentiable at x_0, while $g : f(U) \to Z$ is Fréchet differentiable at $f(x_0)$. Then $g \circ f : U \to Z$ is Fréchet differentiable at x_0 and*

$$D(g \circ f)(x_0) = Dg(f(x_0)) \circ Df(x_0).$$

Proof. Let

$$y_0 = f(x_0),$$
$$L_1 = Df(x_0),$$
$$L_2 = Dg(y_0).$$

There exist $r_1 \in o(X, Y)$, $r_2 \in o(Y, Z)$ such that

$$f(x) = f(x_0) + L_1(x - x_0) + r_1(x - x_0), \quad x \in U,$$
$$g(y) = g(y_0) + L_2(y - y_0) + r_2(y - y_0), \quad y \in f(U).$$

Hence,

$$\begin{aligned}
g(f(x)) &= g(f(x_0)) + L_2(f(x) - y_0) + r_2(f(x) - y_0) \\
&= g(y_0) + L_2(L_1(x - x_0) + r_1(x - x_0)) \\
&\quad + r_2(L_1(x - x_0) + r_1(x - x_0)) \\
&= g(y_0) + L_2(L_1(x - x_0)) + L_2(r_1(x - x_0)) \\
&\quad + r_2(L_1(x - x_0) + r_1(x - x_0)), \quad x \in U.
\end{aligned}$$

Define $r_3 : X \to Z$ as follows:

$$r_3(x) = r_2(L_1(x) + r_1(x)), \quad x \in U.$$

Fix $c > \|L_1\|$ and represent r_1 as follows:

$$r_1(x) = \alpha_1(x)\|x\|, \quad x \in U.$$

We have that $\alpha_1 : X \to Y$, $\alpha_1(0) = 0$, and α_1 is continuous at 0. Then there exists some $\delta > 0$ such that if $\|x\| < \delta$, then

$$\|\alpha_1(x)\| < c - \|L_1\|.$$

Hence, if $\|x\| < \delta$, then

$$\|r_1(x)\| \le (c - \|L_1\|)\|x\|.$$

Then, $\|x\| < \delta$ implies

$$\begin{aligned}\|L_1(x) + r_1(x)\| &\le \|L_1(x)\| + \|r_1(x)\| \\ &\le \|L_1\|\|x\| + (c - \|L_1\|)\|x\| \\ &= c\|x\|.\end{aligned}$$

Then $x \to L_1(x) + r_1(x)$ is stable at 0. Hence, by Theorem B.5, we get $r_3 \in o(X, Z)$. Define $r : X \to Z$ as follows:

$$r = L_1 \circ r_1 + r_3.$$

We have $r \in o(X, Z)$ and

$$g \circ f(r) = g \circ f(x_0) + L_2 \circ L_1(x - x_0) + r(x - x_0), \quad x \in U.$$

Since $L_1 \in \mathcal{L}(X, Y)$, $L_2 \in \mathcal{L}(Y, Z)$, we have $L_2 \circ L_1 \in \mathcal{L}(X, Z)$. Therefore, $g \circ f$ is Fréchet differentiable at x_0 and

$$L_2 \circ L_1 = Dg(y_0) \circ Df(x_0) = Dg(f(x_0)) \circ Df(x_0).$$

This completes the proof. □

Theorem B.11. *Let $f_1, f_2 : U \to \mathbb{R}$ be Fréchet differentiable at x_0. Then $f_1 \cdot f_2$ is Fréchet differentiable at x_0 and*

$$D(f_1 \cdot f_2)(x_0) = f_2(x_0)Df_1(x_0) + f_1(x_0)Df_2(x_0).$$

Proof. Let $r_1, r_2 \in o(X, \mathbb{R})$ be such that

$$\begin{aligned}f_1(x) &= f_1(x_0) + Df_1(x_0)(x - x_0) + r_1(x - x_0), \\ f_2(x) &= f_2(x_0) + Df_2(x_0)(x - x_0) + r_2(x - x_0), \quad x \in U.\end{aligned}$$

Hence,

$$\begin{aligned}f_1(x)f_2(x) = {}& (f_1(x_0) + Df_1(x_0)(x - x_0) + r_1(x - x_0)) \\ & \times (f_2(x_0) + Df_2(x_0)(x - x_0) + r_2(x - x_0)) \\ = {}& f_1(x_0)f_2(x_0) + f_1(x_0)Df_2(x_0)(x - x_0) + f_2(x_0)Df_1(x_0)(x - x_0) \\ & + f_1(x_0)r_2(x - x_0) + Df_1(x_0)(x - x_0)Df_2(x_0)(x - x_0) \\ & + Df_1(x_0)(x - x_0)r_2(x - x_0) + r_1(x - x_0)f_2(x_0) \\ & + Df_2(x_0)(x - x_0)r_1(x - x_0) + r_1(x - x_0)r_2(x - x_0), \quad x \in U.\end{aligned}$$

Let $r : X \to \mathbb{R}$ be defined as follows:

$$r(x) = f_1(x_0)r_2(x) + Df_1(x_0)xDf_2(x_0)x$$
$$+ Df_1(x_0)xr_2(x) + r_1(x)f_2(x_0)$$
$$+ Df_2(x_0)xr_1(x) + r_1(x)r_2(x), \quad x \in U.$$

Then

$$f_1(x)f_2(x) = f_1(x_0)f_2(x_0) + f_1(x_0)Df_2(x_0)(x - x_0)$$
$$+ f_2(x_0)Df_1(x_0)(x - x_0) + r(x - x_0), \quad x \in U.$$

Note that

$$|Df_1(x_0)xDf_2(x_0)x| \le \|Df_1(x_0)\|\|Df_2(x_0)\|\|x\|^2, \quad x \in U.$$

Define $\alpha : X \to \mathbb{R}$ as follows:

$$\alpha(x) = \begin{cases} \frac{Df_1(x_0)xDf_2(x_0)x}{\|x\|}, & x \in U, \, x \neq 0, \\ 0, & x = 0. \end{cases}$$

Then

$$Df_1(x_0)xDf_2(x_0)x = \alpha(x)\|x\|, \quad x \in U,$$
$$|\alpha(x)| = \frac{|Df_1(x_0)xDf_2(x_0)x|}{\|x\|}$$
$$\le \frac{\|Df_1(x_0)\|\|Df_2(x_0)\|\|x\|^2}{\|x\|}$$
$$= \|Df_1(x_0)\|\|Df_2(x_0)\|\|x\|, \quad x \in U, \, x \neq 0.$$

Then

$$\alpha(x) \to 0, \quad \text{as } x \to 0.$$

From here, $r \in o(X, \mathbb{R})$. This completes the proof. □

B.3 The Gâteaux derivative

Let X and Y be normed spaces and U be an open subset of X. Let also, $x_0 \in U$.

Definition B.12. Let $f : U \to Y$. If there is some $T \in \mathcal{L}(X, Y)$ such that

$$\lim_{t \to 0} \frac{f(x_0 + tv) - f(x_0)}{t} = Tv$$

for any $v \in X$, we say that f is Gâteaux differentiable at x_0. We write $f'(x_0) = T$. If f is Gâteaux differentiable at any point of U, then we say that f is Gâteaux differentiable on U.

Example B.13. Let $f : \mathbb{R}^2 \to \mathbb{R}$ be defined as follows:

$$f(x_1, x_2) = \begin{cases} \frac{x_1^4}{x_1^6 + x_2^3} & \text{for } (x_1, x_2) \neq (0, 0), \\ 0 & \text{for } (x_1, x_2) = (0, 0). \end{cases}$$

Let $v = (v_1, v_2) \in \mathbb{R}^2$, $(v_1, v_2) \neq (0, 0)$, be arbitrarily chosen. We have, for $t \neq 0$,

$$f(0 + tv) = \frac{t^4 v_1^4}{t^6 v_1^6 + t^3 v_2^3} = \frac{t v_1^4}{t^3 v_1^6 + v_2^3},$$

$$\lim_{t \to 0} \frac{f(0 + tv) - f(0)}{t} = \lim_{t \to 0} \frac{t v_1^4}{t(t^3 v_1^6 + v_2^3)} = \lim_{t \to 0} \frac{v_1^4}{t^3 v_1^6 + v_2^3} = \frac{v_1^4}{v_2^3}.$$

Therefore,

$$f'(0, 0)(v_1, v_2) = \frac{v_1^4}{v_2^3}, \quad (v_1, v_2) \in \mathbb{R}^2, \quad (v_1, v_2) \neq (0, 0).$$

This ends the example.

Theorem B.14. *If $f : U \to Y$ is Fréchet differentiable at x_0, then it is Gâteaux differentiable at x_0.*

Proof. Since $f : U \to Y$ is Fréchet differentiable at x_0, then there is some $r \in o(X, Y)$ such that

$$f(x) = f(x_0) + Df(x_0)(x - x_0) + r(x - x_0), \quad x \in U,$$

and

$$r(x) = \alpha(x)\|x\|, \quad x \in X,$$

where $\alpha : X \to Y$, $\alpha(0) = 0$, and α is continuous at 0. Then, for $v \in X$ and $t \in \mathbb{R}$, with $|t|$ small enough, we have

$$\frac{f(x_0 + tv) - f(x_0)}{t} = \frac{Df(x_0)(tv) + r(tv)}{t}$$

$$= \frac{tDf(x_0)(v) + |t|\|v\|\alpha(tv)}{t}$$

$$= Df(x_0)(v) + \text{sign}(t)\|v\|\alpha(tv)$$

$$\to Df(x_0)(v) \quad \text{as } t \to 0.$$

This completes the proof. □

C Pötzsche's chain rules

C.1 Measure chains

Let $\mathbb{T}$ be some set of real numbers.

Definition C.1. A triple $(\mathbb{T}, \leq, v)$ is called a measure chain provided it satisfies the following axioms:

(A1) The relation "$\leq$" is, for $r, s, t \in \mathbb{T}$,
1. reflexive, i. e., $t \leq t$,
2. transitive, i. e., if $t \leq r$ and $r \leq s$, then $t \leq s$,
3. antisymmetric, i. e., if $t \leq r$ and $r \leq t$, then $t = r$,
4. total, i. e., either $r \leq s$ or $s \leq r$.

(A2) Any nonvoid subset of $\mathbb{T}$ which is bounded above has a least upper bound, i. e., the measure chain $(T, \leq)$ is conditionally complete.

(A3) The mapping $v : \mathbb{T} \times \mathbb{T} \to \mathbb{R}$ has the following properties, for $r, s, t \in \mathbb{T}$:
1. $v(r, s) + v(s, t) = v(r, t)$ (cocycle property),
2. if $r > s$, then $v(r, s) > 0$ (strong isotony),
3. v is continuous (continuity).

Example C.2. Let $\mathbb{T}$ be any nonvoid closed subset of real numbers, with "$\leq$" being the usual order relation between real numbers and

$$v(r, s) = r - s, \quad r, s \in \mathbb{T}.$$

Definition C.3. The forward jump operator σ and the backward jump operator ρ are defined as follows:

$$\sigma(t) = \inf\{s \in \mathbb{T} : s > t\}, \quad \rho(t) = \sup\{s \in \mathbb{T} : s < t\},$$

where

$$\sigma(t) = t \quad \text{if } t = \max \mathbb{T},$$
$$\rho(t) = t \quad \text{if } t = \min \mathbb{T}.$$

The graininess function is defined as

$$\mu(t) = v(\sigma(t), t), \quad t \in \mathbb{T}.$$

The notions left-scattered, left-dense, right-scattered, right-dense, isolated, and $\mathbb{T}^\kappa$ are defined as in the case of time scales.

https://doi.org/10.1515/9783110787320-015

Definition C.4. Let X be a Banach space with a norm $\| \cdot \|$. We say that $f : \mathbb{T} \to X$ is differentiable at $t \in \mathbb{T}$ if there exists $f^{\Delta}(t) \in X$ such that for any $\varepsilon > 0$ there exists a neighborhood U of t such that

$$\|f(\sigma(t)) - f(s) - f^{\Delta}(t)v(\sigma(t), s)\| \le \varepsilon |v(\sigma(t), s)|$$

for all $s \in U$. In this case $f^{\Delta}(t)$ is said to be the derivative of f at t.

Theorem C.5. *We have*

$$v^{\Delta}(\cdot, t) = 1, \quad t \in \mathbb{T}.$$

Proof. Let $t \in \mathbb{T}$. Let also $\varepsilon > 0$ be arbitrarily chosen and U be a neighborhood of t. Then

$$v(\sigma(t), s) + v(s, t) = v(\sigma(t), t), \quad s \in \mathbb{T},$$

and

$$\left|v(\sigma(t), t) - v(s, t) - v(\sigma(t), s)\right| = \left|v(\sigma(t), t) - v(\sigma(t), t)\right|$$
$$= 0$$
$$\le \varepsilon |v(\sigma(t), s)|,$$

for any $s \in U$. This completes the proof. □

As in the case of time scales, one can prove the following assertion.

Theorem C.6. *Let $f, g : \mathbb{T} \to X$ and $t \in \mathbb{T}$.*
1. *If $t \in \mathbb{T}^{\kappa}$, then f has at most one derivative at t.*
2. *If f is differentiable at t, then f is continuous at t.*
3. *If f is continuous at t and t is right-scattered, then f is differentiable at t and*

$$f^{\Delta}(t) = \frac{f(\sigma(t)) - f(t)}{\mu(t)}.$$

4. *If f and g are differentiable at $t \in \mathbb{T}^{\kappa}$ and $\alpha, \beta \in \mathbb{R}$, then $\alpha f + \beta g$ is differentiable at t and*

$$(\alpha f + \beta g)^{\Delta}(t) = \alpha f^{\Delta}(t) + \beta g^{\Delta}(t).$$

5. *If f and g are differentiable at $t \in \mathbb{T}^{\kappa}$ and "$\cdot$" is bilinear and continuous, then $f \cdot g$ is differentiable at t and*

$$(f \cdot g)^{\Delta}(t) = f^{\Delta}(t) \cdot g(t) + f(\sigma(t)) \cdot g^{\Delta}(t).$$

6. *If f and g are differentiable at $t \in \mathbb{T}^\kappa$ and g is algebraically invertible, then $f \cdot g^{-1}$ is differentiable at t with*

$$(f \cdot g^{-1})^\Delta(t) = (f^\Delta(t) - (f \cdot g^{-1})(t) \cdot g^\Delta(t)) \cdot g^{-1}(\sigma(t)).$$

C.2 Pötzsche's chain rule

Throughout this section we suppose that $(\mathbb{T}, \leq, \nu)$ is a measure chain with forward jump operator σ and graininess μ. Assume that X and Y are Banach spaces, and we will write $\| \cdot \|$ for the norms of X and Y. For a function $f : \mathbb{T} \times X \to Y$ and $x_0 \in X$, we denote the delta derivative of $t \to f(t, x_0)$ by $\Delta_1 f(\cdot, x_0)$, and for a $t_0 \in \mathbb{T}$ we denote the Fréchet derivative of $x \to f(t_0, x)$ by $D_2 f(t_0, \cdot)$, provided these derivatives exist.

Theorem C.7 (Pötzsche's chain rule). *For some fixed $t_0 \in \mathbb{T}^\kappa$, let $g : \mathbb{T} \to X, f : \mathbb{T} \times X \to Y$ be functions such that $g, f(\cdot, g(t_0))$ are differentiable at t_0, and let $U \subseteq \mathbb{T}$ be a neighborhood of t_0 such that $f(t, \cdot)$ is differentiable for $t \in U \cup \{\sigma(t_0)\}$, $D_2 f(\sigma(t_0), \cdot)$ is continuous on the line segment*

$$\{g(t_0) + h\mu(t_0)g^\Delta(t_0) \in X : h \in [0,1]\}$$

and $D_2 f$ is continuous at $(t_0, g(t_0))$. Then the composition function $F : \mathbb{T} \to Y, F(t) = f(t, g(t))$ is differentiable at t_0 with derivative

$$F^\Delta(t_0) = \Delta_1 f(t_0, g(t_0)) + \left(\int_0^1 D_2 f(\sigma(t_0), g(t_0) + h\mu(t_0)g^\Delta(t_0))dh \right) g^\Delta(t_0).$$

Proof. Let $U_0 \subseteq U$ be a neighborhood of t_0 such that

$$\mu(t_0) \leq |\nu(t, \sigma(t_0))| \quad \text{for } t \in U_0.$$

Let

$$\Phi(t, h) = D_2 f(t, g(t_0) + h(g(t) - g(t_0))), \quad t \in U_0, \quad h \in [0, 1].$$

Note that there exists a constant $C > 0$ such that

$$\|\Phi(\sigma(t_0), h) - \Phi(t_0, h)\| \leq C|\nu(t, \sigma(t_0))| \quad \text{for } t \in U_0, \quad h \in [0, 1].$$

Let $\varepsilon > 0$ be arbitrarily chosen. We choose $\varepsilon_1 > 0$, $\varepsilon_2 > 0$ small enough such that

$$\varepsilon_1 \left(1 + C \left\| \int_0^1 \Phi(\sigma(t_0), h)dh \right\| \right) + \varepsilon_2(\varepsilon_1 + 2\|g^\Delta(t_0)\|) \leq \varepsilon.$$

Since g and $f(\cdot, g(t_0))$ are differentiable at t_0, there exists a neighborhood $U_1 \subseteq U_0$ of t_0 such that

$$\|g(t) - g(t_0)\| \leq \varepsilon_1,$$
$$\|g(t) - g(\sigma(t_0)) - v(t, \sigma(t_0))g^\Delta(t_0)\| \leq \varepsilon_1 |v(t, \sigma(t_0))|,$$
$$\|f(t, g(t_0)) - f(\sigma(t_0), g(t_0)) - v(t, \sigma(t_0))\Delta_1 f(t_0, g(t_0))\| \leq \varepsilon_1 |v(t, \sigma(t_0))|$$

for $t \in U_1$. Hence,

$$
\begin{aligned}
\|g(t) - g(t_0)\| &= \|g(t) - g(\sigma(t_0)) - v(t, \sigma(t_0))g^\Delta(t_0) + g^\Delta(t_0)v(t, \sigma(t_0)) \\
&\quad + g(\sigma(t_0)) - g(t_0)\| \\
&\leq \|g(t) - g(\sigma(t_0)) - v(t, \sigma(t_0))g^\Delta(t_0)\| \\
&\quad + \|g^\Delta(t_0)\| |v(t, \sigma(t_0))| + \|g(\sigma(t_0)) - g(t_0)\| \\
&\leq \varepsilon_1 |v(t, \sigma(t_0))| + \|g^\Delta(t_0)\| |v(t, \sigma(t_0))| + \|g^\Delta(t_0)\| \mu(t_0) \\
&= (\varepsilon_1 + \|g^\Delta(t_0)\|) |v(t, \sigma(t_0))| + \|g^\Delta(t_0)\| \mu(t_0) \\
&\leq (\varepsilon_1 + 2\|g^\Delta(t_0)\|) |v(t, \sigma(t_0))|, \quad t \in U_1.
\end{aligned}
$$

Since g is continuous at t_0 and $D_2 f$ is continuous at $(t_0, g(t_0))$, there exists a neighborhood $U_2 \subseteq U$ of t_0 so that

$$\|\Phi(t, h) - \Phi(t_0, h)\| \leq \varepsilon_2 \quad \text{for } t \in U_2, \quad h \in [0, 1].$$

Hence,

$$
\left\| F(t) - F(\sigma(t_0)) - v(t, \sigma(t_0)) \left(\Delta_1 f(t_0, g(t_0)) + \int_0^1 \Phi(\sigma(t_0), h) \, dh \, g^\Delta(t_0) \right) \right\|
$$

$$
= \Big\| f(t, g(t)) - f(\sigma(t_0), g(\sigma(t_0))) - f(\sigma(t_0), g(t_0)) + f(\sigma(t_0), g(t_0))
$$

$$
- f(t, g(t_0)) + f(t, g(t_0)) - v(t, \sigma(t_0))\Delta_1 f(t_0, g(t_0))
$$

$$
- v(t, \sigma(t_0)) \int_0^1 \Phi(\sigma(t_0), h) \, dh \, g^\Delta(t_0) - \int_0^1 \Phi(\sigma(t_0), h) \, dh (g(t) - g(t_0))
$$

$$
+ \int_0^1 \Phi(\sigma(t_0), h) \, dh (g(t) - g(t_0)) \Big\|
$$

$$
\leq \| f(t, g(t_0)) - f(\sigma(t_0), g(t_0)) - v(t, \sigma(t_0))\Delta_1 f(t_0, g(t_0)) \|
$$

$$
+ \left\| \int_0^1 \Phi(\sigma(t_0), h) \, dh (g(t) - g(t_0) - v(t, \sigma(t_0))g^\Delta(t_0)) \right\|
$$

$$+ \left\| f(t,g(t)) - f(t,g(t_0)) - (f(\sigma(t_0),g(\sigma(t_0))) - f(\sigma(t_0),g(t_0))) \right.$$

$$\left. - \int_0^1 \Phi(\sigma(t_0),h)dh(g(t) - g(t_0)) \right\|$$

$$\leq \left\| f(t,g(t_0)) - f(\sigma(t_0),g(t_0)) - v(t,\sigma(t_0))\Delta_1 f(t_0,g(t_0)) \right\|$$

$$+ \left\| \int_0^1 \Phi(\sigma(t_0),h)dh \right\| \left\| g(t) - g(t_0) - v(t,\sigma(t_0))g^\Delta(t_0) \right\|$$

$$+ \left\| \int_0^1 (\Phi(t,h) - \Phi(\sigma(t_0),h))dh(g(t) - g(t_0)) \right\|$$

$$\leq \left\| f(t,g(t_0)) - f(\sigma(t_0),g(t_0)) - v(t,\sigma(t_0))\Delta_1 f(t_0,g(t_0)) \right\|$$

$$+ \left\| \int_0^1 \Phi(\sigma(t_0),h)dh \right\| \left\| g(t) - g(t_0) - v(t,\sigma(t_0))g^\Delta(t_0) \right\|$$

$$+ \left\| \int_0^1 (\Phi(t,h) - \Phi(t_0,h))dh \right\| \left\| g(t) - g(t_0) \right\|$$

$$+ \left\| \int_0^1 (\Phi(t_0,h) - \Phi(\sigma(t_0),h))dh \right\| \left\| g(t) - g(t_0) \right\|$$

$$\leq \varepsilon_1 |v(t,\sigma(t_0))| + \varepsilon_1 |v(t,\sigma(t_0))| \left\| \int_0^1 \Phi(\sigma(t_0),h)dh \right\|$$

$$+ \varepsilon_2(\varepsilon_1 + 2\|g^\Delta(t_0)\|)|v(t,\sigma(t_0))| + \varepsilon_1 C|v(t,\sigma(t_0))|$$

$$= \left(\varepsilon_1 \left(1 + C + \left\| \int_0^1 \Phi(\sigma(t_0),h)dh \right\| \right) + \varepsilon_2(\varepsilon_1 + 2\|g^\Delta(t_0)\|) \right)|v(t,\sigma(t_0))|$$

$$\leq \varepsilon |v(t,\sigma(t_0))|, \quad t \in U_1 \cap U_2.$$

This completes the proof. $\qquad\square$

C.3 A generalization of Pötzsche's chain rule

In this section we state and prove a generalization of Pötzsche's chain rule.

Let $g : \mathbb{T} \times \mathbb{R}^n \to \mathbb{R}$ be a given function. Then for the function $g(t,y_1,\ldots,y_n)$ we denote by $\Delta_1 g(\cdot,y_1,\ldots,y_n)$ its delta derivative.

Theorem C.8. *For some fixed $t_0 \in \mathbb{T}^\kappa$, let $y_j : \mathbb{T} \to \mathbb{R}, j \in \{1,\ldots,n\}, f : \mathbb{T} \times \mathbb{R}^n \to \mathbb{R}$ be continuous functions such that $f(\cdot, y_1(t_0),\ldots,y_n(t_0))$, and $y_j, j \in \{1,\ldots,n\}$, are differentiable at t_0. Let $U \subseteq \mathbb{T}$ be a neighborhood of t_0 such that*
1. *$f(t,\cdot,\ldots,\cdot)$ is continuously-differentiable for $t \in U \cup \{\sigma(t_0)\}$,*
2. *$\Delta_1 f(\cdot, y_1(\cdot),\ldots,y_n(\cdot))$ is continuous at t_0,*
3.

$$\frac{\partial}{\partial y_j} f(\sigma(t_0), y_1(\sigma(t_0)),\ldots,y_{j-1}(\sigma(t_0)),\cdot,y_{j+1}(t),\ldots,y_n(t))$$

is continuous on the line segment

$$\{y_j(t) + h(y_j(\sigma(t_0)) - y_j(t)) \in \mathbb{R} : h \in [0,1]\}, \quad j \in \{1,\ldots,n\}, \ \forall t \in U \cup \{t_0\},$$

4. *$\frac{\partial}{\partial y_j} f$ is continuous at $(t_0, y_1(t_0),\ldots,y_n(t_0))$.*

Then the composition function $F : \mathbb{T} \to \mathbb{R}, F(t) = f(t, y_1(t), y_2(t),\ldots,y_n(t))$, is differentiable at t_0 with derivative

$$F^\Delta(t_0) = \Delta_1 f(t_0, y_1(t_0), y_2(t_0),\ldots,y_n(t_0))$$

$$+ \left(\int_0^1 \frac{\partial}{\partial y_1} f(\sigma(t_0), y_1(t_0) + h\mu(t_0)y_1^\Delta(t_0), y_2(t_0),\ldots,y_n(t_0))dh \right) y_1^\Delta(t_0)$$

$$+ \left(\int_0^1 \frac{\partial}{\partial y_2} f(\sigma(t_0), y_1(\sigma(t_0)), y_2(t_0) + h\mu(t_0)y_2^\Delta(t_0),\ldots,y_n(t_0))dh \right) y_2^\Delta(t_0)$$

$$+ \cdots$$

$$+ \left(\int_0^1 \frac{\partial}{\partial y_n} f(\sigma(t_0), y_1(\sigma(t_0)), y_2(\sigma(t_0)),\ldots,y_{n-1}(\sigma(t_0)), y_n(t_0) \right.$$

$$\left. + h\mu(t_0)y_n^\Delta(t_0))dh \right) y_n^\Delta(t_0).$$

Proof. Let $s \in (t_0 - \delta, t_0 + \delta) \cap \mathbb{T}, s \neq \sigma(t_0)$, for $\delta > 0$ small enough, and $s < \sigma(t_0)$ if $\sigma(t_0) > t_0$. Then

$$F(\sigma(t_0)) - F(s)$$
$$= f(\sigma(t_0), y_1(\sigma(t_0)), y_2(\sigma(t_0)),\ldots,y_n(\sigma(t_0))) - f(s, y_1(s), y_2(s),\ldots,y_n(s))$$
$$= f(\sigma(t_0), y_1(s), y_2(s),\ldots,y_n(s)) - f(s, y_1(s), y_2(s),\ldots,y_n(s))$$
$$+ f(\sigma(t_0), y_1(\sigma(t_0)), y_2(s),\ldots,y_n(s)) - f(\sigma(t_0), y_1(s), y_2(s),\ldots,y_n(s))$$
$$+ f(\sigma(t_0), y_1(\sigma(t_0)), y_2(\sigma(t_0)),\ldots,y_n(s)) - f(\sigma(t_0), y_1(\sigma(t_0)), y_2(s),\ldots,y_n(s))$$
$$+ \cdots$$
$$+ f(\sigma(t_0), y_1(\sigma(t_0)), y_2(\sigma(t_0)),\ldots,y_n(\sigma(t_0))) - f(\sigma(t_0), y_1(\sigma(t_0)), y_2(\sigma(t_0)),\ldots,y_n(s)).$$

Then, we have

$$F(\sigma(t_0)) - F(s)$$
$$= f(\sigma(t_0), y_1(s), y_2(s), \ldots, y_n(s)) - f(s, y_1(s), y_2(s), \ldots, y_n(s))$$
$$+ \left(\int_0^1 \frac{\partial}{\partial y_1} f(\sigma(t_0), y_1(s) + h(y_1(\sigma(t_0)) - y_1(s)), y_2(s), \ldots, y_n(s)) dh \right) (y_1(\sigma(t_0)) - y_1(s))$$
$$+ \left(\int_0^1 \frac{\partial}{\partial y_2} f(\sigma(t_0), y_1(\sigma(t_0)), y_2(s) + h(y_2(\sigma(t_0)) - y_2(s)), \ldots, y_n(s)) dh \right)$$
$$\times (y_2(\sigma(t_0)) - y_2(s))$$
$$+ \cdots$$
$$+ \left(\int_0^1 \frac{\partial}{\partial y_n} f(\sigma(t_0), y_1(\sigma(t_0)), y_2(\sigma(t_0)), \ldots, y_n(s) + h(y_n(\sigma(t_0)) - y_n(s))) dh \right)$$
$$\times (y_n(\sigma(t_0)) - y_n(s)).$$

If $\sigma(t_0) > t_0$, by the mean value theorem there exist $\xi_1, \xi_2 \in [s, \sigma(t_0)) = [s, t_0]$ so that

$$\Delta_1 f(\xi_1, y_1(s), y_2(s), \ldots, y_n(s))(\sigma(t_0) - s) \le f(\sigma(t_0), y_1(s), y_2(s), \ldots, y_n(s)),$$
$$-f(s, y_1(s), y_2(s), \ldots, y_n(s)) \le \Delta_1 f(\xi_2, y_1(s), y_2(s), \ldots, y_n(s))(\sigma(t_0) - s),$$

and

$$\Delta_1 f(t_0, y_1(t_0), y_2(t_0), \ldots, y_n(t_0))$$
$$= \lim_{s \to t_0} \Delta_1 f(\xi_1, y_1(s), y_2(s), \ldots, y_n(s))$$
$$\le \lim_{s \to t_0} \frac{1}{\sigma(t_0) - s} (f(\sigma(t_0), y_1(s), y_2(s), \ldots, y_n(s)) - f(s, y_1(s), y_2(s), \ldots, y_n(s)))$$
$$\le \lim_{s \to t_0} \Delta_1 f(\xi_2, y_1(s), y_2(s), \ldots, y_n(s))$$
$$= \Delta_1 f(t_0, y_1(t_0), y_2(t_0), \ldots, y_n(t_0)).$$

If $\sigma(t_0) = t_0$, by the mean value theorem, there exist ξ_1, ξ_2 between s and t_0 so that

$$\Delta_1 f(\xi_1, y_1(s), y_2(s), \ldots, y_n(s))(t_0 - s) \le f(t_0, y_1(s), y_2(s), \ldots, y_n(s)),$$
$$-f(s, y_1(s), y_2(s), \ldots, y_n(s)) \le \Delta_1 f(\xi_2, y_1(s), y_2(s), \ldots, y_n(s))(t_0 - s).$$

In this case, if $s < t_0$ we have

$$\Delta_1 f(t_0, y_1(t_0), y_2(t_0), \ldots, y_n(t_0))$$
$$= \lim_{s \to t_0-} \Delta_1 f(\xi_1, y_1(s), y_2(s), \ldots, y_n(s))$$
$$\le \lim_{s \to t_0-} \frac{1}{t_0 - s} (f(\sigma(t_0), y_1(s), y_2(s), \ldots, y_n(s)) - f(s, y_1(s), y_2(s), \ldots, y_n(s)))$$

$$\leq \lim_{s \to t_0^-} \Delta_1 f(\xi_2, y_1(s), y_2(s), \ldots, y_n(s))$$

$$= \Delta_1 f(t_0, y_1(t_0), y_2(t_0), \ldots, y_n(t_0)),$$

and if $s > t_0$ we have

$$\Delta_1 f(t_0, y_1(t_0), y_2(t_0), \ldots, y_n(t_0))$$

$$= \lim_{s \to t_0^+} \Delta_1 f(\xi_1, y_1(s), y_2(s), \ldots, y_n(s))$$

$$\geq \lim_{s \to t_0^+} \frac{1}{t_0 - s} (f(\sigma(t_0), y_1(s), y_2(s), \ldots, y_n(s)) - f(s, y_1(s), y_2(s), \ldots, y_n(s)))$$

$$\geq \lim_{s \to t_0^+} \Delta_1 f(\xi_2, y_1(s), y_2(s), \ldots, y_n(s))$$

$$= \Delta_1 f(t_0, y_1(t_0), y_2(t_0), \ldots, y_n(t_0)).$$

Moreover,

$$\lim_{s \to t_0} \left(\left(\int_0^1 \frac{\partial}{\partial y_j} f(\sigma(t_0), y_1(\sigma(t_0)), \ldots, y_{j-1}(\sigma(t_0)), y_j(s) + h(y_j(\sigma(t_0)) - y_j(s)), \right. \right.$$

$$\left. y_{j+1}(s), \ldots, y_n(s)) dh \right) \frac{y_j(\sigma(t_0)) - y_j(t_0)}{\sigma(t_0) - s} \right)$$

$$= \lim_{s \to t_0} \left(\int_0^1 \frac{\partial}{\partial y_j} f(\sigma(t_0), y_1(\sigma(t_0)), \ldots, y_{j-1}(\sigma(t_0)), y_j(s) + h(y_j(\sigma(t_0)) - y_j(s)), \right.$$

$$\left. y_{j+1}(s), \ldots, y_n(s)) dh \right) \lim_{s \to t_0} \frac{y_j(\sigma(t_0)) - y_j(t_0)}{\sigma(t_0) - s}$$

$$= \left(\int_0^1 \frac{\partial}{\partial y_j} f(\sigma(t_0), y_1(\sigma(t_0)), \ldots, y_{j-1}(\sigma(t_0)), y_j(t_0) + h(y_j(\sigma(t_0)) - y_j(t_0)), \right.$$

$$\left. y_{j+1}(t_0), \ldots, y_n(t_0)) dh \right) y_j^\Delta(t_0)$$

$$= \left(\int_0^1 \frac{\partial}{\partial y_j} f(\sigma(t_0), y_1(\sigma(t_0)), \ldots, y_{j-1}(\sigma(t_0)), y_j(t_0) + h\mu(t_0) y_j^\Delta(t_0), y_{j+1}(t_0), \ldots, y_n(t_0)) dh \right)$$

$$\times y_j^\Delta(t_0), \quad j \in \{1, \ldots, n\}.$$

Therefore,

$$\lim_{s \to t_0} \frac{F(\sigma(t_0)) - F(s)}{\sigma(t_0) - s}$$

$$= \lim_{s \to t_0} \frac{f(\sigma(t_0), y_1(s), y_2(s), \ldots, y_n(s)) - f(s, y_1(s), y_2(s), \ldots, y_n(s))}{\sigma(t_0) - s}$$

$$+ \lim_{s \to t_0} \left(\left(\int_0^1 \frac{\partial}{\partial y_1} f(\sigma(t_0), y_1(s) + h(y_1(\sigma(t_0)) - y_1(s)), y_2(s), \dots, y_n(s)) dh \right) \right.$$

$$\left. \times \frac{y_1(\sigma(t_0)) - y_1(s)}{\sigma(t_0) - s} \right)$$

$$+ \lim_{s \to t_0} \left(\left(\int_0^1 \frac{\partial}{\partial y_2} f(\sigma(t_0), y_1(\sigma(t_0)), y_2(s) + h(y_2(\sigma(t_0)) - y_2(s)), \dots, y_n(s)) dh \right) \right.$$

$$\left. \times \frac{y_2(\sigma(t_0)) - y_2(s)}{\sigma(t_0) - s} \right)$$

$$+ \cdots$$

$$+ \lim_{s \to t_0} \left(\left(\int_0^1 \frac{\partial}{\partial y_n} f(\sigma(t_0), y_1(\sigma(t_0)), y_2(\sigma(t_0)), \dots, y_n(s) + h(y_n(\sigma(t_0)) - y_n(s))) dh \right) \right.$$

$$\left. \times \frac{y_n(\sigma(t_0)) - y_n(s)}{\sigma(t_0) - s} \right)$$

$$= \Delta_1 f(t_0, y_1(t_0), y_2(t_0), \dots, y_n(t_0))$$

$$+ \left(\int_0^1 \frac{\partial}{\partial y_1} f(\sigma(t_0), y_1(t_0) + h\mu(t_0) y_1^\Delta(t_0), y_2(t_0), \dots, y_n(t_0)) dh \right) y_1^\Delta(t_0)$$

$$+ \left(\int_0^1 \frac{\partial}{\partial y_2} f(\sigma(t_0), y_1(\sigma(t_0)), y_2(t_0) + h\mu(t_0) y_2^\Delta(t_0), \dots, y_n(t_0)) dh \right) y_2^\Delta(t_0)$$

$$+ \cdots$$

$$+ \left(\int_0^1 \frac{\partial}{\partial y_n} f(\sigma(t_0), y_1(\sigma(t_0)), y_2(\sigma(t_0)), \dots, y_{n-1}(\sigma(t_0)), y_n(t_0) + h\mu(t_0) y_n^\Delta(t_0)) dh \right)$$

$$\times y_n^\Delta(t_0).$$

This completes the proof. $\qquad \qquad \square$

D Lebesgue integration. L^p-spaces. Sobolev spaces

Let $\mathbb{T}$ be a time scale with forward jump operator σ and delta differentiation operator Δ.

D.1 The Lebesgue delta integral

Suppose that $\mathcal{F}_1$ is a family of left-closed and right-open intervals of $\mathbb{T}$ of the form

$$[a, b) = \{t \in \mathbb{T} : a \leq t < b\},$$

$a, b \in \mathbb{T}$, $a \leq b$. The interval $[a, a)$ is understood as the empty set.

Definition D.1. Define $m_1 : \mathcal{F}_1 \to [0, \infty]$ to be the set function that assigns to each interval $[a, b) \in \mathcal{F}_1$ its length, i. e.,

$$m_1([a, b)) = b - a,$$

and satisfies the following properties:
1. $m_1(\emptyset) = 0$,
2. if $\{[a_j, b_j)\}_{j \in A}$ is a finite or countable pairwise disjoint family of intervals of $\mathcal{F}_1$, then

$$m_1\left(\bigcup_{j \in A}([a_j, b_j))\right) = \sum_{j \in A} m_1([a_j, b_j)) = \sum_{j \in A}(b_j - a_j).$$

Here A is an index set.

Definition D.2. Consider the pair $(\mathcal{F}_1, m_1)$. Let E be a subset of $\mathbb{T}$. If there exists at least one finite or countable system of intervals $V_j \in \mathcal{F}_1, j \in \mathbb{N}$, such that $E \subset \bigcup_j V_j$, then we set

$$m_1^*(E) = \inf \sum_j m_1(V_j),$$

where the infimum is taken over all coverings of E by a finite or countable system of intervals $V_j \in \mathcal{F}_1$. If there is no such covering of E, then we set $m_1^*(E) = \infty$.

Note that $m_1^*(\emptyset) = 0$.

Definition D.3. A subset A of $\mathbb{T}$ is said to be m_1^*-measurable, or Δ-measurable, if

$$m_1^*(E) = m_1^*(E \cap A) + m_1^*(E \cap A^c)$$

holds for any $E \subset \mathbb{T}$. Here A^c denotes the complement of A, i. e., $A^c = \mathbb{T} \setminus A$. The family of all m_1^*-measurable subsets of $\mathbb{T}$ will be denoted by $M(m_1^*)$.

https://doi.org/10.1515/9783110787320-016

Exercise D.4. Let $A, B \subset \mathbb{T}$, $A \cap B = \emptyset$. Prove that

$$m_1^*(A \cup B) = m_1^*(A) + m_1^*(B).$$

Exercise D.5. Let $\{A_j\}_{j\in\mathbb{N}} \subset M(m_1^*)$. Prove

1. $(\bigcup_{j=1}^{\infty} A_j)^c = \bigcap_{j=1}^{\infty} A_j^c$,
2. $(\bigcap_{j=1}^{\infty} A_j)^c = \bigcup_{j=1}^{\infty} A_j^c$.

Theorem D.6. *The family $M(m_1^*)$ is a σ-algebra.*

Proof.
1. Let $E \subseteq \mathbb{T}$ be arbitrarily chosen and fixed. Then

$$E \cap \mathbb{T} = E, \quad E \cap \mathbb{T}^c = \emptyset,$$

and

$$m_1^*(E) = m_1^*(E \cap \mathbb{T}) + m_1^*(E \cap \mathbb{T}^c).$$

Therefore, $\mathbb{T} \in M(m_1^*)$.
2. Let $A \in M(m_1^*)$ be arbitrarily chosen and fixed. Take $E \subseteq \mathbb{T}$ arbitrarily. Then, using that $(A^c)^c = A$, we get

$$m_1^*(E) = m_1^*(E \cap A) + m_1^*(E \cap A^c) = m_1^*(E \cap A^c) + m_1^*(E \cap A)$$
$$= m_1^*(E \cap A^c) + m_1^*(E \cap (A^c)^c).$$

Therefore, $A^c \in M(m_1^*)$.
3. Let $\{A_j\}_{j=1}^{\infty} \subset M(m_1^*)$ be arbitrarily chosen and fixed. Take $E \subseteq \mathbb{T}$ arbitrarily. Then

$$m_1^*\left(E \cap \left(\bigcup_{j=1}^{\infty} A_j\right)\right) + m_1^*\left(E \cap \left(\bigcup_{j=1}^{\infty} A_j\right)^c\right)$$

$$= m_1^*\left(E \cap \left(\bigcup_{j=1}^{\infty} A_j\right)\right) + m_1^*\left(E \cap \left(\bigcap_{j=1}^{\infty} A_j^c\right)\right)$$

$$= m_1^*\left(\left(E \cap \left(\bigcup_{j=1}^{\infty} A_j\right)\right) \cup \left(E \cap \left(\bigcap_{j=1}^{\infty} A_j^c\right)\right)\right)$$

$$= m_1^*(E).$$

Therefore, $\bigcup_{j=1}^{\infty} A_j \in M(m_1^*)$. Next,

$$m_1^*\left(E \cap \left(\bigcap_{j=1}^{\infty} A_j\right)\right) + m_1^*\left(E \cap \left(\bigcap_{j=1}^{\infty} A_j\right)^c\right)$$

$$= m_1^*\left(E \cap \left(\bigcap_{j=1}^{\infty} A_j\right)\right) + m_1^*\left(E \cap \left(\bigcup_{j=1}^{\infty} A_j^c\right)\right)$$

$$= m_1^*\left(\left(E \cap \left(\bigcap_{j=1}^{\infty} A_j\right)\right) \cup \left(E \cap \left(\bigcup_{j=1}^{\infty} A_j^c\right)\right)\right)$$

$$= m_1^*(E).$$

Consequently, $\bigcap_{j=1}^{\infty} A_j \in M(m_1^*)$.

This completes the proof. $\square$

The restriction of m_1^* to $M(m_1^*)$ will be denoted by μ_Δ. This μ_Δ (the Lebesgue Δ-measure) is a countable additive measure on $M(m_1^*)$. All intervals of the family $\mathcal{F}_1$, including the empty set, are Δ-measurable. Therefore, $\mathbb{T}$ is Δ-measurable. Assume that $\mathbb{T}$ has a finite maximum τ_0. Then the set $\mathbb{T} \setminus \{\tau_0\}$ can be represented as a finite or countable union of intervals of the family $\mathcal{F}_1$ and therefore it is Δ-measurable. Because the difference of two Δ-measurable sets is a Δ-measurable set, we get that the single-point set $\{\tau_0\}$ is Δ-measurable. Since $\{\tau_0\}$ does not have a finite or countable covering by intervals of $\mathcal{F}_1$, we conclude that the single-point set $\{\tau_0\}$ and any Δ-measurable subset of $\mathbb{T}$, containing the point τ_0, have Δ-measure infinity.

Let $a, b \in \mathbb{T}$, $a < b$.

Lemma D.7. *The set of all right-scattered points of* $\mathbb{T}$ *is at most countable, i. e., there are* $I \subset \mathbb{N}$ *and* $\{t_i\}_{i \in I} \subset \mathbb{T}$ *such that*

$$R = \{t \in \mathbb{T} : t < \sigma(t)\} = \{t_i\}_{i \in I}. \tag{D.1}$$

Proof. Let $g : [a, b] \to \mathbb{R}$ be defined as follows:

$$g(t) = \begin{cases} t & \text{if } t \in \mathbb{T}, \\ \sigma(s) & \text{if } t \in (s, \sigma(s)) \text{ for some } s \in \mathbb{T}. \end{cases}$$

Note that the function g is monotone on $[a, b]$ and continuous on the set

$$[a, b] \setminus \{t \in \mathbb{T} : t < \sigma(t)\}.$$

Because the set of points where a monotone function has discontinuities is at most countable, we get the desired result. This completes the proof. $\square$

Remark D.8. Since

$$\mathbb{T} = \bigcup_{n \in \mathbb{N}} (\mathbb{T} \cap (-n, n)),$$

Lemma D.7 is valid in the case when $\mathbb{T}$ is unbounded.

Theorem D.9. *Let $t_0 \in \mathbb{T} \setminus \{\max \mathbb{T}\}$. Then the single-point set $\{t_0\}$ is Δ-measurable and its Δ-measure is given by*

$$\mu_\Delta(\{t_0\}) = \sigma(t_0) - t_0.$$

Proof.
1. Let t_0 be right-scattered. Then

$$\{t_0\} = [t_0, \sigma(t_0)) \in \mathcal{F}_1.$$

Therefore, $\{t_0\}$ is Δ-measurable and

$$\mu_\Delta(\{t_0\}) = \sigma(t_0) - t_0.$$

2. Let t_0 be right-dense. Then there is a decreasing sequence $\{t_k\}_{k \in \mathbb{N}}$ of points of $\mathbb{T}$ such that $t_k > t_0$, $k \in \mathbb{N}$ and $t_k \to t_0$, as $k \to \infty$. We have

$$[t_0, t_1) \supset [t_0, t_2) \supset \cdots \supset [t_0, t_k) \supset \cdots$$

and

$$\{t_0\} = \bigcap_{k=1}^{\infty} [t_0, t_k).$$

Hence, $\{t_0\}$ is Δ-measurable as a countable intersection of Δ-measurable sets. By the continuity property of μ_Δ, we get

$$\mu_\Delta(\{t_0\}) = \lim_{k \to \infty} \mu_\Delta([t_0, t_k)) = \lim_{k \to \infty} (t_k - t_0) = 0.$$

This completes the proof. □

Theorem D.10. *Let $a, b \in \mathbb{T}$, $a \le b$. Then*

$$\mu_\Delta([a, b)) = b - a,$$
$$\mu_\Delta((a, b)) = b - \sigma(a).$$

Proof. We have

$$\mu_\Delta([a, b)) = m_1([a, b)) = b - a.$$

Next observe that

$$[a, b) = \{a\} \cup (a, b).$$

Since

$$\{a\} \cap (a, b) = \emptyset,$$

we get

$$
\begin{aligned}
b - a &= \mu_\Delta([a, b)) \\
&= \mu_\Delta(\{a\} \cup (a, b)) \\
&= \mu_\Delta(\{a\}) + \mu_\Delta((a, b)) \\
&= \sigma(a) - a + \mu_\Delta((a, b)).
\end{aligned}
$$

Hence,

$$\mu_\Delta((a, b)) = \sigma(a) - a - b + a = \sigma(a) - b.$$

This completes the proof. □

Example D.11. Let $\mathbb{T} = 2\mathbb{Z}$. We will find

$$\mu_\Delta([0, 10)) \quad \text{and} \quad \mu_\Delta((0, 10)).$$

We have

$$\sigma(t) = t + 2, \quad t \in \mathbb{T}.$$

Then

$$\mu_\Delta([0, 10)) = 10 - 0 = 10,$$
$$\mu_\Delta((0, 10)) = 10 - \sigma(0) = 10 - 2 = 8.$$

Example D.12. Let $\mathbb{T} = 2^{\mathbb{N}_0} \cup \{0\}$. We will find

$$\mu_\Delta([1, 16)) \quad \text{and} \quad \mu_\Delta((2, 32)).$$

We have

$$\sigma(0) = 1, \quad \sigma(t) = 2t, \quad t \in 2^{\mathbb{N}_0}.$$

Then

$$\mu_\Delta([1, 16)) = 16 - 1 = 15,$$

$$\mu_\Delta((2, 32)) = 32 - \sigma(2) = 32 - 4 = 28.$$

Exercise D.13. Let $\mathbb{T} = 3^{\mathbb{N}_0} \cup \{0\}$. Find
1. $\mu_\Delta([0, 27))$,
2. $\mu_\Delta([1, 81))$,
3. $\mu_\Delta((0, 81))$,
4. $\mu_\Delta((0, 243))$,
5. $\mu_\Delta([1, 243))$.

Theorem D.14. *Let $a, b \in \mathbb{T} \setminus \{\max \mathbb{T}\}$, $a \le b$. Then*

$$\mu_\Delta((a, b]) = \sigma(b) - \sigma(a), \quad \mu_\Delta([a, b]) = \sigma(b) - a.$$

Proof. We have

$$(a, b] = (a, b) \cup \{b\}.$$

Then, using that

$$(a, b) \cap \{b\} = \emptyset,$$

we obtain

$$\begin{aligned}
\mu_\Delta((a, b]) &= \mu_\Delta((a, b) \cup \{b\}) \\
&= \mu_\Delta((a, b)) + \mu_\Delta(\{b\}) \\
&= b - \sigma(a) + \sigma(b) - b \\
&= \sigma(b) - \sigma(a).
\end{aligned}$$

Next,

$$[a, b] = \{a\} \cup (a, b].$$

Since

$$\{a\} \cap (a, b] = \emptyset,$$

we obtain

$$\begin{aligned}
\mu_\Delta([a, b]) &= \mu_\Delta(\{a\} \cup (a, b]) \\
&= \mu_\Delta(\{a\}) + \mu_\Delta((a, b]) \\
&= \sigma(a) - a + \sigma(b) - \sigma(a) \\
&= \sigma(b) - a.
\end{aligned}$$

This completes the proof. □

Example D.15. Let $\mathbb{T} = \mathbb{Z}$. We will find

$$\mu_\Delta((0,10]) \quad \text{and} \quad \mu_\Delta([-2,8]).$$

We have

$$\sigma(t) = t + 1, \quad t \in \mathbb{T}.$$

Then

$$\mu_\Delta((0,10]) = \sigma(10) - \sigma(0) = 11 - 1 = 10,$$
$$\mu_\Delta([-2,8]) = \sigma(8) - (-2) = 9 + 2 = 11.$$

Example D.16. Let $\mathbb{T} = 4^{\mathbb{N}_0} \cup \{0\}$. We will find

$$\mu_\Delta((0,4]) \quad \text{and} \quad \mu_\Delta([1,64]).$$

We have

$$\sigma(0) = 1,$$
$$\sigma(t) = 4t, \quad t \in 4^{\mathbb{N}_0}.$$

Then

$$\mu_\Delta((0,4]) = \sigma(4) - \sigma(0) = 16 - 1 = 15,$$
$$\mu_\Delta([1,64]) = \sigma(64) - 1 = 256 - 1 = 255.$$

Exercise D.17. Let $\mathbb{T} = 8\mathbb{Z}$. Find
1. $\mu_\Delta((0,8])$,
2. $\mu_\Delta([-8,8])$,
3. $\mu_\Delta([-16,32])$,
4. $\mu_\Delta([-16,0])$,
5. $\mu_\Delta((-16,8])$.

For a set $E \subset [a,b]$, define

$$I_E = \{i \in I : t_i \in E \cap R\},$$

where $I \subset \mathbb{N}$ and R is given by (D.1).

Theorem D.18. *If $E \subset [a, b]$, then the following properties are satisfied:*

1. $\mu_\Delta(E) \le m_1^*(E)$.
2. *If $b \notin E$ and E has no right-scattered points, then*

$$\mu_\Delta(E) = m_1^*(E).$$

3. *If R is defined by (D.1) and $I \setminus R$ is Lebesgue measurable, then*

$$\mu(R) = 0,$$

   where $\mu(\cdot)$ is the classical Lebesgue measure.

4.

$$\mu_\Delta(E \cap R) = \sum_{i \in I_E} (\sigma(t_i) - t_i) \le b - a = \mu_\Delta([a, b)).$$

5. *If $b \notin E$, then*

$$m_1^*(E) = \sum_{i \in I_E} (\sigma(t_i) - t_i) + \mu(E).$$

6. $m_1^*(E) = \mu_\Delta(E)$ *if and only if $b \notin E$ and E has no right-scattered points.*

Proof.

1. Assertions 1, 2, 3, and 4 follow directly from the definitions of m_1^* and μ_Δ.
2. Now we will prove Assertion 5. Suppose that $b \notin E$. We have

$$\begin{aligned}
\mu_\Delta(E) &= \mu_\Delta(E \cap \mathbb{T}) \\
&= \mu_\Delta(E \cap (R \cup (\mathbb{T} \setminus R))) \\
&= \mu_\Delta(E \cap R) + \mu_\Delta(E \cap (\mathbb{T} \setminus R)) \\
&= \mu_\Delta(E \cap (\mathbb{T} \setminus R)).
\end{aligned}$$

Because $b \notin E \cap (\mathbb{T} \setminus R)$ and $E \cap (\mathbb{T} \setminus R)$ has no right-scattered points, by Assertion 2, we have

$$\mu_\Delta(E) = \mu_\Delta(E \cap (\mathbb{T} \setminus R)) = m_1^*(E \cap (\mathbb{T} \setminus R)).$$

Thus,

$$\begin{aligned}
m_1^*(E) &= m_1^*(E \cap R) + m_1^*(E \cap (\mathbb{T} \setminus R)) \\
&= \sum_{i \in I_E} (\sigma(t_i) - t_i) + \mu_\Delta(E).
\end{aligned}$$

3. Now we will prove Assertion 6.

(a) Let $b \notin E$ and suppose E has no right-scattered points. Then

$$\sum_{i \in I_E} (\sigma(t_i) - t_i) = 0$$

and

$$m_1^*(E) = \mu_\Delta(E).$$

(b) Let $m_1^*(E) = \mu_\Delta(E)$. Then, using Assertion 5, we get

$$\sum_{i \in I_E} (\sigma(t_i) - t_i) = 0.$$

This completes the proof. ☐

Theorem D.19. *Let $A \subset [a, b]$. Then A is Δ-measurable if and only if A is Lebesgue measurable. In such a case, the following properties hold for any Δ-measurable set A:*
(i) *If $b \notin A$, then*

$$\mu_\Delta(A) = \sum_{i \in I_A} (\sigma(t_i) - t_i) + \mu(A).$$

(ii) *$\mu_\Delta(A) = \mu(A)$ if and only if $b \notin A$ and A has no right-scattered points.*

Proof.
1. Let A be Δ-measurable.
    (a) Let $b \notin A$. Take $E \subset [a, b]$ arbitrarily.
        i. Suppose that $b \notin E$. Then, using that

        $$[a, b] \setminus A = (\mathbb{T} \setminus A) \cup ([a, b] \setminus \mathbb{T}),$$

        A is Δ-measurable, and $\mathbb{T}$ is Lebesgue measurable, we obtain

        $$\begin{aligned}
        \mu(E) &\leq \mu(E \cap A) + \mu(E \cap ([a, b] \setminus A)) \\
        &\leq \mu(E \cap A) + \mu(E \cap (\mathbb{T} \setminus A)) + \mu(E \cap ([a, b] \setminus \mathbb{T})) \\
        &= m_1^*(E \cap A) + m_1^*(E \cap (\mathbb{T} \setminus A)) \\
        &\quad - \sum_{i \in I_{E \cap \mathbb{T}}} (\sigma(t_i) - t_i) + \mu(E \cap ([a, b] \setminus \mathbb{T})) \\
        &= m_1^*(E \cap \mathbb{T}) - \sum_{i \in I_{E \cap \mathbb{T}}} (\sigma(t_i) - t_i) + \mu(E \cap ([a, b] \setminus \mathbb{T})) \\
        &= \mu(E \cap \mathbb{T}) + \mu(E \cap ([a, b] \setminus \mathbb{T})) \\
        &= \mu(E).
        \end{aligned}$$

Thus,

$$\mu(E) = \mu(E \cap A) + \mu(E \cap ([a, b] \setminus A)).$$

ii. Let $b \in E$. Then

$$\mu(\{b\}) = 0$$

and

$$
\begin{aligned}
\mu(E) &\leq \mu(E \cap A) + \mu(E \cap ([a, b] \setminus A)) \\
&\leq \mu((E \cap [a, b)) \cap A) + \mu((E \cap [a, b)) \cap ([a, b] \setminus A)) \\
&= \mu(E \cap [a, b)) \\
&\leq \mu(E),
\end{aligned}
$$

i. e.,

$$\mu(E) = \mu(E \cap A) + \mu(E \cap ([a, b] \setminus A)).$$

Consequently, A is Lebesgue measurable.

(b) Let $b \in A$. Then $A \setminus \{b\}$ is Δ-measurable and, by the previous case, it follows that $A \setminus \{b\}$ is Lebesgue measurable. Since $\{b\}$ is Lebesgue measurable and the union of two Lebesgue measurable sets is a Lebesgue measurable set, we conclude that the set A is a Lebesgue measurable set.

2. The fact that if A is a Lebesgue measurable set, then it is a Δ-measurable set follows similarly, and we leave its proof to the reader as an exercise.

Note that (i) and (ii) follow by Assertions 5 and 6 of Theorem D.18. This completes the proof. □

Definition D.20. The Lebesgue integral associated with the measure μ_Δ we call the Lebesgue Δ-integral on $\mathbb{T}$. For a set $E \subseteq \mathbb{T}$ and a (measurable) function $f : E \to \mathbb{R}$, the corresponding integral of f over E we denote by

$$\int_E f(t) \Delta t.$$

So, all theorems of the general Lebesgue integration theory, including the Lebesgue dominated convergence theorem, will hold for the Lebesgue delta integral on $\mathbb{T}$. Below is a comparison of the Lebesgue Δ-integral with the Riemann Δ-integral.

Theorem D.21. *Let $[a, b)$ be a half-closed bounded interval in $\mathbb{T}$ and let f be a bounded real-valued function on $[a, b)$. If f is Riemann Δ-integrable from a to b, then f is Lebesgue Δ-integrable on $[a, b)$, and*

$$R \int_a^b f(t)\Delta t = L \int_a^b f(t)\Delta t, \tag{D.2}$$

where R and L indicate the Riemann and Lebesgue integrals, respectively.

Proof. Let f be a Riemann Δ-integrable from a to b. Then for each positive integer k, we can choose a $\delta_k > 0$, $\delta_k \to 0$, as $k \to \infty$, and a partition

$$P_k : a = t_0^{(k)} < t_1^{(k)} < \cdots < t_{n(k)}^k = b$$

of the interval $[a, b)$ such that $P_k \in \mathcal{P}_{\delta_k}$ and

$$U(f, P_k) - L(f, P_k) < \frac{1}{k}.$$

Here $\mathcal{P}_{\delta_k}$ is the set of all partitions

$$P : a = t_0 < t_1 < \cdots < t_n = b$$

such that either $t_j - t_{j-1} < \delta$ or $t_j - t_{j-1} > \delta$ and $\rho(t_j) = t_{j-1}$, $U(f, P_k)$ and $L(f, P_k)$ are the upper and lower Darboux Δ-sums of f with respect to P_k, respectively. Then

$$\lim_{k \to \infty} L(f, P_k) = \lim_{k \to \infty} U(f, P_k) = R \int_a^b f(t)\Delta t.$$

By replacing partitions P_k with finer partitions, if necessary, we can assume that, for each k, partition P_{k+1} is a refinement of partition P_k. Set

$$m_j^{(k)} = \inf\{f(t) : t \in [t_{j-1}^{(k)}, t_j^{(k)})\},$$
$$M_j^{(k)} = \sup\{f(t) : t \in [t_{j-1}^{(k)}, t_j^{(k)})\}, \quad j = 1, 2, \ldots, n(k).$$

Define the sequences $\{\phi_k\}$ and $\{\Phi_k\}$ of functions on $[a, b)$ such that

$$\phi_k(t) = m_j^{(k)} \quad \text{and} \quad \Phi_k(t) = M_j^{(k)}, \quad t \in [t_{j-1}^{(k)}, t_j^{(k)}),$$

$j = 1, 2, \ldots, n(k)$. We have that $\{\phi_k\}$ is a nondecreasing sequence and $\{\Phi_k\}$ is a nonincreasing sequence. Also, for each positive integer k, we have

$$\phi_k \le \phi_{k+1},$$
$$\Phi_k \ge \Phi_{k+1},$$
$$\phi_k \le f \le \Phi_k,$$
$$L \int_{[a,b)} \phi_k(t)\Delta t = L(f, P_k),$$

$$L \int_{[a,b)} \Phi_k(t)\Delta t = U(f, P_k).$$

Because f is bounded, we have that the sequences $\{\phi_k\}$ and $\{\Phi_k\}$ are bounded. Therefore,

$$\phi(t) = \lim_{k\to\infty} \phi_k(t),$$

$$\Phi(t) = \lim_{k\to\infty} \Phi_k(t), \quad t \in [a, b).$$

We have

$$\phi(t) \le f(t) \le \Phi(t), \quad t \in [a, b),$$

and ϕ and Φ are Δ-measurable functions on $[a, b)$. By the Lebesgue dominated theorem, we obtain

$$\lim_{k\to\infty} L \int_{[a,b)} \phi_k(t)\Delta t = L \int_{[a,b)} \phi(t)\Delta t,$$

$$\lim_{k\to\infty} L \int_{[a,b)} \Phi_k(t)\Delta t = L \int_{[a,b)} \Phi(t)\Delta t.$$

Therefore,

$$\begin{aligned}
L \int_{[a,b)} \phi(t)\Delta t &= \lim_{k\to\infty} \int_{[a,b)} \phi_k(t)\Delta t \\
&= \lim_{k\to\infty} L(f, P_k) \\
&= R \int_a^b f(t)\Delta t \\
&= \lim_{k\to\infty} U(f, P_k) \\
&= \lim_{k\to\infty} L \int_{[a,b)} \Phi_k(t)\Delta t \\
&= L \int_{[a,b)} \Phi(t)\Delta t.
\end{aligned}$$

Hence,

$$L \int_{[a,b)} (\Phi(t) - \phi(t))\Delta t = 0$$

and, using that

$$\phi(t) = \Phi(t), \quad t \in [a,b),$$

we get

$$\phi(t) = \Phi(t)$$

for Δ-almost every $t \in [a,b)$. Consequently,

$$\phi(t) = f(t) = \Phi(t)$$

for Δ-almost every $t \in [a,b)$. Therefore, f is Lebesgue Δ-integrable and (D.2) holds. This completes the proof. □

Theorem D.22. *Let f be a bounded function defined on the half-closed bounded interval $[a,b)$ of $\mathbb{T}$. Then f is Riemann Δ-integrable from a to b if and only if the set of all right-dense points of $[a,b)$ at which f is discontinuous is a set of Δ-measure zero.*

Proof.

1. Suppose that f is Riemann Δ-integrable from a to b. For each positive integer k, let P_k, ϕ_k, Φ_k, ϕ, and Φ be defined as in the proof of Theorem D.21. Let

$$\Lambda = \bigcup_{k=1}^{\infty} P_k,$$
$$\Lambda_{rd} = \{t \in [a,b) : t \in \Lambda \text{ and } t \text{ is right-dense}\},$$
$$G = \{t \in [a,b) : f \text{ is discontinuous at } t\},$$
$$G_{rd} = \{t \in G : t \text{ is right-dense}\},$$
$$A = \{t \in [a,b) : \phi(t) \neq \Phi(t)\}.$$

Let $t \in [a,b)$ be such that

$$\phi(t) = f(t) = \Phi(t)$$

and $t \notin \Lambda$. Assume that f is not continuous at t. Then there exist an $\varepsilon > 0$ and a sequence $\{t_j\}_{j \in \mathbb{N}}$ such that $t_j \to t$, as $j \to \infty$, and

$$|f(t_j) - f(t)| > \varepsilon$$

for any $j \in \mathbb{N}$. Hence,

$$f(t_j) > \varepsilon + f(t)$$

and

$$\Phi(t) \geq \varepsilon + \phi(t),$$

which is a contradiction. Therefore, f is continuous at t. Observe that all right-scattered points of $[a, b)$ belong to Λ. Hence, for each right-scattered point t of $[a, b)$ and all sufficiently large k, we have

$$\phi_k(t) = \Phi_k(t) = f(t)$$

and from here

$$\phi(t) = \Phi(t) = f(t).$$

Therefore,

$$G_{rd} \subset A \cup \Lambda_{rd}.$$

By the proof of Theorem D.21, it follows that

$$\phi(t) = \Phi(t)$$

for Δ-almost every t in $[a, b)$. Therefore,

$$\mu_\Delta(A) = \mu_\Delta(\Lambda_{rd}) = 0,$$

and hence,

$$\mu_\Delta(G_{rd}) = 0.$$

2. Suppose that the set of all right-dense points of $[a, b)$ at which f is discontinuous is of Δ-measure zero. Then $\mu_\Delta(G_{rd}) = 0$. For each positive integer k, we choose $\delta_k > 0$, $\delta_k \to 0$, as $k \to \infty$, and a partition

$$P_k : a = t_0^{(k)} < t_1^{(k)} < \cdots < t_{n(k)}^{(k)} = b$$

of $[a, b)$ such that $P_k \in \mathcal{P}_{\delta_k}$ and P_{k+1} is a refinement of P_k. Let ϕ_k, Φ_k, ϕ, and Φ be defined as in the proof of Theorem D.21. Suppose that $t \in [a, b)$ is right-dense and f is continuous at t. Then for a given $\varepsilon > 0$, there exists a $\delta > 0$ such that

$$\sup f - \inf f < \varepsilon,$$

where the supremum and the infimum are taken over the interval $(t - \delta, t + \delta)$. For all k sufficiently large, a subinterval of P_k containing t will be in $(t - \delta, t + \delta)$ and then

$$\Phi_k(t) - \phi_k(t) < \varepsilon.$$

Since $\varepsilon > 0$ was arbitrarily chosen, we conclude that

$$\phi(t) = \Phi(t).$$

Next, at each right-scattered point t of $[a, b)$, using the first part of the proof, we have

$$\phi(t) = \Phi(t).$$

Consequently, $A \subset G_{rd}$ and, using that $\mu_\Delta(G_{rd}) = 0$, we conclude that $\mu_\Delta(A) = 0$. From here,

$$\phi(t) = \Phi(t)$$

Δ-almost everywhere on $[a, b)$ and

$$L \int_{[a,b)} \phi(t)\Delta t = L \int_{[a,b)} \Phi(t)\Delta t.$$

Hence, using the proof of Theorem D.21, we obtain

$$\lim_{k \to \infty} L(f, P_k) = \lim_{k \to \infty} U(f, P_k)$$

and thus, f is Riemann Δ-integrable on $[a, b)$.

This completes the proof. $\qquad\qquad\qquad\qquad\qquad\qquad\qquad\qquad\qquad\qquad$ □

D.2 The spaces $\mathbb{L}^p(\mathbb{T})$

Having defined the space $\mathbb{L}^1(\mathbb{T})$, it is usual to define $\mathbb{L}^p(\mathbb{T})$ for any $p \geq 1$ in the following way:

$$\mathbb{L}^p(\mathbb{T}) = \{u \in \mathbb{L}^1(\mathbb{T}) : |u|^p \in \mathbb{L}^1(\mathbb{T})\}.$$

For any $p \geq 1$, we provide $\mathbb{L}^p(\mathbb{T})$ with a norm

$$\|u\|_{\mathbb{L}^p(\mathbb{T})} = \left(\int_a^b |u|^p \Delta \right)^{\frac{1}{p}}, \quad u \in \mathbb{L}^p(\mathbb{T}). \tag{D.3}$$

Exercise D.23. Prove that (D.3) satisfies all the axioms for a norm.

Theorem D.24. *For any $p \geq 1$, we have*

$$\mathbb{C}(\mathbb{T}) \subset \mathbb{C}_{rd}(\mathbb{T}) \subset \mathbb{L}^p(\mathbb{T}).$$

Proof. Let $u \in C_{rd}(\mathbb{T})$ be arbitrarily chosen. Then $|u|^p \circ E$ is bounded on $[a, b]$ and it is continuous on $[a, b]$ except possibly at right-scattered points. Since the set of such points is countable, it has Lebesgue measure zero. Thus, $|u|^p \circ E$ is Lebesgue measurable, and hence, integrable. Therefore, $u \in L^p(\mathbb{T})$. Because $u \in C_{rd}(\mathbb{T})$ was arbitrarily chosen and we get that $u \in L^p(\mathbb{T})$, we conclude that

$$C_{rd}(\mathbb{T}) \subset L^p(\mathbb{T}).$$

This completes the proof. □

Theorem D.25. *For any $p \in [1, \infty)$, we have that $C(\mathbb{T})$ is dense in $L^p(\mathbb{T})$.*

Proof. Let $u \in L^p(\mathbb{T})$ be arbitrarily chosen. Consider $\mathbb{T}$ as a topological space with measure $\mu_\mathbb{T}$. By the standard Lebesgue integration theory, it follows that there exists a sequence $\{u_n\}_{n \in \mathbb{N}}$ of elements of $C(\mathbb{T})$ such that

$$\|u - u_n\|_{L^1(\mathbb{T})} \to 0,$$

as $n \to \infty$. Thus, the result is proved for $p = 1$. In addition, by the standard Lebesgue integration theory, it follows that we can suppose that there exists a constant $C > 0$ for which $|u(t)| \le C$, $|u_n(t)| \le C$ for any $n \in \mathbb{N}$ and for any $t \in \mathbb{T}$. Hence,

$$\begin{aligned}
\|u - u_n\|_{L^p(\mathbb{T})}^p &= \int_a^b |u - u_n|^p \Delta \\
&= \int_a^b |u - u_n|^{p-1}|u - u_n|\Delta \\
&\le \int_a^b (|u| + |u_n|)^{p-1}|u - u_n|\Delta \\
&\le (2C)^{p-1} \int_a^b |u - u_n|\Delta \\
&= (2C)^{p-1}\|u - u_n\|_{L^1(\mathbb{T})} \\
&\to 0,
\end{aligned}$$

as $n \to \infty$. This completes the proof. □

Theorem D.26. *Suppose that $\{u_n\}_{n \in \mathbb{N}}$ is a sequence in $L^p(\mathbb{T})$ for some $p \ge 1$. If $\|u - u_n\|_{L^p(\mathbb{T})} \to 0$, as $n \to \infty$, for some $u \in L^p(\mathbb{T})$, and if $t \in \mathbb{T}$ is right-scattered, then $u_n(t) \to u(t)$, as $n \to \infty$.*

Proof. Let $\|\cdot\|_{\mathbb{L}^p(a,b)}$ denote the standard $\mathbb{L}^p(a,b)$-norm on the real interval $[a,b]$. Then

$$0 = \lim_{n\to\infty} \|(u - u_n) \circ E\|_{\mathbb{L}^p(a,b)}^p$$

$$= \lim_{n\to\infty} \int_a^b |(u - u_n) \circ E|^p dx$$

$$\geq \lim_{n\to\infty} \int_t^{\sigma(t)} |(u - u_n) \circ E|^p dx$$

$$= \lim_{n\to\infty} \int_t^{\sigma(t)} |u - u_n|^p \Delta$$

$$= \lim_{n\to\infty} |u(t) - u_n(t)|^p (\sigma(t) - t)$$

$$\geq 0.$$

Consequently, $u_n(t) \to u(t)$, as $n \to \infty$. This completes the proof. $\square$

Theorem D.27. *If $\{u_n\}_{n\in\mathbb{N}}$ is a Cauchy sequence in $\mathbb{L}^p(\mathbb{T})$, then there exists a unique $u \in \mathbb{L}^p(\mathbb{T})$ such that $\|u - u_n\|_{\mathbb{L}^p(\mathbb{T})} \to 0$, as $n \to \infty$.*

Proof. Note that the sequence $\{u_n \circ E\}_{n\in\mathbb{N}}$ is a Cauchy sequence in $\mathbb{L}^p(a,b)$. Then, by the standard Lebesgue theory, it follows that there is a unique $z \in \mathbb{L}^p(a,b)$ such that

$$\|z - u_n \circ E\|_{\mathbb{L}^p((a,b))} \to 0,$$

as $n \to \infty$. Suppose that $t \in \mathbb{T}$ is right-scattered. Then

$$\|(u_n - u_m) \circ E\|_{\mathbb{L}^p(\mathbb{T})}^p = \int_a^b |(u_n - u_m) \circ E|^p dx$$

$$\geq \int_t^{\sigma(t)} |(u_n - u_m) \circ E|^p dx$$

$$= \int_t^{\sigma(t)} |u_n - u_m|^p \Delta$$

$$= |u_n(t) - u_m(t)|^p (\sigma(t) - t).$$

Therefore, the sequence $\{u_n(t)\}_{n\in\mathbb{N}}$ is a Cauchy sequence. Thus, it converges to $c(t)$ and the function z must equal to $c(t)$ almost everywhere on the interval $[t, \sigma(t)]$, and so we may suppose that this equality holds everywhere on $[t, \sigma(t))$. Since the set of

---

right-scattered points $t \in \mathbb{T}$ is at most countable, we may suppose that this is true for all right-scattered points $t \in \mathbb{T}$. Defining u to be the restriction of z to $\mathbb{T}$, it follows that $u \in \mathbb{L}^p(\mathbb{T})$, $z = u \circ E$, and $\|u - u_n\|_{\mathbb{L}^p(\mathbb{T})} \to 0$, as $n \to \infty$. This completes the proof. $\square$

Remark D.28.
1. By Theorem D.27, it follows that $\mathbb{L}^p(\mathbb{T})$, $p \geq 1$, is a Banach space with respect to the norm $\|\cdot\|_{\mathbb{L}^p(\mathbb{T})}$.
2. We can define a natural inner product on $\mathbb{L}^2(\mathbb{T})$ by

$$\langle u, v \rangle_{\mathbb{T}} = \int_a^b uv\Delta, \quad u, v \in \mathbb{L}^2(\mathbb{T}). \tag{D.4}$$

   With respect to this inner product, the space $\mathbb{L}^2(\mathbb{T})$ is a Hilbert space.
3. To simplify the notation, from now on we will use the notation $\|\cdot\|_{\mathbb{T}}$ for the norm $\|\cdot\|_{\mathbb{L}^2(\mathbb{T})}$.
4. The notations $\|\cdot\|_{\mathbb{T}}$ and $\langle\cdot,\cdot\rangle_{\mathbb{T}}$ for the above norm and inner product on $\mathbb{L}^2(\mathbb{T})$ indicate that their values depend on the entire time scale $\mathbb{T}$, even if u is not defined at b.

Exercise D.29. Prove that (D.4) satisfies all the axioms for an inner product.

D.3 Sobolev-type spaces and generalized derivatives

Definition D.30. For $u \in \mathbb{C}^1_{rd}(\mathbb{T})$, we define

$$\|u\|^2_{1,\mathbb{T}} = \|u\|^2_{\mathbb{T}} + \|u^\Delta\|^2_{\mathbb{T}}, \tag{D.5}$$

and define the space $\mathbb{H}^1(\mathbb{T}) \subset \mathbb{L}^2(\mathbb{T})$ to be the completion of $\mathbb{C}^1(\mathbb{T})$ with respect to the norm $\|\cdot\|_{1,\mathbb{T}}$. The space $\mathbb{H}^1(\mathbb{T})$ will be called a Sobolev space over $\mathbb{T}$.

Exercise D.31. Prove that (D.5) satisfies all the axioms for a norm.

Theorem D.32. *Function $u \in \mathbb{H}^1(\mathbb{T})$ if and only if there exists a function $u^{\Delta_g} \in \mathbb{L}^2(\mathbb{T})$ such that the following condition holds: there exists a sequence $\{u_n\}_{n\in\mathbb{N}}$ in $\mathbb{C}^1(\mathbb{T})$ such that $u_n \to u$ and $u_n^\Delta \to u^{\Delta_g}$, as $n \to \infty$, in $\mathbb{L}^2(\mathbb{T})$. If $u \in \mathbb{H}^1(\mathbb{T})$, then u^{Δ_g} is unique in the $\mathbb{L}^2(\mathbb{T})$ sense. If $u \in \mathbb{C}^1(\mathbb{T})$, then $u^{\Delta_g} = u^\Delta$.*

Proof.
1. Let $u \in \mathbb{H}^1(\mathbb{T})$. By the definition of the space $\mathbb{H}^1(\mathbb{T})$, it follows that there exists a sequence $\{u_n\}_{n\in\mathbb{N}}$ of elements of $\mathbb{C}^1(\mathbb{T})$ such that $\|u_n-u\|_{1,\mathbb{T}} \to 0$, as $n \to \infty$. Hence, $\|u_n - u_m\|_{1,\mathbb{T}} \to 0$, as $m, n \to \infty$. Consequently, $\{u_n^\Delta\}_{n\in\mathbb{N}}$ is a Cauchy sequence in

$\mathbb{L}^2(\mathbb{T})$. Because $\mathbb{L}^2(\mathbb{T})$ is a Banach space, there exists a unique $u^{\Delta_g} \in \mathbb{L}^2(\mathbb{T})$ such that $u_n^\Delta \to u^{\Delta_g}$, as $n \to \infty$, in $\mathbb{L}^2(\mathbb{T})$. In particular, if $u \in \mathbb{C}^1(\mathbb{T})$, then $u^\Delta = u^{\Delta_g}$.

2. Let $\{u_n\}_{n \in \mathbb{N}}$ be a sequence of elements of $\mathbb{C}^1(\mathbb{T})$ such that $u_n \to u$ and $u_n^\Delta \to u^{\Delta_g}$, as $n \to \infty$, in $\mathbb{L}^2(\mathbb{T})$, for some $u^{\Delta_g} \in \mathbb{L}^2(\mathbb{T})$. Hence, using (D.5), we conclude that $\{u_n\}_{n \in \mathbb{N}}$ is a convergent sequence in $\mathbb{H}^1(\mathbb{T})$. By the definition of $\mathbb{H}^1(\mathbb{T})$, it follows that there exists $u_1 \in \mathbb{H}^1(\mathbb{T})$ such that $u_n \to u_1$, as $n \to \infty$, in $\mathbb{H}^1(\mathbb{T})$. Hence, $u_n \to u_1$, as $n \to \infty$, in $\mathbb{L}^2(\mathbb{T})$. Because $u_n \to u$, as $n \to \infty$, in $\mathbb{L}^2(\mathbb{T})$, we conclude that $u = u_1$ and $u \in \mathbb{H}^1(\mathbb{T})$. This completes the proof. □

Definition D.33. For any $u \in \mathbb{H}^1(\mathbb{T})$, the function u^{Δ_g} in Theorem D.32 will be called a generalized derivative of u.

Theorem D.34. *If $u \in \mathbb{H}^1(\mathbb{T})$, then $u \in \mathbb{C}(\mathbb{T})$. There exists a constant $C > 0$ such that*

$$|u|_{0,\mathbb{T}} \le C\|u\|_{1,\mathbb{T}}, \quad u \in \mathbb{H}^1(\mathbb{T}). \tag{D.6}$$

Furthermore,

$$u(t) - u(s) = \int_s^t u^{\Delta_g} \Delta, \quad s, t \in \mathbb{T}. \tag{D.7}$$

Proof. Let $u \in \mathbb{C}^1(\mathbb{T})$ and $s, t \in \mathbb{T}$, $s \le t$. By Theorem D.22, we get

$$u(t) - u(s) = \int_s^t u^\Delta \Delta. \tag{D.8}$$

Hence,

$$|u(t)| = \left| u(s) + \int_s^t u^\Delta \Delta \right|$$

$$\le |u(s)| + \left| \int_s^t u^\Delta \Delta \right|$$

$$\le |u(s)| + \int_s^t |u^\Delta| \Delta$$

$$\le |u(s)| + (t - s)^{\frac{1}{2}} \left(\int_s^t |u^\Delta|^2 \Delta \right)^{\frac{1}{2}}$$

$$\le |u(s)| + (b - a)^{\frac{1}{2}} \|u^\Delta\|_{\mathbb{T}}.$$

Now we integrate the latter inequality over $\mathbb{T}$ with respect to s and get

$$\int_a^b |u(t)|\Delta \le \int_a^b |u(s)|\Delta + (b-a)^{\frac{1}{2}} \int_a^b \|u^\Delta\|_{\mathbb{T}}\Delta$$

$$\le (b-a)^{\frac{1}{2}}\left(\int_a^b |u(s)|^2\Delta\right)^{\frac{1}{2}} + (b-a)^{\frac{1}{2}}\|u^\Delta\|_{\mathbb{T}}$$

$$= (b-a)^{\frac{1}{2}}\|u\|_{\mathbb{T}} + (b-a)^{\frac{3}{2}}\|u^\Delta\|_{\mathbb{T}},$$

or

$$(b-a)|u(t)| \le (b-a)^{\frac{1}{2}}\|u\|_{\mathbb{T}} + (b-a)^{\frac{3}{2}}\|u^\Delta\|_{\mathbb{T}}.$$

Hence,

$$|u(t)| \le \frac{1}{(b-a)^{\frac{1}{2}}}\|u\|_{\mathbb{T}} + (b-a)^{\frac{1}{2}}\|u^\Delta\|_{\mathbb{T}}$$

and

$$\sup_{t\in\mathbb{T}}|u(t)| \le \frac{1}{(b-a)^{\frac{1}{2}}}\|u\|_{\mathbb{T}} + (b-a)^{\frac{1}{2}}\|u^\Delta\|_{\mathbb{T}},$$

or

$$|u|_{0,\mathbb{T}} \le \frac{1}{(b-a)^{\frac{1}{2}}}\|u\|_{\mathbb{T}} + (b-a)^{\frac{1}{2}}\|u^\Delta\|_{\mathbb{T}}.$$

Let

$$C = \max\left\{\frac{1}{(b-a)^{\frac{1}{2}}}, (b-a)^{\frac{1}{2}}\right\}.$$

Then

$$|u|_{0,\mathbb{T}} \le C(\|u\|_{\mathbb{T}} + \|u^\Delta\|_{\mathbb{T}}) = C\|u\|_{1,\mathbb{T}}.$$

By Theorem D.32, we have that $u^\Delta = u^{\Delta g}$. Hence, by (D.8), we get (D.7). Let now $u \in \mathbb{H}^1(\mathbb{T})$. Then there exists a sequence $\{u_n\}_{n\in\mathbb{N}}$ of elements of $\mathbb{C}^1(\mathbb{T})$ such that $u_n \to u$, as $n \to \infty$, in $\mathbb{H}^1(\mathbb{T})$. For this sequence, we have

$$|u_n - u_m|_{0,\mathbb{T}} \le C\|u_n - u_m\|_{1,\mathbb{T}}, \quad m,n \in \mathbb{N}. \tag{D.9}$$

Note that $u_n \to u$, as $n \to \infty$, in $\mathbb{L}^2(\mathbb{T})$. Because $\|u_n - u_m\|_{1,\mathbb{T}} \to 0$, as $m,n \to \infty$, using (D.9), we get

$$|u_n - u_m|_{0,\mathbb{T}} \to 0,$$

as $m, n \to \infty$. Therefore, $\{u_n\}_{n \in \mathbb{N}}$ is a Cauchy sequence in $\mathbb{C}(\mathbb{T})$. Because $\mathbb{C}(\mathbb{T})$ is a Banach space, we conclude that the sequence $\{u_n\}_{n \in \mathbb{N}}$ is a convergent sequence in $\mathbb{C}(\mathbb{T})$. Let $u^1 \in \mathbb{C}(\mathbb{T})$ be its limit, i. e., $u_n \to u^1$, as $n \to \infty$, in $\mathbb{C}(\mathbb{T})$. Then

$$\|u_n - u^1\|_{\mathbb{T}}^2 = \int_a^b |u_n - u^1|^2 \Delta$$

$$\leq (b - a)|u_n - u^1|_{0,\mathbb{T}}^2$$

$$\to 0,$$

as $n \to \infty$. Therefore, $u_n \to u^1$, as $n \to \infty$, in $\mathbb{L}^2(\mathbb{T})$. Because $u_n \to u$, as $n \to \infty$, in $\mathbb{L}^2(\mathbb{T})$, we conclude that $u = u^1$ and $u \in \mathbb{C}(\mathbb{T})$. From here, using

$$|u_n|_{0,\mathbb{T}} \leq C\|u_n\|_{1,\mathbb{T}},$$

we obtain that u satisfies the inequality (D.6). Since $u \in \mathbb{H}^1(\mathbb{T})$, using Theorem D.32, there exist a unique function $u^{\Delta_g} \in \mathbb{L}^2(\mathbb{T})$ and a sequence $\{v_n\}_{n \in \mathbb{N}}$ of elements of $\mathbb{C}^1(\mathbb{T})$ such that $v_n \to u$ and $v_n^\Delta \to u^{\Delta_g}$, as $n \to \infty$, in $\mathbb{L}^2(\mathbb{T})$. Also, we have

$$v_n(t) - v_n(s) = \int_s^t v_n^{\Delta_g} \Delta, s, t \in \mathbb{T}, \quad s \leq t. \tag{D.10}$$

Note that

$$\left| \int_s^t v_n^{\Delta_g} \Delta - \int_s^t u^{\Delta_g} \Delta \right| = \left| \int_s^t v_n^\Delta \Delta - \int_s^t u^{\Delta_g} \Delta \right|$$

$$= \left| \int_s^t (v_n^\Delta - u^{\Delta_g}) \Delta \right|$$

$$\leq \int_s^t |v_n^\Delta - u^{\Delta_g}| \Delta$$

$$\leq (t - s)^{\frac{1}{2}} \left(\int_s^t |v_n^\Delta - u^{\Delta_g}|^2 \Delta \right)^{\frac{1}{2}}$$

$$\leq (b - a)^{\frac{1}{2}} \|v_n^\Delta - u^{\Delta_g}\|_{\mathbb{T}}$$

$$\to 0, \quad s \leq t,$$

as $n \to \infty$. Hence, using that $v_n \to u$, as $n \to \infty$, in $\mathbb{C}(\mathbb{T})$, and (D.10), we conclude that u satisfies (D.7). This completes the proof. □

Theorem D.35. *Let $u \in \mathbb{H}^1(\mathbb{T})$. If $u^{\Delta_g} = 0$, then u is a constant on $\mathbb{T}$.*

Proof. By (D.7), we get

$$u(t) = u(s)$$

for any $s, t \in \mathbb{T}$. This completes the proof. □

Theorem D.36. *Let $u \in \mathbb{H}^1(\mathbb{T})$. If $t \in \mathbb{T}$ is a right-scattered point, then*

$$u^{\Delta g}(t) = \frac{u(\sigma(t)) - u(t)}{\sigma(t) - t}.$$

Proof. Let $t \in \mathbb{T}$ be a right-scattered point. Since $u \in \mathbb{H}^1(\mathbb{T})$, we have that $u \in \mathbb{L}^2(\mathbb{T})$ and there exist a sequence $\{u_n\}_{n \in \mathbb{N}}$ of elements of $\mathbb{C}^1(\mathbb{T})$ and an unique $u^{\Delta g} \in \mathbb{L}^2(\mathbb{T})$ such that $u_n \to u$ and $u_n^\Delta \to u^{\Delta g}$, as $n \to \infty$, in $\mathbb{L}^2(\mathbb{T})$. Hence, by Theorem D.26, we get that $u_n(t) \to u(t)$ and $u_n^\Delta(t) \to u^{\Delta g}(t)$, as $n \to \infty$. By (D.7), we have

$$u(\sigma(t)) = u(t) + \int_t^{\sigma(t)} u^{\Delta g} \Delta$$

and

$$u_n(\sigma(t)) = u_n(t) + \int_t^{\sigma(t)} u_n^\Delta \Delta = u_n(t) + u_n^\Delta(t)(\sigma(t) - t).$$

Hence,

$$\lim_{n \to \infty} u_n(\sigma(t)) = \lim_{n \to \infty} (u_n(t) + u_n^\Delta(t)(\sigma(t) - t)) = u(t) + \int_t^{\sigma(t)} u^{\Delta g} \Delta = u(\sigma(t)).$$

Therefore,

$$u^{\Delta g}(t) = \lim_{n \to \infty} u_n^\Delta(t)$$

$$= \lim_{n \to \infty} \frac{u_n(\sigma(t)) - u_n(t)}{\sigma(t) - t}$$

$$= \frac{u(\sigma(t)) - u(t)}{\sigma(t) - t}.$$

This completes the proof. □

Theorem D.37. *Let $u, v \in \mathbb{H}^1(\mathbb{T})$ and $\alpha, \beta \in \mathbb{R}$. Then $\alpha u + \beta v \in \mathbb{H}^1(\mathbb{T})$ and*

$$(\alpha u + \beta v)^{\Delta g} = \alpha u^{\Delta g} + \beta v^{\Delta g}.$$

Proof. Since $u, v \in \mathbb{H}^1(\mathbb{T})$, there exist sequences $\{u_n\}_{n \in \mathbb{N}}$, $\{v_n\}_{n \in \mathbb{N}}$ of elements of $\mathbb{C}^1(\mathbb{T})$ and unique $u^{\Delta_g}, v^{\Delta_g} \in \mathbb{L}^2(\mathbb{T})$ such that

$$u_n \to u, \quad v_n \to v, \quad u_n^\Delta \to u^{\Delta_g}, \quad v_n^\Delta \to v^{\Delta_g},$$

as $n \to \infty$, in $\mathbb{L}^2(\mathbb{T})$. Hence,

$$\alpha u_n + \beta v_n \to \alpha u + \beta v, \quad \alpha u_n^\Delta + \beta v_n^\Delta \to \alpha u^{\Delta_g} + \beta v^{\Delta_g},$$

as $n \to \infty$, in $\mathbb{L}^2(\mathbb{T})$. Hence, by Theorem D.32, it follows that $\alpha u + \beta v \in \mathbb{H}^1(\mathbb{T})$. Applying again Theorem D.32, we have that there exists a unique $(\alpha u + \beta v)^{\Delta_g} \in \mathbb{L}^2(\mathbb{T})$ such that

$$(\alpha u_n + \beta v_n)^\Delta \to (\alpha u + \beta v)^{\Delta_g},$$

as $n \to \infty$, in $\mathbb{L}^2(\mathbb{T})$. Since

$$(\alpha u_n + \beta v_n)^\Delta = \alpha u_n^\Delta + \beta v_n^\Delta \to \alpha u^{\Delta_g} + \beta v^{\Delta_g},$$

as $n \to \infty$, in $\mathbb{L}^2(\mathbb{T})$, we conclude that

$$(\alpha u + \beta v)^{\Delta_g} = \alpha u^{\Delta_g} + \beta v^{\Delta_g}.$$

This completes the proof. $\qquad\qquad\qquad\qquad\qquad\qquad\qquad\qquad\qquad\qquad\qquad\quad$ □

Theorem D.38. *Let $u, v \in \mathbb{H}^1(\mathbb{T})$. Then $uv \in \mathbb{H}^1(\mathbb{T})$ and*

$$(uv)^{\Delta_g} = u^{\Delta_g} v + u^\sigma v^{\Delta_g} = uv^{\Delta_g} + u^{\Delta_g} v^\sigma, \qquad (D.11)$$

and

$$\int_s^t u^{\Delta_g} v \Delta + \int_s^t u^\sigma v^{\Delta_g} \Delta = \int_s^t uv^{\Delta_g} \Delta + \int_s^t u^{\Delta_g} v^\sigma \Delta$$

$$= (uv)(t) - (uv)(s), \quad s, t \in \mathbb{T}.$$

Proof. Since $u, v \in \mathbb{H}^1(\mathbb{T})$, there exist sequences $\{u_n\}_{n \in \mathbb{N}}$, $\{v_n\}_{n \in \mathbb{N}}$ of elements of $\mathbb{C}^1(\mathbb{T})$ and unique $u^{\Delta_g}, v^{\Delta_g} \in \mathbb{L}^2(\mathbb{T})$ such that

$$u_n \to u, \quad v_n \to v, \quad u_n^\Delta \to u^{\Delta_g}, \quad v_n^\Delta \to v^{\Delta_g},$$

as $n \to \infty$, in $\mathbb{L}^2(\mathbb{T})$. Then

$$u_n v_n \to uv,$$

as $n \to \infty$, in $\mathbb{L}^2(\mathbb{T})$,

$$u^{\Delta_g} v, \quad u^\sigma v^{\Delta_g}, \quad u v^{\Delta_g}, \quad u^{\Delta_g} v^\sigma \in \mathbb{L}^2(\mathbb{T}),$$
$$u_n^\sigma \to u^\sigma, \quad v_n^\sigma \to v^\sigma,$$

as $n \to \infty$, in $\mathbb{L}^2(\mathbb{T})$. Hence,

$$(u_n v_n)^\Delta = u_n^\Delta v_n + u_n^\sigma v_n^\Delta \to u^{\Delta_g} v + u^\sigma v^{\Delta_g}, \tag{D.12}$$
$$(u_n v_n)^\Delta = u_n v_n^\Delta + u_n^\Delta v_n^\sigma \to u v^{\Delta_g} + u^{\Delta_g} v^\sigma, \tag{D.13}$$

as $n \to \infty$, in $\mathbb{L}^2(\mathbb{T})$. Therefore, $uv \in \mathbb{H}^1(\mathbb{T})$. Hence, using Theorem D.32, there exists a unique $(uv)^{\Delta_g} \in \mathbb{L}^2(\mathbb{T})$ such that

$$(u_n v_n)^\Delta \to (uv)^{\Delta_g},$$

as $n \to \infty$, in $\mathbb{L}^2(\mathbb{T})$. From here and (D.12), (D.13), we obtain (D.11). Hence, by (D.7), we get

$$\int_s^t u^{\Delta_g} v \Delta + \int_s^t u^\sigma v^{\Delta_g} \Delta = \int_s^t (u^{\Delta_g} v + u^\sigma v^{\Delta_g}) \Delta$$

$$= \int_s^t (uv)^{\Delta_g} \Delta$$

$$= (uv)(t) - (uv)(s)$$

and

$$\int_s^t u v^{\Delta_g} \Delta + \int_s^t u^{\Delta_g} v^\sigma \Delta = \int_s^t (u v^{\Delta_g} + u^{\Delta_g} v^\sigma) \Delta$$

$$= \int_s^t (uv)^{\Delta_g} \Delta$$

$$= (uv)(t) - (uv)(s), \quad s, t \in \mathbb{T}.$$

This completes the proof. $\qquad\square$

Theorem D.39. *Let $u \in \mathbb{L}^2(\mathbb{T})$ and $U(t) = \int_a^t u \Delta, t \in \mathbb{T}$. Then $U \in \mathbb{H}^1(\mathbb{T})$, $U^{\Delta_g} = u$, and, if $V \in H^1(\mathbb{T})$ satisfies $V^{\Delta_g} = u$, then $U - V$ is a constant. In addition, there exists a constant $C > 0$ such that*

$$\|U\|_{1,\mathbb{T}} \le C \|u\|_{\mathbb{T}}.$$

Proof. Let $u \in \mathbb{L}^2(\mathbb{T})$. By Theorem D.25, it follows that there exists a sequence $\{u_n\}_{n \in \mathbb{N}}$ of elements of $\mathbb{C}(\mathbb{T})$ such that $u_n \to u$, as $n \to \infty$, in $\mathbb{L}^2(\mathbb{T})$. Since $u_n \in \mathbb{C}(\mathbb{T})$, $n \in \mathbb{N}$, we have that

$$U_n(t) = \int_a^t u_n \Delta \in \mathbb{C}^1(\mathbb{T}) \quad \text{and} \quad U_n^\Delta(t) = u_n(t), \quad t \in \mathbb{T}.$$

For $n \in \mathbb{N}$, we have

$$\|U_n - U\|_{\mathbb{T}}^2 = \int_a^b \left| \int_a^t u_n \Delta - \int_a^t u \Delta \right|^2 \Delta$$

$$= \int_a^b \left| \int_a^t (u_n - u) \Delta \right|^2 \Delta$$

$$\leq \int_a^b \left(\int_a^t (u_n - u)^2 \Delta \right) (t - a) \Delta$$

$$\leq (b - a) \int_a^b \left(\int_a^b (u_n - u)^2 \Delta \right) \Delta$$

$$= (b - a)^2 \|u_n - u\|_{\mathbb{T}}^2$$

$$\to 0,$$

$$\|U_n^\Delta - u\|_{\mathbb{T}} = \|u_n - u\|_{\mathbb{T}} \to 0,$$

as $n \to \infty$, i.e., $U_n \to U$, $U_n^\Delta \to u$, as $n \to \infty$, in $\mathbb{L}^2(\mathbb{T})$. Hence, by Theorem D.32, it follows that $U \in \mathbb{H}^1(\mathbb{T})$ and $U^{\Delta_g} = u$. Next,

$$\|U\|_{\mathbb{T}}^2 = \int_a^b \left| \int_a^t u \Delta \right|^2 \Delta$$

$$\leq \int_a^b (t - a) \left(\int_a^t u^2 \Delta \right) \Delta$$

$$\leq (b - a) \int_a^b \left(\int_a^b u^2 \Delta \right) \Delta$$

$$= (b - a)^2 \|u\|_{\mathbb{T}}^2,$$

$$\|U^{\Delta_g}\|_{\mathbb{T}}^2 = \|u\|_{\mathbb{T}}^2.$$

Hence,

$$\|U\|_{1,\mathbb{T}} = \left(\|U\|_{\mathbb{T}}^2 + \|U^{\Delta_g}\|_{\mathbb{T}}^2 \right)^{\frac{1}{2}}$$

$$\leq \left((b - a)^2 \|u\|_{\mathbb{T}}^2 + \|u\|_{\mathbb{T}}^2 \right)^{\frac{1}{2}}$$

$$= \left(1 + (b - a)^2 \right)^{\frac{1}{2}} \|u\|_{\mathbb{T}}.$$

Let $V \in \mathbb{H}^1(\mathbb{T})$ be such that $V^{\Delta_g} = u$. Then, using Theorem D.37, we obtain

$$(U - V)^{\Delta_g} = U^{\Delta_g} - V^{\Delta_g} = u - u = 0.$$

Hence, by Theorem D.35, we conclude that $U - V$ is a constant. This completes the proof. □

Definition D.40. The function U in Theorem D.39 will be called the antiderivative of u.

Theorem D.41. *We have*

$$\mathbb{C}^1_{rd}(\mathbb{T}) \subset \mathbb{H}^1(\mathbb{T}). \tag{D.14}$$

Proof. Let $v \in \mathbb{C}^1_{rd}(\mathbb{T})$ be arbitrarily chosen and $u = v^\Delta \in \mathbb{C}_{rd}(\mathbb{T})$. By Theorem D.24, it follows that $u \in \mathbb{L}^2(\mathbb{T})$. Hence, by Theorem D.39, we have

$$U = \int_a^t u\Delta \in \mathbb{H}^1(\mathbb{T}) \quad \text{and} \quad U^{\Delta_g} = u.$$

Because $v^\Delta = v^{\Delta_g} = u$, by Theorem D.39, it follows that there exists a constant C such that $v = U + C \in \mathbb{H}^1(\mathbb{T})$. Because $v \in \mathbb{C}^1_{rd}(\mathbb{T})$ was arbitrarily chosen and for it we get that it is an element of $\mathbb{H}^1(\mathbb{T})$, we obtain the inclusion (D.14). This completes the proof. □

Definition D.42. Define the space

$$\mathbb{H}^2(\mathbb{T}) = \{u \in \mathbb{C}^1(\mathbb{T}) : u^\Delta \in \mathbb{H}^1(\mathbb{T}^\kappa)\}$$

with the norm

$$\|u\|^2_{2,\mathbb{T}} = \|u\|^2_{\mathbb{T}} + \|u^\Delta\|^2_{1,\mathbb{T}}. \tag{D.15}$$

Exercise D.43. Prove that (D.15) satisfies all the axioms for a norm.

Theorem D.44. *If $\{u_n\}_{n \in \mathbb{N}}$ is a bounded sequence in $\mathbb{H}^1(\mathbb{T})$, then $\{u_n\}_{n \in \mathbb{N}}$ has a subsequence that converges in $\mathbb{C}(\mathbb{T})$.*

Proof. Since $\{u_n\}_{n \in \mathbb{N}}$ is a bounded sequence in $\mathbb{H}^1(\mathbb{T})$, there exists a constant $M > 0$ such that

$$\|u_n\|_{1,\mathbb{T}} \leq M$$

for any $n \in \mathbb{N}$. Hence, by Theorem D.34, it follows that

$$|u_n|_{0,\mathbb{T}} \leq C\|u_n\|_{1,\mathbb{T}} \leq CM$$

with $C = \max\{\frac{1}{(b-a)^{\frac{1}{2}}}, (b-a)^{\frac{1}{2}}\}$, for any $n \in \mathbb{N}$. Therefore, the sequence $\{u_n\}_{n\in\mathbb{N}}$ is a bounded sequence in $\mathbb{C}(\mathbb{T})$. Also, using Cauchy–Schwartz inequality and (D.7), we have

$$|u_n(s) - u_n(t)| = \left| \int_s^t u_n^{\Delta_g} \Delta \right|$$

$$\leq \int_s^t |u_n^{\Delta_g}| \Delta$$

$$\leq (t-s)^{\frac{1}{2}} \left(\int_s^t |u_n^{\Delta_g}|^2 \Delta \right)^{\frac{1}{2}}$$

$$\leq (t-s)^{\frac{1}{2}} \|u_n\|_{1,\mathbb{T}}$$

$$\leq M(t-s)^{\frac{1}{2}}, \quad s,t \in \mathbb{T}, \quad s \leq t.$$

Repeating as above,

$$|u_n(t) - u_n(s)| \leq M(s-t)^{\frac{1}{2}}, \quad s,t \in \mathbb{T}, \quad s \geq t.$$

Therefore, the sequence $\{u_n\}_{n\in\mathbb{N}}$ is equicontinuous on $\mathbb{T}$. Hence, using the Arzela–Ascoli theorem, there exists a subsequence $\{u_{n_m}\}_{m\in\mathbb{N}}$ of the sequence $\{u_n\}_{n\in\mathbb{N}}$ that converges in $\mathbb{C}(\mathbb{T})$. This completes the proof. □

Theorem D.45. *The embeddings*

$$\mathbb{H}^{i+1}(\mathbb{T}) \hookrightarrow \mathbb{C}(\mathbb{T}), \quad i = 0,1,$$

are compact.

Proof. Let $i = 0$. By Theorem D.34, we have

$$|u|_{0,\mathbb{T}} \leq C\|u\|_{1,\mathbb{T}},$$

with $C = \max\{\frac{1}{(b-a)^{\frac{1}{2}}}, (b-a)^{\frac{1}{2}}\}$. Therefore,

$$\mathbb{H}^1(\mathbb{T}) \hookrightarrow \mathbb{C}(\mathbb{T}). \tag{D.16}$$

By Theorem D.44, we have that every bounded sequence in $\mathbb{H}^1(\mathbb{T})$ has a subsequence that converges in $\mathbb{C}(\mathbb{T})$. Therefore, the embedding (D.16) is compact. Let $i = 1$. Suppose that $u \in \mathbb{H}^2(\mathbb{T})$. By the definition of $\mathbb{H}^2(\mathbb{T})$, it follows that $u \in \mathbb{C}^1(\mathbb{T})$. Therefore,

$$\mathbb{H}^2(\mathbb{T}) \hookrightarrow \mathbb{C}^1(\mathbb{T}). \tag{D.17}$$

Let $\{u_n\}_{n\in\mathbb{N}}$ be a bounded sequence in $\mathbb{H}^2(\mathbb{T})$. Then $\{u_n^\Delta\}_{n\in\mathbb{N}}$ is a bounded sequence in $\mathbb{H}^1(\mathbb{T})$. Hence, by Theorem D.44, it follows that there exists a subsequence $\{u_{n_m}^\Delta\}_{m\in\mathbb{N}}$ of the sequence $\{u_n^\Delta\}_{n\in\mathbb{N}}$ that converges in $\mathbb{C}(\mathbb{T})$. From here, it follows that $\{u_{n_m}^\Delta\}_{m\in\mathbb{N}}$ is a bounded sequence in $\mathbb{C}(\mathbb{T})$ and then there exists a positive constant M such that

$$|u_{n_m}^\Delta(t)| \leq M$$

for any $t \in \mathbb{T}$ and for any $m \in \mathbb{N}$. Note that

$$
\begin{aligned}
\left|u_{n_m}(t) - u_{n_m}(s)\right| &= \left|\int_s^t u_{n_m}^\Delta \Delta\right| \\
&\leq \int_s^t |u_{n_m}^\Delta| \Delta \\
&\leq M(t - s), \quad s, t \in \mathbb{T}, \quad s \leq t.
\end{aligned}
$$

Repeating as above,

$$\left|u_{n_m}(t) - u_{n_m}(s)\right| \leq M(s - t), \quad s, t \in \mathbb{T}, \quad s \geq t.$$

Therefore, $\{u_{n_m}\}_{m\in\mathbb{N}}$ is equicontinuous on $\mathbb{T}$. Since $u_{n_m} \in \mathbb{C}^1(\mathbb{T})$, we have that $u_{n_m} \in \mathbb{H}^1(\mathbb{T})$ for any $m \in \mathbb{N}$. By Theorem D.34, it follows that

$$|u_{n_m}|_{0,\mathbb{T}} \leq C\|u_{n_m}\|_{1,\mathbb{T}} \leq C\|u_{n_m}\|_{2,\mathbb{T}}$$

with $C = \max\{\frac{1}{(b-a)^{\frac{1}{2}}}, (b - a)^{\frac{1}{2}}\}$, for any $m \in \mathbb{N}$. Since $\{u_n\}_{n\in\mathbb{N}}$ is a bounded sequence in $\mathbb{H}^2(\mathbb{T})$ and $\{u_{n_m}\}_{m\in\mathbb{N}}$ is its subsequence, we conclude that $\{u_{n_m}\}_{m\in\mathbb{N}}$ is a bounded sequence in $\mathbb{C}(\mathbb{T})$. Hence and the Arzela–Ascoli theorem, it follows that there exists a subsequence $\{u_{n_{m_k}}\}_{k\in\mathbb{N}}$ of the sequence $\{u_{n_m}\}_{m\in\mathbb{N}}$ that converges in $\mathbb{C}(\mathbb{T})$. Because $\{u_{n_m}^\Delta\}_{m\in\mathbb{N}}$ converges in $\mathbb{C}(\mathbb{T})$, we have that $\{u_{n_{m_k}}^\Delta\}_{k\in\mathbb{N}}$ converges in $\mathbb{C}(\mathbb{T})$. Consequently, the sequence $\{u_{n_{m_k}}\}_{k\in\mathbb{N}}$ converges in $\mathbb{C}^1(\mathbb{T})$. Therefore, the embedding (D.17) is compact. This completes the proof. $\qquad\square$

Theorem D.46. *For any $\varepsilon > 0$, there exists $C(\varepsilon) > 0$ such that*

$$|u|_{0,\mathbb{T}} \leq \varepsilon\|u^{\Delta_g}\|_{\mathbb{T}} + C(\varepsilon)(\|u\|_{\mathbb{T}} + \|u^\sigma\|_{\mathbb{T}}), \quad u \in \mathbb{H}^1(\mathbb{T}). \tag{D.18}$$

Proof. Let $u \in \mathbb{H}^1(\mathbb{T})$. Then there exists a sequence $\{u_n\}_{n\in\mathbb{N}}$ of elements of $\mathbb{C}^1(\mathbb{T})$ such that $u_n \to u$, $u_n^\Delta \to u^{\Delta_g}$, as $n \to \infty$, in $\mathbb{L}^2(\mathbb{T})$. Take $n \in \mathbb{N}$ arbitrarily. Consider an arbitrary $t_0 \in \mathbb{T}$. Suppose that

$$\mathbb{T}_{0,\varepsilon} = \left[t_0 - \frac{\varepsilon^2}{2}, t_0 + \frac{\varepsilon^2}{2}\right] \cap \mathbb{T} \neq \{t_0\},$$

i. e., t_0 is not isolated in $\mathbb{T}$. Let

$$i_0 = \inf \mathbb{T}_{0,\varepsilon}, \quad s_0 = \sup \mathbb{T}_{0,\varepsilon}.$$

We have

$$0 < s_0 - i_0 \le \varepsilon^2.$$

Then

$$\begin{aligned}
|u_n(t)| &= \left| u_n(s) + \int_s^t u_n^\Delta \Delta \right| \\
&\le |u_n(s)| + \left| \int_s^t u_n^\Delta \Delta \right| \\
&\le |u_n(s)| + \int_s^t |u_n^\Delta| \Delta \\
&\le |u_n(s)| + (t-s)^{\frac{1}{2}} \left(\int_s^t |u_n^\Delta|^2 \Delta \right)^{\frac{1}{2}} \\
&\le |u_n(s)| + \varepsilon \|u_n^\Delta\|_{\mathbb{T}}, \quad s, t \in \mathbb{T}_{0,\varepsilon}, \quad s \le t.
\end{aligned}$$

Repeating as above,

$$|u_n(t)| \le |u_n(s)| + \varepsilon \|u_n^\Delta\|_{\mathbb{T}}, \quad s, t \in \mathbb{T}_{0,\varepsilon}, \quad s \ge t.$$

Integrating over $\mathbb{T}_{0,\varepsilon}$ with respect to s, we get

$$\begin{aligned}
(s_0 - i_0)|u_n(t)| &\le \int_{t_0 - \frac{\varepsilon^2}{2}}^{t_0 + \frac{\varepsilon^2}{2}} |u_n(s)| \Delta + \varepsilon(s_0 - i_0)\|u_n^\Delta\|_{\mathbb{T}} \\
&\le (s_0 - i_0)^{\frac{1}{2}} \left(\int_{t_0 - \frac{\varepsilon^2}{2}}^{t_0 + \frac{\varepsilon^2}{2}} |u_n(s)|^2 \Delta \right)^{\frac{1}{2}} + \varepsilon(s_0 - i_0)\|u_n^\Delta\|_{\mathbb{T}} \\
&\le (s_0 - i_0)^{\frac{1}{2}} \|u_n\|_{\mathbb{T}} + \varepsilon(s_0 - i_0)\|u_n^\Delta\|_{\mathbb{T}}, \quad t \in \mathbb{T}_{0,\varepsilon}.
\end{aligned}$$

Hence,

$$|u_n(t)| \le \varepsilon(s_0 - i_0)^{-\frac{1}{2}} \|u_n\|_{\mathbb{T}} + \varepsilon \|u_n^\Delta\|_{\mathbb{T}}, \quad t \in \mathbb{T}_{0,\varepsilon}. \tag{D.19}$$

Now, we suppose that $\mathbb{T}_{0,\varepsilon} = \{t_0\}$, i. e., t_0 is isolated in $\mathbb{T}$. If $\sigma(t_0) > t_0$, then

$$
|u_n(t_0)| = (\sigma(t_0) - t_0)^{-1} \int_{t_0}^{\sigma(t_0)} |u_n| \Delta
$$

$$
\leq (\sigma(t_0) - t_0)^{-\frac{1}{2}} \left(\int_{t_0}^{\sigma(t_0)} |u_n|^2 \Delta \right)^{\frac{1}{2}}
$$

$$
\leq (\sigma(t_0) - t_0)^{-\frac{1}{2}} \|u_n\|_{\mathbb{T}}. \tag{D.20}
$$

If $t_0 > \rho(t_0)$, then

$$
|u_n(t_0)| = (t_0 - \rho(t_0))^{-1} \int_{\rho(t_0)}^{t_0} |u_n^\sigma| \Delta
$$

$$
\leq (t_0 - \rho(t_0))^{-\frac{1}{2}} \left(\int_{\rho(t_0)}^{t_0} |u_n^\sigma|^2 \Delta \right)^{\frac{1}{2}}
$$

$$
\leq (t_0 - \rho(t_0))^{-\frac{1}{2}} \|u_n^\sigma\|_{\mathbb{T}}. \tag{D.21}
$$

Suppose that there exists $\varepsilon > 0$ such that for each $m \in \mathbb{N}$ and $t_m \in \mathbb{T}$ we have

$$
|u_n(t_m)| > \varepsilon \|u_n^\Delta\|_{\mathbb{T}} + m(\|u_n\|_{\mathbb{T}} + \|u_n^\sigma\|_{\mathbb{T}}), \quad t_m \to t_0. \tag{D.22}
$$

If t_0 is isolated, (D.22) contradicts (D.20) and (D.21). If t_0 is not isolated, then (D.22) contradicts (D.19). Consequently, for each $\varepsilon > 0$, there exists $C(\varepsilon) > 0$ such that

$$
|u_n|_{0,\mathbb{T}} \leq \varepsilon \|u_n^\Delta\|_{\mathbb{T}} + C(\varepsilon)(\|u_n\|_{\mathbb{T}} + \|u_n^\sigma\|_{\mathbb{T}}). \tag{D.23}
$$

Because $n \in \mathbb{N}$ was arbitrarily chosen and

$$
|u_n|_{0,\mathbb{T}} \to |u|_{0,\mathbb{T}}, \quad \|u_n^\Delta\|_{\mathbb{T}} \to \|u^{\Delta_g}\|_{\mathbb{T}},
$$
$$
\|u_n\|_{\mathbb{T}} \to \|u\|_{\mathbb{T}}, \quad \|u_n^\sigma\|_{\mathbb{T}} \to \|u^\sigma\|_{\mathbb{T}},
$$

as $n \to \infty$, by (D.23), we get (D.18). This completes the proof. □

Theorem D.47. *For any $\varepsilon > 0$, there exists $C(\varepsilon) > 0$ such that*

$$
|u|_{0,\mathbb{T}_\kappa} \leq \varepsilon \|u^{\Delta_g}\|_{\mathbb{T}} + C(\varepsilon)\|u^\sigma\|_{\mathbb{T}}, \quad u \in \mathbb{H}^1(\mathbb{T}). \tag{D.24}
$$

Proof. Let $u \in \mathbb{H}^1(\mathbb{T})$. Then there exists a sequence $\{u_n\}_{n \in \mathbb{N}}$ of elements of $\mathbb{C}^1(\mathbb{T})$ such that $u_n \to u$, $u_n^\Delta \to u^{\Delta_g}$, as $n \to \infty$, in $\mathbb{L}^2(\mathbb{T})$. Take $n \in \mathbb{N}$ arbitrarily. Consider an arbitrary $t_0 \in \mathbb{T}_\kappa$. Suppose that

$$\mathbb{T}_{0,\varepsilon} = \left[t_0 - \frac{\varepsilon^2}{2}, t_0 + \frac{\varepsilon^2}{2} \right] \cap \mathbb{T}_\kappa \neq \{t_0\},$$

i. e., t_0 is not isolated in $\mathbb{T}$. Let

$$i_0 = \inf \mathbb{T}_{0,\varepsilon}, \quad s_0 = \sup \mathbb{T}_{0,\varepsilon}.$$

We have

$$0 < s_0 - i_0 \leq \varepsilon^2.$$

Then

$$|u_n(t)| = \left| u_n^\sigma(s) + \int_{\sigma(s)}^t u_n^\Delta \Delta \right|$$

$$\leq |u_n^\sigma(s)| + \left| \int_{\sigma(s)}^t u_n^\Delta \Delta \right|$$

$$\leq |u_n^\sigma(s)| + \int_{\sigma(s)}^t |u_n^\Delta| \Delta$$

$$\leq |u_n^\sigma(s)| + (t - \sigma(s))^{\frac{1}{2}} \left(\int_{\sigma(s)}^t |u_n^\Delta|^2 \Delta \right)^{\frac{1}{2}}$$

$$\leq |u_n^\sigma(s)| + \varepsilon \|u_n^\Delta\|_{\mathbb{T}}, \quad s, \sigma(s), t \in \mathbb{T}_{0,\varepsilon}, \quad \sigma(s) \leq t.$$

Repeating as above,

$$|u_n(t)| \leq |u_n^\sigma(s)| + \varepsilon \|u_n^\Delta\|_{\mathbb{T}}, \quad s, \sigma(s), t \in \mathbb{T}_{0,\varepsilon}, \quad \sigma(s) \geq t.$$

Integrating over $\mathbb{T}_{0,\varepsilon}$ with respect to s, we get

$$(s_0 - i_0)|u_n(t)| \leq \int_{t_0 - \frac{\varepsilon^2}{2}}^{t_0 + \frac{\varepsilon^2}{2}} |u_n^\sigma(s)| \Delta + \varepsilon(s_0 - i_0) \|u_n^\Delta\|_{\mathbb{T}}$$

$$\leq (s_0 - i_0)^{\frac{1}{2}} \left(\int_{t_0 - \frac{\varepsilon^2}{2}}^{t_0 + \frac{\varepsilon^2}{2}} |u_n^\sigma(s)|^2 \Delta \right)^{\frac{1}{2}} + \varepsilon(s_0 - i_0) \|u_n^\Delta\|_{\mathbb{T}}$$

$$\leq (s_0 - i_0)^{\frac{1}{2}} \|u_n\|_{\mathbb{T}} + \varepsilon(s_0 - i_0) \|u_n^\Delta\|_{\mathbb{T}}, \quad t \in \mathbb{T}_{0,\varepsilon}.$$

Hence,

$$|u_n(t)| \le \varepsilon(s_0 - i_0)^{-\frac{1}{2}}\|u_n^\sigma\|_{\mathbb{T}} + \varepsilon\|u_n^\Delta\|_{\mathbb{T}}, \quad t \in \mathbb{T}_{0,\varepsilon}. \tag{D.25}$$

Now, we suppose that $\mathbb{T}_{0,\varepsilon} = \{t_0\}$, i. e., t_0 is isolated in $\mathbb{T}$. Then

$$|u_n(t_0)| = (t_0 - \rho(t_0))^{-1} \int_{\rho(t_0)}^{t_0} |u_n^\sigma|\Delta$$

$$\le (t_0 - \rho(t_0))^{-\frac{1}{2}} \left(\int_{\rho(t_0)}^{t_0} |u_n^\sigma|^2 \Delta \right)^{\frac{1}{2}}$$

$$\le (t_0 - \rho(t_0))^{-\frac{1}{2}} \|u_n^\sigma\|_{\mathbb{T}}. \tag{D.26}$$

Suppose that there exists $\varepsilon > 0$ such that for each $m \in \mathbb{N}$ and $t_m \in \mathbb{T}_\kappa$ we have

$$|u_n(t_m)| > \varepsilon\|u_n^\Delta\|_{\mathbb{T}} + m\|u_n^\sigma\|_{\mathbb{T}}, \quad t_m \to t_0. \tag{D.27}$$

If t_0 is isolated, then (D.27) contradicts (D.26). If t_0 is not isolated, then (D.27) contradicts (D.25). Consequently, for each $\varepsilon > 0$, there exists $C(\varepsilon) > 0$ such that

$$|u_n|_{0,\mathbb{T}} \le \varepsilon\|u_n^\Delta\|_{\mathbb{T}} + C(\varepsilon)\|u_n^\sigma\|_{\mathbb{T}}. \tag{D.28}$$

Because $n \in \mathbb{N}$ was arbitrarily chosen and

$$|u_n|_{0,\mathbb{T}} \to |u|_{0,\mathbb{T}}, \quad \|u_n^\Delta\|_{\mathbb{T}} \to \|u^{\Delta_g}\|_{\mathbb{T}},$$
$$\|u_n^\sigma\|_{\mathbb{T}} \to \|u^\sigma\|_{\mathbb{T}},$$

as $n \to \infty$, by (D.28), we get (D.24). This completes the proof. □

D.4 Weak solutions of dynamical systems

Definition D.48. Let $f : \mathbb{T} \to \mathbb{R}$ be a Lebesgue measurable function. If

$$\int_K |f(t)|\Delta t < \infty$$

on all compact subsets K of $\mathbb{T}$, then we say that f is locally integrable. The set of all such functions we will denote by $\mathbb{L}^1_{\mathrm{loc}}(\mathbb{T})$.

Define

$$\Gamma_1(\mathbb{T}) = \{p \in \mathbb{L}^1_{\mathrm{loc}}(\mathbb{T}) : 1 + \mu(t)p(t) \ne 0\}.$$

Note that for any $p \in \Gamma_1(\mathbb{T})$ and $a, b \in \mathbb{R}$ fixed, there exist positive constants a_1 and a_2 such that

$$a_1 \le |1 + \mu(t)p(t)| \le a_2, \quad t \in [a, b] \cap \mathbb{T},$$

and the set

$$\{t \in \mathbb{T} : 1 + \mu(t)p(t) < 0\} \subset \mathbb{T}$$

is finite.

Definition D.49. For $p \in \Gamma_1(\mathbb{T})$ and $s, t \in \mathbb{T}$, define

$$\xi_{\mu(t)}(p(t)) = \begin{cases} \frac{\text{Log}(1+\mu(t)p(t))}{\mu(t)} & \text{if } \mu(t) \ne 0, \\ p(t) & \text{if } \mu(t) = 0, \end{cases}$$

and

$$e_p(t, s) = e^{\int_{[s,t)} \xi_{\mu(\tau)}(p(\tau))\Delta\tau}.$$

Exercise D.50. Let $p \in \Gamma_1(\mathbb{T})$. Prove that

$$\xi_{\mu(\cdot)}(p(\cdot)) \in \mathbb{L}^1_{\text{loc}}(\mathbb{T}).$$

Definition D.51. For $p, q \in \Gamma_1(\mathbb{T})$, define

$$p \oplus q = p + q + \mu pq,$$
$$\ominus p = -\frac{p}{1 + \mu p},$$
$$p \ominus q = \frac{p - q}{1 + \mu q}.$$

Exercise D.52. Let $p, q \in \Gamma_1(\mathbb{T})$. Prove that

$$p \oplus q, \quad p \ominus q, \quad \ominus p \in \Gamma_1(\mathbb{T}).$$

Exercise D.53. Let $p, q \in \Gamma_1(\mathbb{T})$ and $s, t, r \in \mathbb{T}$. Prove that

1. $e_0(t, s) = 1$,
2. $e_p(t, t) = 1$,
3. $e_p(t, s)e_p(s, r) = e_p(t, r)$,
4. $e_p(\sigma(t), s) = (1 + \mu(t)p(t))e_p(t, s)$,
5. $e_p(t, s) = \frac{1}{e_p(s,t)} = e_{\ominus p}(s, t)$,
6. $e_p(t, s)e_q(t, s) = e_{p \oplus q}(t, s)$,
7. $\frac{e_p(t,s)}{e_q(t,s)} = e_{p \ominus q}(t, s)$,

8. $e_p(\cdot, s) \in C_{rd}(\mathbb{T})$,
9. $(e_p(\cdot, s))^\Delta = p(\cdot)e_p(\cdot, s)$ Δ-a.e. on $\mathbb{T}$,
10. $(e_p(s, \cdot))^\Delta = -p(\cdot)e_p(s, \sigma(\cdot))$ Δ-a.e. on $\mathbb{T}$.

Theorem D.54. *Let $x \in C_{rd}(\mathbb{T})$, $p, f \in \mathbb{L}^1(\mathbb{T})$, $p \geq 0$ on $\mathbb{T}$, $a \in \mathbb{T}$, and*

$$x^\Delta(t) \leq p(t)x(t) + f(t) \quad \Delta\text{-a.e. on } \mathbb{T}.$$

Then

$$x(t) \leq e_p(t, a)x(a) + \int_{[a,t)} e_p(t, \sigma(\tau))f(\tau)\Delta\tau, \quad t \in \mathbb{T}.$$

Proof. Since $p \in \mathbb{L}^1(\mathbb{T})$ and $p \geq 0$ on $\mathbb{T}$, we have that $p \in \Gamma_1(\mathbb{T})$ and

$$1 + \mu(t)p(t) > 0, \quad t \in \mathbb{T}.$$

Therefore, for any $s, t \in \mathbb{T}$, we have

$$e_p(t, s) > 0, \quad e_{\ominus p}(t, s) > 0.$$

Next,

$$\begin{aligned}(x(\cdot)e_{\ominus p}(\cdot, s))^\Delta(t) &= x^\Delta(t)e_{\ominus p}(\sigma(t), a) + x(t)(\ominus p)(t)e_{\ominus p}(t, a) \\ &= \frac{x^\Delta(t)}{1 + \mu(t)p(t)}e_{\ominus p}(t, a) - \frac{p(t)x(t)}{1 + \mu(t)p(t)}e_{\ominus p}(t, a) \\ &= \frac{x^\Delta(t) - p(t)x(t)}{1 + \mu(t)p(t)}e_{\ominus p}(t, a) \\ &= (x^\Delta(t) - p(t)x(t))e_{\ominus p}(\sigma(t), a) \quad \Delta\text{-a.e. on } \mathbb{T}.\end{aligned}$$

Therefore,

$$\begin{aligned}x(t)e_{\ominus p}(t, a) - x(a) &= \int_{[a,t)} (x^\Delta(\tau) - p(\tau)x(\tau))e_{\ominus p}(\sigma(\tau), a)\Delta\tau \\ &\leq \int_{[a,t)} f(\tau)e_{\ominus p}(\sigma(\tau), a)\Delta\tau, \quad t \in \mathbb{T}.\end{aligned}$$

Hence,

$$x(t)e_{\ominus p}(t, a) \leq x(a) + \int_{[a,t)} f(\tau)e_{\ominus p}(\sigma(\tau), a)\Delta\tau, \quad t \in \mathbb{T},$$

and

$$x(t) \leq x(a)e_p(t,a) + \int_{[a,t)} f(\tau)e_{\ominus p}(\sigma(\tau),a)e_{\ominus p}(a,t)\Delta\tau$$

$$= x(a)e_p(t,a) + \int_{[a,t)} f(\tau)e_{\ominus p}(\sigma(\tau),t)\Delta\tau$$

$$= x(a)e_p(t,a) + \int_{[a,t)} f(\tau)e_p(t,\sigma(\tau))\Delta\tau, \quad t \in \mathbb{T}.$$

This completes the proof. ☐

Theorem D.55. *Let* $x \in C_{rd}(\mathbb{T})$, $p, g \in \mathbb{L}^1(\mathbb{T})$, $x \geq 0$, $g \geq 0$ *on* $\mathbb{T}$, $\alpha \geq 0$, $\lambda \in (0,1)$, $a, b \in \mathbb{T}$, $a < b$,

$$x(t) \leq \alpha + \int_{[a,t)} p(\tau)x(\tau)\Delta\tau + \int_{[a,t)} g(\tau)(x(\sigma(\tau)))^\lambda \Delta\tau, \quad t \in [a,b].$$

Then there exists a positive constant M such that

$$x(t) \leq M, \quad t \in [a,b].$$

Proof. Let

$$y(t) = \alpha + \int_{[a,t)} p(\tau)x(\tau)\Delta\tau + \int_{[a,t)} g(\tau)(x(\sigma(\tau)))^\lambda \Delta\tau, \quad t \in [a,b].$$

Then

$$x(t) \leq y(t), \quad t \in [a,b],$$

and *y* is differentiable Δ-a.e. on $[a,b]$, $y(a) = \alpha$. We have

$$y^\Delta(t) = p(t)x(t) + g(t)(x(\sigma(t)))^\lambda$$
$$\leq p(t)y(t) + g(t)(y(\sigma(t)))^\lambda \quad \text{Δ-a.e. on } [a,b].$$

Hence, by Theorem D.54, it follows that

$$y(t) \leq \alpha e_p(t,a) + \int_{[a,t)} e_p(t,\sigma(\tau))g(\tau)(y(\sigma(\tau)))^\lambda \Delta\tau$$

$$\leq \alpha e_p(b,a) + e_p(b,a) \int_{[a,t)} e_p(a,\sigma(\tau))g(\tau)(y(\sigma(\tau)))^\lambda \Delta\tau$$

$$\leq \alpha e_p(b, a) + e_p(b, a) \int_{[a,t)} g(\tau)(y(\sigma(\tau)))^\lambda \Delta\tau$$

$$\leq \alpha e_p(b, a) + e_p(b, a) \int_{[a,b)} g(\tau)(y(\sigma(\tau)))^\lambda \Delta\tau, \quad t \in [a, b].$$

Define

$$q(t) = \alpha e_p(b, a) + e_p(b, a) \int_{[a,b)} g(\tau)(y(\sigma(\tau)))^\lambda \Delta\tau$$

$$+ e_p(b, a) \int_{[a,t)} g(\tau)(y(\sigma(\tau)))^\lambda \Delta\tau, \quad t \in [a, b].$$

Then q is monotone increasing on $[a, b]$ and

$$y(t) \leq q(a),$$

$$y(\sigma(t)) \leq q(a), \quad t \in [a, b],$$

$$q(b) = \alpha e_p(b, a) + 2e_p(b, a) \int_{[a,b)} g(\tau)(y(\sigma(\tau)))^\lambda \Delta\tau$$

$$= -\alpha e_p(b, a) + 2q(a),$$

and

$$q^\Delta(t) = e_p(b, a)g(t)(y(\sigma(t)))^\lambda$$

$$\leq e_p(b, a)g(t)(q(a))^\lambda$$

$$\leq e_p(b, a)g(t)(q(t))^\lambda \quad \Delta\text{-a.e. on } [a, b].$$

Hence,

$$q(t) - q(a) \leq e_p(b, a) \int_{[a,t)} g(\tau)(q(\tau))^\lambda \Delta\tau$$

$$\leq (q(t))^\lambda e_p(b, a) \int_{[a,t)} g(\tau)\Delta\tau, \quad t \in [a, b].$$

From here,

$$(q(t))^{1-\lambda} - (q(a))^{1-\lambda} \leq (q(t))^{1-\lambda} - (q(t))^{-\lambda}q(a)$$

$$\leq e_p(b, a) \int_{[a,t)} g(\tau)\Delta\tau, \quad t \in [a, b].$$

Hence,

$$(q(b))^{1-\lambda} - (q(a))^{1-\lambda} \le e_p(b,a) \int_{[a,b)} g(\tau)\Delta\tau,$$

or

$$(2q(a) - \alpha e_p(b,a))^{1-\lambda} - (q(a))^{1-\lambda} \le e_p(b,a) \int_{[a,b)} g(\tau)\Delta\tau.$$

Let

$$h(z) = (2z - \alpha e_p(b,a))^{1-\lambda} - z^{1-\lambda}, \quad z \in \mathbb{R}.$$

We have

$$\lim_{z \to \infty} h(z) = \lim_{z \to \infty} \frac{h(z)}{z^{1-\lambda}} z^{1-\lambda}$$

$$= \lim_{z \to \infty} \left(\left(2 - \frac{\alpha e_p(b,a)}{z} \right)^{1-\lambda} - 1 \right) z^{1-\lambda}$$

$$= \infty.$$

Hence, there is a positive constant M such that

$$q(a) \le M$$

and

$$x(t) \le y(t) \le q(a) \le M, \quad t \in [a,b].$$

This completes the proof. □

Consider the following IVP:

$$x^{\Delta}(t) + p(t)x^{\sigma}(t) = f(t),$$
$$x(a) = x_0, \quad \text{(D.29)}$$

where $p \in \Gamma_1(\mathbb{T}), f \in \mathbb{L}^1(\mathbb{T}), x_0 \in \mathbb{R}$ is given, $a \in \mathbb{T}$. Note that the integral

$$\int_{[a,\cdot)} e_{\ominus p}(\cdot, \tau)f(\tau)\Delta\tau$$

is well-defined.

Definition D.56. The function $x \in C_{rd}(\mathbb{T})$ given by

$$x(t) = e_{\ominus p}(t,a)x_0 + \int_{[a,t)} e_{\ominus p}(t,\tau)f(\tau)\Delta\tau, \quad t \in \mathbb{T}, \quad \text{(D.30)}$$

is said to be a weak solution of the IVP (D.29).

If $x \in C_{rd}(\mathbb{T})$ is a weak solution of the IVP (D.29) and if it is given by the expression (D.30), we have

$$x^\Delta(t) + p(t)x^\sigma(t) = f(t) \quad \Delta\text{-a.e. on } \mathbb{T}.$$

Now we consider the IVP

$$x^\Delta(t) + p(t)x^\sigma(t) = f(t, x(t), x^\sigma(t)), \quad a \leq t < b,$$
$$x(a) = x_0, \tag{D.31}$$

where $a, b \in \mathbb{T}$, $a < b$, $p \in \Gamma_1(\mathbb{T})$,

(H1) $f : \mathbb{T} \times \mathbb{R} \times \mathbb{R} \to \mathbb{R}$ is Δ-measurable in $t \in \mathbb{T}$ and

$$\left| f(t, x_1, y_1) - f(t, x_2, y_2) \right| \leq L\left(|x_1 - x_2| + |y_1 - y_2| \right)$$

for all $t \in \mathbb{T}$, $x_1, x_2, y_1, y_2 \in \mathbb{R}$, and for some positive constant L,

(H2) there exist a constant $\lambda \in (0, 1)$ and a function $q \in \mathbb{L}^1(\mathbb{T})$, such that

$$1 - 2 \sup_{t,s \in [a,b)} \left| e_{\ominus p}(t, s) \right| \int_{[a,b)} q(\tau)\Delta\tau > 0,$$

and

$$\left| f(t, x, y) \right| \leq q(t)\left(1 + |x| + |y|^\lambda \right)$$

for all $t \in \mathbb{T}$, $x, y \in \mathbb{R}$.

Definition D.57. A function $x \in C_{rd}(\mathbb{T})$ is said to be a weak solution of the IVP (D.31), if x satisfies the following integral equation:

$$x(t) = e_{\ominus p}(t, a)x_0 + \int_{[a,t)} e_{\ominus p}(t, \tau) f(\tau, x(\tau), x^\sigma(\tau))\Delta\tau, \quad t \in \mathbb{T}.$$

Theorem D.58. $p \in \Gamma_1(\mathbb{T})$. *Suppose* (H1) *and* (H2) *hold. Then the IVP* (D.31) *has a unique weak solution* $x \in C_{rd}([a, b))$ *such that*

$$x^\Delta(t) + p(t)x^\sigma(t) = f(t, x(t), x^\sigma(t)) \quad \Delta\text{-a.e. on } \mathbb{T}.$$

Proof. For $x \in C_{rd}([a, b))$, define the operator

$$(Qx)(t) = e_{\ominus p}(t, a)x_0 + \int_{[a,t)} e_{\ominus p}(t, \tau) f(\tau, x(\tau), x^\sigma(\tau))\Delta\tau.$$

We have $Qx \in C_{rd}([a, b))$. Let $\rho > 0$ be arbitrarily chosen and

$$M = \sup_{t,s \in [a,b)} |e_{\ominus p}(t,s)|.$$

Take $x, y \in \mathcal{C}_{rd}([a,b))$ such that

$$\|x\|_{\mathcal{C}_{rd}([a,b))} \leq \rho, \quad \|y\|_{\mathcal{C}_{rd}([a,b))} \leq \rho,$$

where

$$\| \cdot \|_{\mathcal{C}_{rd}([a,b))} = \sup_{[a,b)} | \cdot |.$$

We have

$$\begin{aligned}
|(Qx)(t) - (Qy)(t)| &= \left| e_{\ominus p}(t,a)x_0 + \int_{[a,t)} e_{\ominus p}(t,\tau)f(\tau,x(\tau),x^\sigma(\tau))\Delta\tau \right. \\
&\quad \left. - e_{\ominus p}(t,a)x_0 - \int_{[a,t)} e_{\ominus p}(t,\tau)f(\tau,y(\tau),y^\sigma(\tau))\Delta\tau \right| \\
&= \left| \int_{[a,t)} e_{\ominus p}(t,\tau)(f(\tau,x(\tau),x^\sigma(\tau)) - f(\tau,y(\tau),y^\sigma(\tau)))\Delta\tau \right| \\
&\leq \int_{[a,t)} |e_{\ominus p}(t,\tau)||f(\tau,x(\tau),x^\sigma(\tau)) - f(\tau,y(\tau),y^\sigma(\tau))|\Delta\tau \\
&\leq \int_{[a,b)} |e_{\ominus p}(t,\tau)||f(\tau,x(\tau),x^\sigma(\tau)) - f(\tau,y(\tau),y^\sigma(\tau))|\Delta\tau \\
&\leq LM \int_{[a,b)} (|x(\tau) - y(\tau)| + |x^\sigma(\tau) - y^\sigma(\tau)|)\Delta\tau \\
&\leq 2LM(b-a)\|x-y\|_{\mathcal{C}_{rd}([a,b))}, \quad t \in [a,b).
\end{aligned}$$

Hence,

$$\|Qx - Qy\|_{\mathcal{C}_{rd}([a,b))} \leq 2LM(b-a)\|x-y\|_{\mathcal{C}_{rd}([a,b))}.$$

Therefore,

$$Q : \mathcal{C}_{rd}([a,b)) \to \mathcal{C}_{rd}([a,b))$$

is a continuous operator. Define

$$W = \{x \in \mathcal{C}_{rd}([a,b)) : \|x\|_{\mathcal{C}_{rd}([a,b))} \leq \rho\}.$$

Note that, for $x \in W$, we have

$$\left|f(t, x(t), x^\sigma(t))\right| \leq q(t)\left(1 + |x(t)| + |x^\sigma(t)|^\lambda\right)$$
$$\leq q(t)\left(1 + \|x\|_{\mathcal{C}_{rd}([a,b))} + \|x\|^\lambda_{\mathcal{C}_{rd}([a,b))}\right)$$
$$\leq q(t)(1 + \rho + \rho^\lambda), \quad t \in [a, b),$$

and

$$\left|(Qx)(t)\right| = \left|e_{\ominus p}(t, a)x_0 + \int_{[a,t)} e_{\ominus p}(t, \tau)f(\tau, x(\tau), x^\sigma(\tau))\Delta\tau\right|$$
$$\leq \left|e_{\ominus p}(t, a)x_0\right| + \int_{[a,t)} \left|e_{\ominus p}(t, \tau)\right|\left|f(\tau, x(\tau), x^\sigma(\tau))\right|\Delta\tau$$
$$\leq M|x_0| + M(1 + \rho + \rho^\lambda) \int_{[a,b)} q(\tau)\Delta\tau$$
$$= M\left(|x_0| + (1 + \rho + \rho^\lambda) \int_{[a,b)} q(\tau)\Delta\tau\right), \quad t \in [a, b),$$

and

$$\|Qx\|_{\mathcal{C}_{rd}([a,b))} \leq M\left(|x_0| + (1 + \rho + \rho^\lambda) \int_{[a,b)} q(\tau)\Delta\tau\right).$$

Therefore, $QW \subseteq \mathcal{C}_{rd}([a, b))$ is bounded. Let $t_1, t_2 \in [a, b)$, $t_2 > t_1$. Then

$$\left|(Qx)(t_2) - (Qx)(t_1)\right| = \left|e_{\ominus p}(t_2, a)x_0 + \int_{[a,t_2)} e_{\ominus p}(t_2, \tau)f(\tau, x(\tau), x^\sigma(\tau))\Delta\tau\right.$$
$$\left. - e_{\ominus p}(t_1, a)x_0 - \int_{[a,t_1)} e_{\ominus p}(t_1, \tau)f(\tau, x(\tau), x^\sigma(\tau))\Delta\tau\right|$$
$$\leq \left|e_{\ominus p}(t_2, a) - e_{\ominus p}(t_1, a)\right||x_0|$$
$$+ \left|\int_{[a,t_2)} e_{\ominus p}(t_2, \tau)f(\tau, x(\tau), x^\sigma(\tau))\Delta\tau\right.$$
$$\left. - \int_{[a,t_1)} e_{\ominus p}(t_1, \tau)f(\tau, x(\tau), x^\sigma(\tau))\Delta\tau\right|$$
$$\leq \left|e_{\ominus p}(t_1, a)\right|\left|e_{\ominus p}(t_2, t_1) - 1\right||x_0|$$
$$+ \int_{[a,t_1)} \left|e_{\ominus p}(t_2, \tau) - e_{\ominus p}(t_1, \tau)\right|\left|f(\tau, x(\tau), x^\sigma(\tau))\right|\Delta\tau$$
$$+ \int_{[t_1,t_2)} \left|e_{\ominus p}(t_2, \tau)\right|\left|f(\tau, x(\tau), x^\sigma(\tau))\right|\Delta\tau$$

$$\leq M|x_0||e_{\ominus p}(t_2,t_1)-1|$$
$$+\int_{[a,t_1)}|e_{\ominus p}(t_2,t_1)-1||e_{\ominus p}(t_1,\tau)||f(\tau,x(\tau),x^\sigma(\tau))|\Delta\tau$$
$$+\int_{[t_1,t_2)}|e_{\ominus p}(t_2,\tau)||f(\tau,x(\tau),x^\sigma(\tau))|\Delta\tau$$
$$\leq M|x_0||e_{\ominus p}(t_2,t_1)-1|$$
$$+M(1+\rho+\rho^\lambda)|e_{\ominus p}(t_2,t_1)-1|\int_{a,t_1)}q(\tau)\Delta\tau$$
$$+M(1+\rho+\rho^\lambda)\int_{[t_1,t_2)}q(\tau)\Delta\tau$$
$$\leq M(|x_0|+1+\rho+\rho^\lambda+\|q\|_{\mathbb{L}^1([a,b))})|e_{\ominus p}(t_2,t_1)-1|$$
$$+M(1+\rho+\rho^\lambda)\int_{[t_1,t_2)}q(\tau)\Delta\tau.$$

Therefore, QW is rd-equicontinuous. By the Arzela–Ascoli theorem, it follows that Q is a compact operator in $C_{rd}([a,b))$. Define
$$Y=\{x\in C_{rd}([a,b)):x=\delta Qx,\quad \delta\in[0,1]\}.$$

Let
$$y=\frac{1}{\delta}x,\quad \delta\neq 0,$$

otherwise $y=0$ for $x\in Y$. Then
$$|y(t)|=|(Q(\delta y))(t)|$$
$$=\left|e_{\ominus p}(t,a)x_0+\int_{[a,t)}e_{\ominus p}(t,\tau)f(\tau,\delta y(\tau),\delta y^\sigma(\tau))\Delta\tau\right|$$
$$\leq|e_{\ominus p}(t,a)x_0|+\int_{[a,t)}|e_{\ominus p}(t,\tau)||f(\tau,\delta y(\tau),\delta y^\sigma(\tau))|\Delta\tau$$
$$\leq M|x_0|+M\int_{[a,t)}q(\tau)(1+\delta|y(\tau)|+\delta^\lambda|y(\tau)|^\lambda)\Delta\tau$$
$$=M|x_0|+M\int_{[a,t)}q(\tau)\Delta\tau+M\delta\int_{[a,t)}q(\tau)|y(\tau)|\Delta\tau$$
$$+M\delta^\lambda\int_{[a,t)}q(\tau)|y^\sigma(\tau)|^\lambda\Delta\tau,\quad t\in[a,b).$$

Hence, by Theorem D.55, it follows that there exists a positive constant r such that

$$|y(t)| \le r, \quad t \in [a,b).$$

We take $r > 0$ large enough and b close enough to a so that

$$M|x_0| + M(1 + r + r^\Delta) \int_{[a,b)} q(\tau)\Delta\tau \le r.$$

Thus Y is a bounded set. Hence, by the Leray–Schauder fixed point theorem, it follows that Q has a fixed point $x \in C_{rd}([a,b))$. We have

$$x(t) = e_{\ominus p}(t,a)x_0 + \int_{[a,t)} e_{\ominus p}(t,\tau)f(\tau,x(\tau),x^\sigma(\tau))\Delta\tau, \quad t \in [a,b).$$

Consequently, x is a weak solution of the IVP (D.31) and

$$x^\Delta(t) + p(t)x^\sigma(t) = f(t,x(t),x^\sigma(t)) \quad \Delta\text{-a.e. on } [a,b).$$

For $x \in C_{rd}([a,b))$, define

$$\|x\|_\beta = \sup_{t \in [a,b)} \frac{|x(t)|}{e_\beta(t,a)}, \tag{D.32}$$

where $\beta > 0$ is chosen so that

$$\int_{[a,b)} \frac{\Delta\tau}{e_\beta(b,\sigma(\tau))} \le \frac{1}{4LM}.$$

Note that $C_{rd}([a,b))$ is a Banach space with respect to the norm $\|\cdot\|_\beta$. Define

$$B = \{x \in C_{rd}([a,b)) : \|x\|_{C_{rd}([a,b))} \le r\}.$$

Note that B is a Banach space with respect to the norm (D.32). For $x \in B$, define the operator

$$Hx(t) = e_{\ominus p}(t,a)x_0 + \int_{[a,t)} e_{\ominus p}(t,\tau)f(\tau,x(\tau),x^\sigma(\tau))\Delta\tau, \quad t \in [a,b).$$

For $x \in B$, we get

$$|Hx(t)| = \left| e_{\ominus p}(t, a)x_0 + \int_{[a,t)} e_{\ominus p}(t, \tau)f(\tau, x(\tau), x^\sigma(\tau))\Delta\tau \right|$$

$$\leq |e_{\ominus p}(t, a)||x_0| + \int_{[a,t)} |e_{\ominus p}(t, \tau)||f(\tau, x(\tau), x^\sigma(\tau))|\Delta\tau$$

$$\leq M|x_0| + M \int_{[a,t)} q(\tau)(1 + |x(\tau)| + |x^\sigma(\tau)|^\lambda)\Delta\tau$$

$$\leq M|x_0| + M(1 + r + r^\lambda) \int_{[a,t)} q(\tau)\Delta\tau$$

$$\leq M|x_0| + M(1 + r + r^\lambda) \int_{[a,b)} q(\tau)\Delta\tau$$

$$\leq r, \quad t \in [a, b).$$

Therefore, $HB \subseteq B$. Next, for $x, y \in B$, we have

$$\|Hx - Hy\|_\beta = \sup_{t\in[a,b)} \frac{1}{e_\beta(t, a)} \left| e_{\ominus p}(t, a)x_0 + \int_{[a,t)} e_{\ominus p}(t, \tau)f(\tau, x(\tau), x^\sigma(\tau))\Delta\tau \right.$$

$$\left. - e_{\ominus p}(t, a)x_0 - \int_{[a,t)} e_{\ominus p}(t, \tau)f(\tau, y(\tau), y^\sigma(\tau))\Delta\tau \right|$$

$$\leq \sup_{t\in[a,b)} \frac{1}{e_\beta(t, a)} \int_{[a,t)} |e_{\ominus p}(t, \tau)||f(\tau, x(\tau), x^\sigma(\tau)) - f(\tau, y(\tau), y^\sigma(\tau))|\Delta\tau$$

$$\leq \sup_{t\in[a,b)} \frac{ML}{e_\beta(t, a)} \int_{[a,t)} (|x(\tau) - y(\tau)| + |x^\sigma(\tau) - y^\sigma(\tau)|)\Delta\tau$$

$$\leq \|x - y\|_\beta \left(ML \sup_{t\in[a,b)} \frac{1}{e_\beta(t, a)} \int_{[a,t)} (e_\beta(\tau, a) + e_\beta(\sigma(\tau), a))\Delta\tau \right)$$

$$= \|x - y\|_\beta \left(ML \sup_{t\in[a,b)} \int_{[a,t)} \left(\frac{1}{e_\beta(t, \tau)} + \frac{1}{e_\beta(t, \sigma(\tau))} \right)\Delta\tau \right)$$

$$= \|x - y\|_\beta \left(ML \sup_{t\in[a,b)} \int_{[a,t)} \frac{1}{e_\beta(t, \sigma(\tau))} \left(1 + \frac{1}{1 + \beta\mu(\tau)} \right)\Delta\tau \right)$$

$$\leq \|x - y\|_\beta \left(2ML \sup_{t\in[a,b)} \int_{[a,t)} \frac{\Delta\tau}{e_\beta(t, \sigma(\tau))} \right)$$

$$\leq \frac{1}{2}\|x - y\|_\beta.$$

Therefore, H has a unique weak solution x in $(B, \|\cdot\|_\beta)$. Hence, the IVP (D.31) has a unique weak solution. This completes the proof. $\qquad\square$

D.5 A Gronwall-type inequality

Let $a, b \in \mathbb{T}$, $a < b$. In this section, we state and prove a result which is a sort of the Gronwall inequality.

Theorem D.59 (Gronwall-type inequality). *Let $X : \mathbb{T} \to \mathbb{R}$ and $k : \mathbb{T} \to [0, \infty)$, $k \in \mathbb{L}^1([a, b))$. Suppose*

$$\left|x^{\Delta}(t)\right| \leq \gamma_1 |x(t)| + k(t) \quad \Delta\text{-}a.\,e. \quad t \in [a, b],$$

for some positive constant γ. Then

1.

$$|x(t) - x(a)| \leq \gamma_1 |x(a)| \int_{[a,t)} e^{\gamma_1(t-s)} \Delta s + \int_{[a,t)} k(s) e^{\gamma_1(t-s)} \Delta s, \quad t \in \mathbb{T}.$$

2.

$$|x(t)| \leq (\gamma_1 e^{\gamma_1(b-a)}(b-a) + 1)|x(a)| + e^{\gamma_1(b-a)} \int_{[a,b)} k(s) \Delta s, \quad t \in \mathbb{T}.$$

Proof. Define $z : \mathbb{T} \to [0, \infty)$ as follows:

$$z(t) = |x(t) - x(a)|$$

and take $t_* \in [a, b)$ such that z and x are Δ-differentiable at t_*. If $\sigma(t_*) > t_*$, then

$$
\begin{aligned}
z^{\Delta}(t_*) &= \frac{z(\sigma(t_*)) - z(t_*)}{\mu(t_*)} \\
&= \frac{|x(\sigma(t_*)) - x(a)| - |x(t_*) - x(a)|}{\mu(t_*)} \\
&\leq \frac{|x(\sigma(t_*)) - x(t_*)|}{\mu(t_*)} \\
&= |x^{\Delta}(t_*)|.
\end{aligned}
$$

If $\sigma(t_*) = t_*$, let $\{s_j\}_{j \in \mathbb{N}} \subset \mathbb{T}$ be a decreasing sequence such that $s_j \to t_*$, as $j \to \infty$. Since

$$\frac{z(s_j) - z(t_*)}{s_j - t_*} = \frac{|x(s_j) - x(a)| - |x(t_*) - x(a)|}{s_j - t_*} \leq \frac{|x(s_j) - x(t_*)|}{s_j - t_*},$$

we arrive at the inequality

$$z^{\Delta}(t_*) \leq |x^{\Delta}(t_*)|.$$

Therefore,

$$z^\Delta(t) \le |x^\Delta(t)| \quad \Delta\text{-a. e.} \quad t \in [a, b).$$

Hence,

$$
\begin{aligned}
z^\Delta(t) &\le |x^\Delta(t)| \\
&\le \gamma_1 |x(t)| + k(t) \\
&\le \gamma_1 |x(t) - x(a)| + \gamma_1 |x(a)| + k(t) \\
&= \gamma_1 z(t) + \gamma_1 |x(a)| + k(t) \quad \Delta\text{-a. e.} \quad t \in [a, b). \tag{D.33}
\end{aligned}
$$

Define the functions $m : \mathbb{T} \to \mathbb{R}$ and $m_c : \mathbb{R} \to \mathbb{R}$ as follows:

$$
\begin{aligned}
m(t) &= e^{-\gamma_1 t}, \quad t \in \mathbb{T}, \\
m_c(t) &= e^{-\gamma_1 t}, \quad t \in \mathbb{R}.
\end{aligned}
$$

Let

$$\psi(t) = z(t)m(t), \quad t \in \mathbb{T}.$$

Take $t_* \in [a, b)$ such that z and m are Δ-differentiable at t_* and (D.33) holds. We have

$$\psi^\Delta(t_*) = m^\Delta(t_*)z(t_*) + m(\sigma(t_*))z^\Delta(t_*).$$

If $\sigma(t_*) = t_*$, then we take a sequence $\{s_j\}_{j\in\mathbb{N}} \subset \mathbb{T}$ such that $s_j \to t_*$ and

$$
\begin{aligned}
m^\Delta(t_*) &= \lim_{s_j \to t_*} \frac{m(t_*) - m(s_j)}{t_* - s_j} \\
&= \lim_{s_j \to t_*} \frac{m_c(t_*) - m_c(s_j)}{t_* - s_j} \\
&= m_c'(t_*) \\
&= -\gamma_1 e^{-\gamma_1 t_*}.
\end{aligned}
$$

So,

$$
\begin{aligned}
\psi^\Delta(t_*) &= e^{-\gamma_1 t_*} z^\Delta(t_*) - \gamma_1 e^{-\gamma_1 t_*} z(t_*) \\
&= (z^\Delta(t_*) - \gamma_1 z(t_*))e^{-\gamma_1 t_*}.
\end{aligned}
$$

If $\sigma(t_*) > t_*$, then, applying the mean value theorem, there exists a $\theta \in (t_*, \sigma(t_*))$ such that

$$
\begin{aligned}
m^\Delta(t_*) &= \frac{m(\sigma(t_*)) - m(t_*)}{\mu(t_*)} \\
&= \frac{m_c(\sigma(t_*)) - m_c(t_*)}{\mu(t_*)}
\end{aligned}
$$

$$= m'_c(\theta)$$
$$= -\gamma_1 e^{-\gamma_1 \theta}$$
$$\leq -\gamma_1 e^{-\gamma_1 \sigma(t_*)}.$$

Hence,

$$\psi^\Delta(t_*) \leq -\gamma_1 z(t_*) e^{-\gamma_1 \sigma(t_*)} + e^{-\gamma_1 \sigma(t_*)} z^\Delta(t_*)$$
$$= (z^\Delta(t_*) - \gamma_1 z(t_*)) e^{-\gamma_1 \sigma(t_*)}.$$

Let

$$\lambda(t_*) = z^\Delta(t_*) - \gamma_1 z(t_*).$$

If $\lambda(t_*) \leq 0$, then

$$\psi^\Delta(t_*) \leq \lambda(t_*) e^{-\gamma_1 \sigma(t_*)}$$
$$\leq 0$$
$$\leq (\gamma_1 |x(a)| + k(t_*)) e^{-\gamma_1 t_*}.$$

If $\lambda(t_*) > 0$, then

$$\psi^\Delta(t_*) \leq \lambda(t_*) e^{-\gamma_1 \sigma(t_*)}$$
$$\leq \lambda(t_*) e^{-\gamma_1 t_*}$$
$$= (z^\Delta(t_*) - \gamma_1 z(t_*)) e^{-\gamma_1 t_*}$$
$$\leq (|\gamma_1||x(a)| + k(t_*)) e^{-\gamma_1 t_*}.$$

So,

$$\psi^\Delta(t) \leq (\gamma_1 |x(a)| + k(t)) e^{-\gamma_1 t} \quad \Delta\text{-a. e.} \quad t \in [a, b).$$

Now, for each $t \in \mathbb{T}$, we have

$$\psi(t) - \psi(a) = \int_{[a,t)} \psi^\Delta(s) \Delta s$$
$$\leq \int_{[a,t)} (\gamma_1 |x(a)| + k(s)) e^{-\gamma_1 s} \Delta s, \quad t \in \mathbb{T}. \tag{D.34}$$

Then

$$|x(t) - x(a)| \le e^{\gamma_1 t} \int_{[a,t)} (\gamma_1 |x(a)| e^{-\gamma_1 s} + k(s) e^{-\gamma_1 s}) \Delta s$$

$$= \gamma_1 |x(a)| \int_{[a,t)} e^{\gamma_1 (t-s)} \Delta s + \int_{[a,t)} k(s) e^{\gamma_1 (t-s)} \Delta s, \quad t \in \mathbb{T}.$$

Now, by (D.34), we obtain

$$\psi(t) \le \gamma_1 |x(a)| e^{-\gamma_1 a} (b - a) + e^{-\gamma_1 a} \int_{[a,b)} k(s) \Delta s, \quad t \in \mathbb{T}.$$

Hence,

$$|x(t)| - |x(a)| \le |x(t) - x(a)|$$

$$\le \gamma_1 |x(a)| e^{\gamma_1 t} e^{-\gamma_1 a} (b - a) + e^{\gamma_1 t} e^{-\gamma_1 a} \int_{[a,b)} k(s) \Delta s$$

$$\le \gamma_1 |x(a)| e^{\gamma_1 (b-a)} (b - a) + e^{\gamma_1 (b-a)} \int_{[a,b)} k(s) \Delta s, \quad t \in \mathbb{T},$$

whereupon

$$|x(t)| \le (\gamma_1 e^{\gamma_1 (b-a)} (b - a) + 1) |x(a)| + e^{\gamma_1 (b-a)} \int_{[a,b)} k(s) \Delta s, \quad t \in \mathbb{T}.$$

This completes the proof. $\qquad\qquad\qquad\qquad\qquad\qquad\qquad\qquad\qquad\qquad$ □

E Mazur's theorem

Definition E.1. Let **X** be a normed linear space. A linear functional $\mathbb{T}$ on **X** is said to be bounded if there is an $M \geq 0$ such that

$$|\mathbb{T}(f)| \leq M\|f\| \quad \text{for any } f \in \mathbf{X}.$$

The infimum of all such M is called the norm of $\mathbb{T}$ and it is denoted by $\|\mathbb{T}\|_*$. The collection of bounded linear functionals on **X** is denoted by $\mathbf{X}^*$ and is called the dual space of **X** which is a linear space.

Definition E.2. The linear operator $\mathbb{J} : \mathbf{X} \to (\mathbf{X}^*)^*$ defined by

$$\mathbb{J}(x)[\psi] = \psi(x) \quad \text{for all } x \in \mathbf{X}, \psi \in \mathbf{X}^*,$$

is called the natural embedding of **X** into $(\mathbf{X}^*)^*$. Also, the space **X** is said to be reflexive when $\mathbb{J}(\mathbf{X}) = (\mathbf{X}^*)^*$. It is customary to denote $(\mathbf{X}^*)^*$ by $\mathbf{X}^{**}$ and call $\mathbf{X}^{**}$ the bidual of **X**.

Definition E.3. A normed linear space **X** is said to be separable when there is a countable subset of X that is dense in **X**.

Remark E.4. If a set **E** is measurable and $1 \leq p < \infty$, the normed linear space $L^p(\mathbf{E})$ is separable.

Definition E.5. A Banach space is a normed linear space that is a complete metric space with respect to the metric derived from its norm.

Let Y be a real Banach space and Y^* be its dual space.

Definition E.6. An operator $B : Y \to Y^*$ is said to be bounded if it maps bounded sets of Y into bounded subsets of Y^*.

Definition E.7. A sequence $\{x_n\}$ in a normed space X is said to be strongly convergent if there is an $x \in X$ such that

$$\lim_{n\to\infty} \|x_n - x\| = 0.$$

Definition E.8. A sequence $\{x_n\}$ in a normed space X is said to be weakly convergent if there is an $x \in X$ such that

$$\lim_{n\to\infty} T(x_n) = T(x),$$

for every continuous linear functional T in X^*.

Theorem E.9 (Mazur's theorem). *Let $\{u_n\}_{n\in\mathbb{N}}$ be a sequence in Y that converges weakly to some $u_0 \in Y$. Then there exists a function $N : \mathbb{N} \to \mathbb{N}$ such that the sequence $\{\bar{u}_n\}_{n\in\mathbb{N}}$ defined by*

https://doi.org/10.1515/9783110787320-017

$$\bar{u}_n = \sum_{k=n}^{N(n)} \lambda_k u_k$$

converges strongly in Y to u_0, where $\lambda_k \geq 0$, $k = n, \ldots, N(n)$, $\sum_{k=n}^{N(n)} \lambda_k = 1$.

Bibliography

[1] M. Bohner and A. Peterson. Advances in Dynamic Equations on Time Scales, Birkhäuser Boston, Inc., Boston, MA, 2003.

[2] M. Bohner and S. Georgiev. Multivariable Dynamic Calculus on Time Scales, Springer, 2016.

[3] M. Bohner, I. Erhan and S. Georgiev. The Euler Method for Dynamic Equations on Time Scales, Nonlinear Studies, Vol. 27, No. 2, pp. 415–431, 2020.

[4] J. Butcher. The Numerical Analysis of Ordinary Differential Equations. Runge-Kutta and General Linear Methods, John Wiley & Sons, Chichester, 1987.

[5] E. Cheney. Introduction to Approximation Theory, McGraw-Hill, New York, 1966.

[6] P. Davis and P. Rabinowitz. Methods of Numerical Integration, Second Edition, Academic Press, Orlando, FL, 1984.

[7] H. Engels. Numerical Quadrature and Cubature, Computational Mathematics and Applications, Academic Press, London, 1980.

[8] S. Georgiev. Integral Equations on Time Scales, Atlantis Press, 2016.

[9] S. Georgiev. Fractional Dynamic Calculus and Fractional Dynamic Equations on Time Scales, Springer, 2018.

[10] S. Georgiev and I. Erhan. Nonlinear Integral Equations on Time Scales, Nova Science Publishers, 2019.

[11] S. Georgiev and I. Erhan. The Taylor Series Method and Trapezoidal Rule on Time Scales, Applied Mathematics and Computation, Vol. 378, 125200, 2020.

[12] S. Georgiev and I. Erhan. Series Solution Method for Cauchy Problems with Fractional Δ-Derivative on Time Scales, Fractional Differential Calculus, Vol. 9, No. 2, pp. 243–261, 2019.

[13] S. Georgiev and I. Erhan. Adomian Polynomials Method for Dynamic Equations on Time Scales, Advances in the Theory of Nonlinear Analysis and its Applications, Vol. 5, No. 3, pp. 300–315, 2020.

[14] F. Hildebrand. Introduction to Numerical Analysis, McGraw-Hill, New York, 1956.

[15] J. Lambert. Numerical Methods for Ordinary Differential Systems, John Wiley & Sons, Chichester, 1991.

[16] M. Powell. Approximation Theory and Methods. Cambridge University Press, Cambridge, 1996.

https://doi.org/10.1515/9783110787320-018

Index

https://doi.org/10.1515/9783110787320-019

www.ingramcontent.com/pod-product-compliance
Lightning Source LLC
Chambersburg PA
CBHW080705220326
41598CB00033B/5311